The IMA Volumes in Mathematics and its Applications

Volume 82

Series Editors
Avner Friedman Robert Gulliver

Springer
New York
Berlin
Heidelberg
Barcelona
Budapest
Hong Kong
London
Milan
Paris
Santa Clara
Singapore
Tokyo

Institute for Mathematics and its Applications IMA

The **Institute for Mathematics and its Applications** was established by a grant from the National Science Foundation to the University of Minnesota in 1982. The IMA seeks to encourage the development and study of fresh mathematical concepts and questions of concern to the other sciences by bringing together mathematicians and scientists from diverse fields in an atmosphere that will stimulate discussion and collaboration.

The IMA Volumes are intended to involve the broader scientific community in this process.

Avner Friedman, Director

Robert Gulliver, Associate Director

* * * * * * * * * * *

IMA ANNUAL PROGRAMS

1982–1983	Statistical and Continuum Approaches to Phase Transition
1983–1984	Mathematical Models for the Economics of Decentralized Resource Allocation
1984–1985	Continuum Physics and Partial Differential Equations
1985–1986	Stochastic Differential Equations and Their Applications
1986–1987	Scientific Computation
1987–1988	Applied Combinatorics
1988–1989	Nonlinear Waves
1989–1990	Dynamical Systems and Their Applications
1990–1991	Phase Transitions and Free Boundaries
1991–1992	Applied Linear Algebra
1992–1993	Control Theory and its Applications
1993–1994	Emerging Applications of Probability
1994–1995	Waves and Scattering
1995–1996	Mathematical Methods in Material Science
1996–1997	Mathematics of High Performance Computing
1997–1998	Emerging Applications of Dynamical Systems
1998–1999	Mathematics in Biology

Continued at the back

Jill P. Mesirov Klaus Schulten
De Witt Sumners
Editors

Mathematical Approaches to Biomolecular Structure and Dynamics

With 52 Illustrations

Springer

Jill P. Mesirov
Computer Science Department
Boston University
111 Cummington Street
Boston, MA 02215 USA

Klaus Schulten
Beckman Institute
Theoretical Biophysics Group
University of Illinois
405 N. Mathews Avenue
Urbana, IL 61801 USA

De Witt Sumners
Department of Mathematics
Florida State University
Tallahassee, FL 32306-3027 USA

Series Editors:
Avner Friedman
Robert Gulliver
Institute for Mathematics and its Applications
University of Minnesota
Minneapolis, MN 55455 USA

Mathematics Subject Classifications (1991): 92-06, 92C05, 92C40, 92E10, 92-04, 92-08, 82C21, 53A04, 57M25, 60J15, 82D60

Library of Congress Cataloging-in-Publication Data
Mathematical approaches to biomolecular structure and dynamics/ [edited by]
Jill P. Mesirov, Klaus Schulten, De Witt Sumners
p. cm. – (The IMA volumes in mathematics and its applications; v. 82)
Includes bibliographical references.

ISBN-13: 978-1-4612-8485-7 e-ISBN-13: 978-1-4612-4066-2
DOI: 10.1007/978-1-4612-4066-2

1. Molecular dynamics—Mathematical models. 2. Biomolecules—Structure—Mathematical models. I. Mesirov, Jill P. II. Schulten, K. (Klaus) III. Sumners, De Witt L. IV. Series.
QP517.M65M38 1996
574.8'8–dc20 96-24164

Printed on acid-free paper.

Softcover reprint of the hardcover 1st edition 1996

Production managed by Hal Henglein; manufacturing supervised by Johanna Tschebull.
Camera-ready copy prepared by the IMA.

9 8 7 6 5 4 3 2 1

FOREWORD

This IMA Volume in Mathematics and its Applications

MATHEMATICAL APPROACHES TO BIOMOLECULAR STRUCTURE AND DYNAMICS

is one of the two volumes based on the proceedings of the 1994 IMA Summer Program on "Molecular Biology" and comprises Weeks 3 and 4 of the four-week program. Weeks 1 and 2 appeared as Volume 81: *Genetic Mapping and DNA Sequencing.* We thank Jill P. Mesirov, Klaus Schulten, and De Witt Sumners for organizing Weeks 3 and 4 of the workshop and for editing the proceedings. We also take this opportunity to thank the National Institutes of Health (NIH) (National Center for Human Genome Research), the National Science Foundation (NSF) (Biological Instrumentation and Resources), and the Department of Energy (DOE), whose financial support made the summer program possible.

Avner Friedman

Robert Gulliver

PREFACE

The revolutionary progress in molecular biology within the last 30 years opens the way to full understanding of the molecular structures and mechanisms of living organisms. Interdisciplinary research in mathematics and molecular biology is driven by ever growing experimental, theoretical and computational power. The mathematical sciences accompany and support much of the progress achieved by experiment and computation as well as provide insight into geometric and topological properties of biomolecular structure and processes.

This volume consists of a representative sample of the papers presented during the last two weeks of the month-long Institute for Mathematics and Its Applications Summer 1994 Program in Molecular Biology. The papers in this volume cover the spectrum from experiment to computation to simulation to theory. Week 3 (July 18–22, 1994) of the Program was Protein Structure and Dynamics, organized by Jill P. Mesirov and Klaus Schulten. Week 4 (July 25–29, 1994) was Topology and Geometry of DNA and RNA, organized by De Witt Sumners.

Protein structure and dynamics is a broad field with a large array of interesting topics. It is of key importance in answering basic scientific questions about the nature of all living organisms and has practical biomedical applications. The major subareas of structure prediction and classification, techniques and heuristics for the simulation of protein folding, and molecular dynamics provide a rich problem domain where mathematics can be helpful in analysis, modeling, and simulation. One would like to infer information about the gross geometric conformation of a protein from its primary amino acid sequence (secondary structure prediction); a robust and biologically significant geometric classification scheme is required to do this. The ab initio simulation of protein folding is beyond current technology. However, groups are experimenting with different techniques and heuristics for simulating folding, for example using lattice models and searching conformational space, or exploring certain "folding pathways". Molecular dynamics is an area where recent progress in n-body algorithms has already had a direct beneficial effect.

Experimental scientists have become increasingly aware that the powerful analytical techniques of geometry and topology can be used in the interpretation and design of experiments. Chemists have long been interested in developing techniques that will allow them to synthesize molecules with interesting three-dimensional structure. Polymer scientists continue to study the chemical and physical ramifications of random topological entanglement in large molecules. One of the important problems in molecular biology is the three-dimensional structure of proteins, DNA and RNA in the cell, and the relationship between structure and function. We need to develop greater understanding of the topology of cellular DNA, RNA and proteins and the various life-sustaining mechanisms used by the cell which modify this molecular topology. Models for molecular structure and metabolic mechanism must be built and understood; simulation, computation and proof of molecular spatial structure and metabolically induced change in structure must be produced. The mathematical tools used in this analysis include statistical and continuum mechanics, partial differential equations, differential geometry, geometric topology, and Monte Carlo simulation, both on lattices and in the continuum.

Jill P. Mesirov

Klaus Schulten

De Witt Sumners

CONTENTS

III. Biomolecular Structure and Dynamics: Computation

TANGLE COMPLEXITY AND THE TOPOLOGY OF THE CHINESE RINGS

LOUIS H. KAUFFMAN*

Abstract. This paper discusses a question in complexity of untanglement of graph embeddings in relation to the well-known Chinese Rings puzzle. A series of examples is described in which the complexity of untanglement is conjectured to rise exponentially as a function of the structural complexity of the associated graphs. Possible relations with biological and physical systems are discussed.

1. Introduction. The Chinese Rings is a mechanical puzzle involving a recursive sequence of actions for its solution. The actions involve the movement of a shuttle that is topologically unlinked from the body of the puzzle, but mechanically entangled with it. The number of moves needed to solve the puzzle can be determined as a function of the starting position by using a reformulation of the mechanical structure via the Gray code. See section 1 of this paper for a description of this classical aspect of the Chinese Rings.

In section 2 of the paper, we formulate a well-defined topological problem in relation to the Chinese Rings. This problem asks for the least number of crossing exchanges needed to disentangle a loop from a graph, given that the crossings must be between the loop and a specified edge of the graph. The original Chinese Rings puzzle can be reformulated in terms of a flexible loop and a rigid graph. By adding a special edge to this rigid graph and removing its rigidity we obtain the specific class of exchange problems discussed here. This addition of an edge to the graph in the Chinese Rings produces a topological problem that is very closely related to the mechanism of the classical Rings. In particular we can prove that the number of movements of the shuttle across the first ring is an upper bound for the topological exchange number in the corresponding problem about graph embeddings. This number can be computed using the Gray code representation. We conjecture that the topological and mechanical exchange numbers are equal. This is the *Ring Conjecture* of section 2. Section 3 briefly discusses possible relations with biological and physical systems. In particular, if the Ring Conjecture is true then we have examples of crossing exchange numbers in a topological problem that grow exponentially as a function of the complexity of the underlying graph.

2. The Chinese Rings. In its most popular form the Chinese rings puzzle is a tricky mechanical exercise that depends upon the linking of rigid objects. Since we are going to obtain a genuine topological problem from this primarily mechanical puzzle, it is best to proceed with a capsule

* Department of Mathematics, Statistics and Computer Science, University of Illinois at Chicago, 851 South Morgan Street, Chicago, IL 60607-7045.

description of the puzzle in its usual incarnation.

There are rings, and each ring is attached to an *arm* on which it can freely pivot. Each arm penetrates a hole in the *baseboard* of the puzzle and is kept from slipping completely from the hole by a knob or thickening at the end of the arm opposite to the ring. The holes in the baseboard are lined up and each is penetrated by its arm. *Each ring is linked around the arm of the neighboring ring to its right.* (See Figure 1.) A *shuttle* consisting of a long metal loop attached to a handle is linked about some or all of the arms. The problem is to remove the shuttle from the arms by motions that do not bend or break the parts of the little machine.

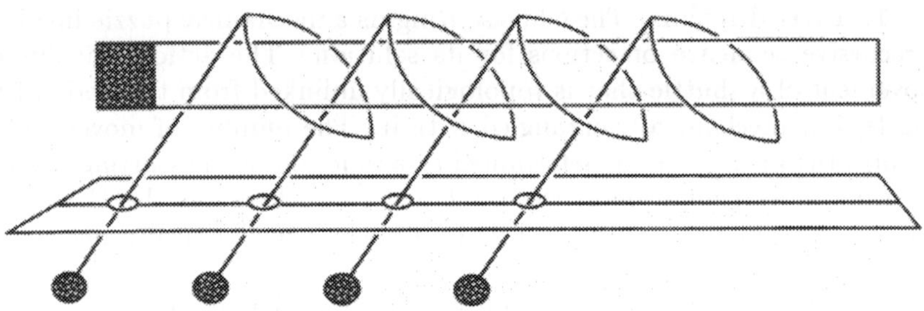

FIG. 1.

It can be done! But as the number of arms goes up the number of movements needed to extricate the shuttle goes up exponentially. The key to the solution of this puzzle lies in a property of the situation where the shuttle is no longer linked with the first k-1 arms. (We count arms from right to left as shown in Figures 1 and 2, with the rightmost arm labelled k=1.) When the shuttle is unlinked from the first k-1 arms, then there is a movement, shown in Figure 2, that allows the linkage or unlinkage of the arm k+1 from the shuttle. This movement does not affect arm k. View Figure 2 for an example of this movement when k=3. The sequence of events depicted in this Figure begins with arm 4 linked to the shuttle while arms 1 and 2 are unlinked. In the end, arm 4 is unlinked leaving only arm 3 linked to the shuttle.

Since arm 1 can always be linked or unlinked with the shuttle, the type of special movement illustrated in Figure 2 can be used repeatedly to solve the puzzle. This method of solving the puzzle can be schematized by using the following notation: A *state* of the shuttle is designated by a row of n zeroes and ones when the puzzle has n arms. In our notation the rightmost digit (0 or 1) denotes the state of the shallowest ring. The shallowest ring is the ring that can always be removed or added to the shuttle. This convention of left-to-right corresponds to the pictorial conventions of Figures 1 and 2. Thus the state 0000 denotes the state where the shuttle is removed from the rings while 1111 denotes the state where the shuttle is engaged around each arm and through each ring. This is the state depicted in Figure 1. The initial state in Figure 2 is 1100 and the final state is 0100.

The principle behind the special move is then the following: ...11000...00 is interchangeable with ...01000...00. This statement is a symbolic expression of the principle that an arm that is linked at stage k, but unlinked at all earlier (shallower) stages can be used to link or unlink the shuttle at stage k+1. The translation is: *In a given state, locate the* 1 *that has only zeroes to its right. The digit just to the left of this special* 1 *can be changed without affecting the rest of the state.*

For example, the following represents the sequence of changes solving the puzzle with n=4, starting from the state 1111.

1111

1101

1100

0100

0101

0111

0110

0010

0011

0001

0000.

This solution is accomplished in 10 moves. This method of moving from one string of binary digits to another by changing only one digit is known as the *Gray code* [2],[1]. We have shown the (well-known) fact that the

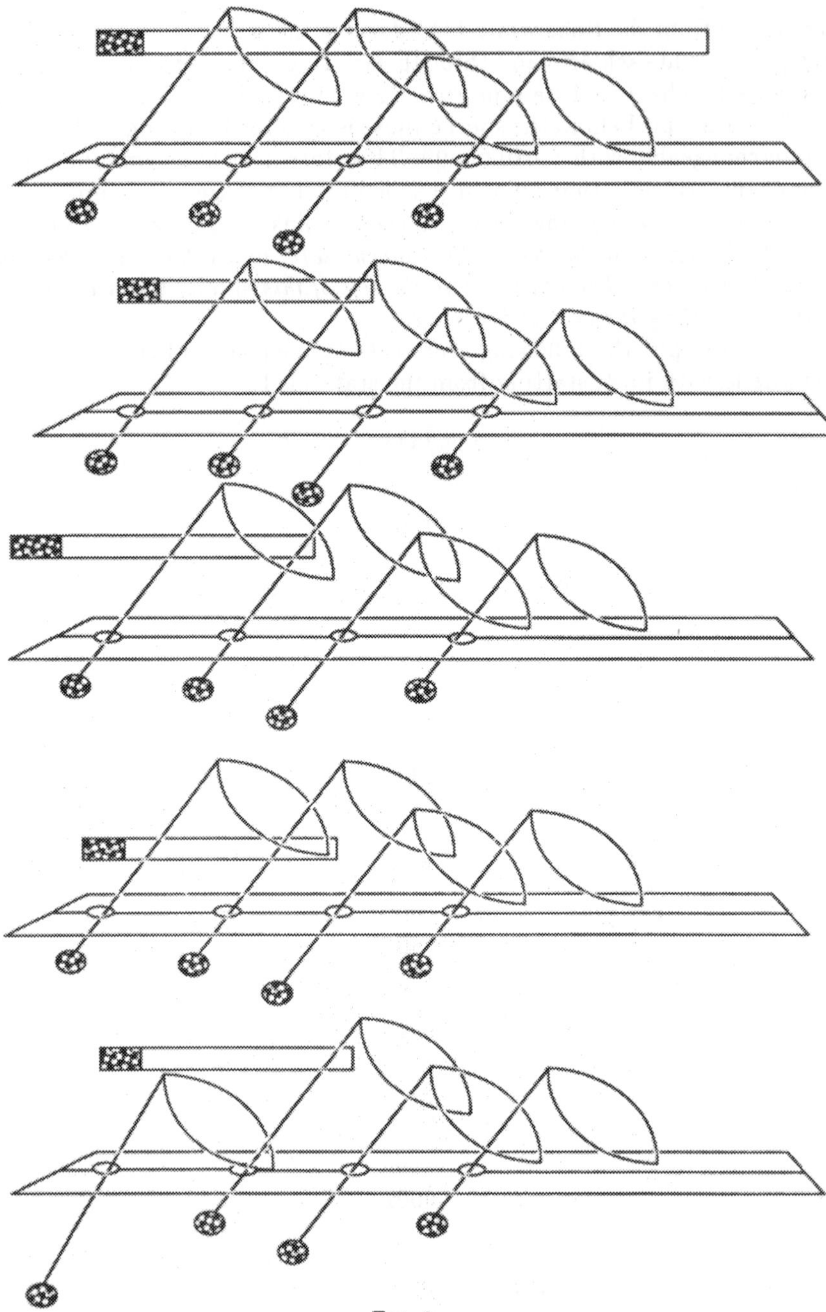

FIG. 2.

Chinese Rings is solved in the pattern of the Gray code. For history of the Chinese Rings see [6] and [5].

Let $c(n)$ denote the number of moves needed to solve the Chinese Rings with n rings and starting position 111...11. I leave it as an exercise for the reader to verify that

$$c(n) = (2^{n+1} - 1)/3 \text{ if n is odd}$$

$$c(n) = (2^{n+1} - 2)/3 \text{ if n is even.}$$

Note that in the course of the solution, the shuttle exchanges the first ring at every other step. Thus in the example above there are 5 first ring exchanges out of the 10 moves. In general, for n even the number of first ring exchanges is one half the total number of moves. For n odd, the number of first ring exchanges is one half of one more than the total number of moves. *Let $e(n)$ denote the number of first ring exchanges in the Gray code solution starting from the state* 111...11. Then the remarks above imply the formula

$$e(n) = (2^n + (-1)^{n+1})/3.$$

We shall reinterpret this formula and its generalizations in topological terms in the next section.

Definition. Let S be a binary string, regarded as a state of a Chinese Rings puzzle with ring cardinality equal to the number of bits in the string S. Let $E[S]$ denote the number of first ring exchanges in the optimal Gray code solution to the Chinese Rings starting from the configuration indicated by S. Call $E[S]$ the *exchange number* of the string S. Let $C[S]$ denote the total number of moves in the Chinese Rings solution starting from the state S.

Thus we have $e(n) = E[111...11]$ where there are n 1's in the string 111...11, and $c(n) = C[111...11]$.

Another interesting example of the exchange number occurs for the initial state $T = 100...00$. If the string T has n digits, let $f(n) = E[T]$. Then it is easy to prove that

$$C[T] = 2^n - 1$$

and

$$f(n) = E[T] = 2^{n-1}.$$

For example, here is the sequence for T with n=4:

1000

1001

1011

1010

1110

1111

1101

1100

0100

0101

0111

0110

0010

0011

0001

0000.

This completes our description of the combinatorics of the Chinese Rings.

3. Topology of the Chinese Rings. In Figure 3 we illustrate an equivalent formulation of the Chinese Rings puzzle. Versions of the device are often manufactured in this form. The interlaced rings form a rigid wire frame, and the shuttle is replaced by a loop of rope laced around the frame in such a way that the loop is not topologically linked with the wire frame. The puzzle is to manipulate the loop (which is flexible and of sufficient length) to remove it from the frame without bending the frame. Among the possible configurations for the loop there is a subset in precise

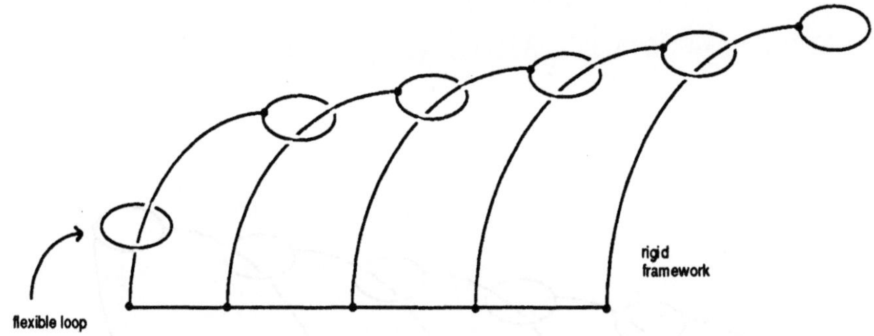

FIG. 3. Chinese Rings

correspondence with the shuttle states of the "classical" Chinese Rings of section 1. In the Figure we have placed the loop in the state 1000.

If both the wire frame and the loop are topologically flexible then the rings puzzle seems to lose all interest since the the rings can be successively disentangled and the loop is easily removed. Note however, that the rings cannot be disentangled in an arbitrary way. The first ring must be slipped through the second ring. This allows the second ring to be slipped through the third and so on. There is a hierarchical structure to this disentanglement.

Is it possible to reformulate the Chinese Rings so that it is a completely topological problem? Here is one answer to this question: *Yes!* Add the extra arc as shown in Figure 4 from ring number 1 to the base of the frame. Call this the *blue* arc. Call the loop in the puzzle the *red* loop. Call the wire-frame with blue arc and red loop the *Chinese Graph.* Call the wire-frame with blue arc the *augmented frame.* We now have the

Chinese Graph Problem. What is the least number of red-blue crossing exchanges needed to topologically disentangle the red loop from the augmented flexible frame? Call this number the *exchange number* of the Chinese Graph.

A *crossing exchange* is a local replacement in the graph embedding that replaces a crossing of one type with a crossing of the opposite type without changing the rest of the embedding. See the illustration below.

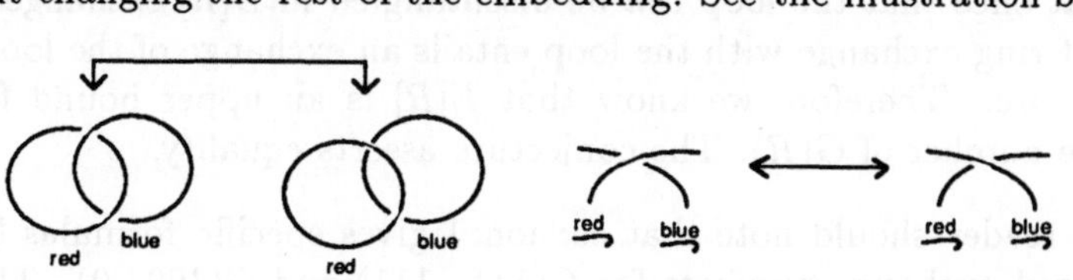

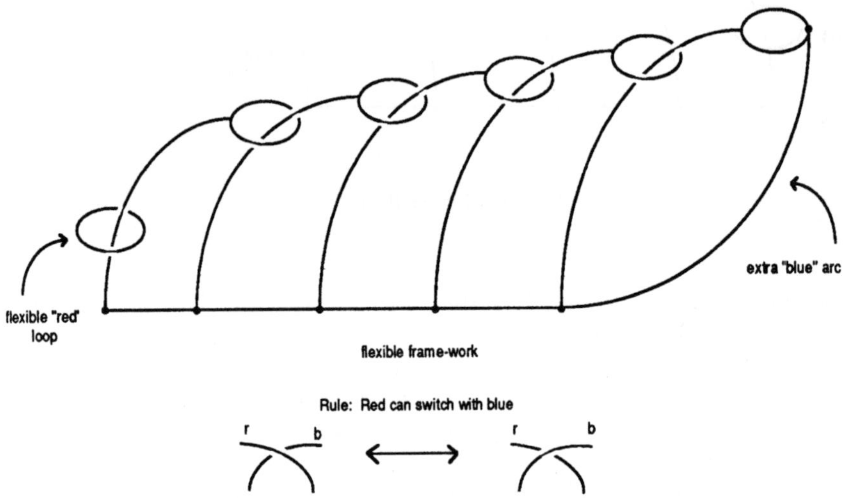

FIG. 4. Chinese Rings

The Chinese Graph Problem is a special uncrossing number problem. Rather than allowing arbitrary crossing exchanges, we only allow exchanges between red and blue. A result of this restriction is the following

Ring Conjecture. Let $G(R)$ denote a Chinese graph - ring and loop configuration corresponding to the binary string R as described in section 1. That is $G(R)$ is the augmented wire-frame and loop corresponding to the classical Chinese Rings device with shuttle state the binary string R. *Then the (topological) exchange number for $G(R)$ is equal to the (mechanical) exchange number $E[R]$.*

This conjecture is the crux of this paper. By the previous discussion it follows at once that the loop *can* be disentangled in $E[R]$ exchanges since each first ring exchange with the loop entails an exchange of the loop with the blue arc. Therefore we know that $E[R]$ is an upper bound for the exchange number of $G(R)$. The conjecture asserts equality.

The reader should note that section 1 gives specific formulas for the conjectured exchange numbers for $G(111...111)$ and $G(100...0)$. Thus we

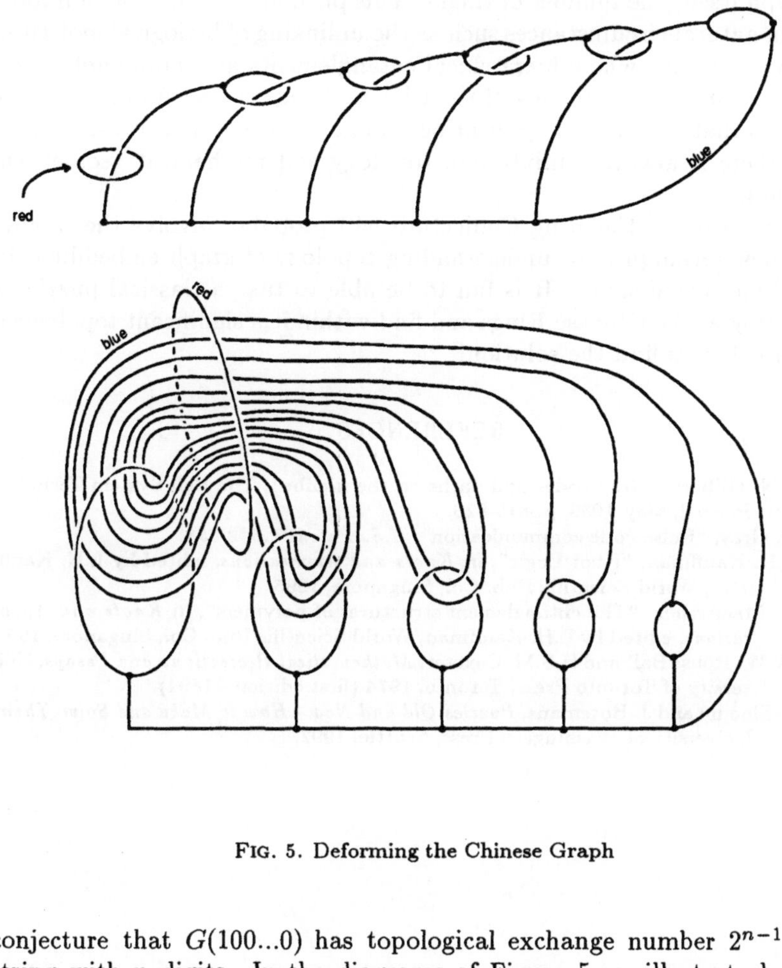

FIG. 5. Deforming the Chinese Graph

conjecture that $G(100...0)$ has topological exchange number 2^{n-1} for a string with n digits. In the diagrams of Figure 5 we illustrate how the exchange number of $G(10000)$ is less than or equal to 16 by deforming the graph embedding (compare [3]). The blue arc in the deformed graph becomes the trace of the shuttle operations of section 1. Here the upper bound is apparent through topological deformation for all the graphs $G(100...00)$.

In the case of $G(111...111)$ the conjecture states that the exchange number for the graph is equal to $(2^n + (-1)^{n+1})/3$. The reader may enjoy seeing the topological evidence for this one via graph deformations! In this case and those of other initial binary strings, the solution to the classical Chinese Rings puzzle is essential in obtaining the conjectured exchange number and rigorous upper bound.

4. Discussion. If the Ring Conjecture is true then we have produced examples of families of crossing exchange problems whose complexity (exchange number) grows exponentially as a function of the complexity of the graph itself (the number of rings). This phenomenon may be of importance in natural circumstances such as the unlinking of biological molecules or polymer chains where hierarchical entanglements are commonplace [4]. This problem in the topology of the Chinese Rings is a useful test case for questions that can arise in applications of knot theory to natural structures where there is always a mixture of topology and mechanical/geometrical modelling.

A solution to the Ring Conjecture will probably involve the discovery of new techniques for understanding topology of graph embeddings in three-dimensional space. It is fun to be able to take a classical puzzle as fascinating as the Chinese Rings and find within it a significant topological problem. Let us find the solution!

REFERENCES

[1] E.N. Gilbert, "Gray codes and paths on the n-cube", The Bell System Technical Journal, May 1958, p. 815-826.

[2] F. Gray, "Pulse code communication", *U.S. Patent 2.632.058.*

[3] L.H. Kauffman, "Knot Logic", in *Knots and Applications*, edited by L.H. Kauffman, World Scientific Pub. Co., Singapore, 1995.

[4] A. MacArthur, "The entanglement structures of polymers", in *Knots and Applications*, edited by L.H. Kauffman, World Scientific Pub. Co., Singapore, 1995.

[5] W.W. Rouse Ball and H.S.M. Coxeter, *Mathematical Recreations and Essays*, University of Toronto Press, Toronto, 1974 (first edition - 1894).

[6] J. Slocum and J. Botemans, *Puzzles Old and New - How to Make and Solve Them*, University of Washington Press, Seattle, 1992.

LATTICE INVARIANTS FOR KNOTS

E.J. JANSE VAN RENSBURG*

Abstract. The geometry of polygonal knots in the cubic lattice may be used to define some knot invariants. One such invariant is the minimal edge number, which is the minimum number of edges necessary (and sufficient) to construct a lattice knot of given type. In addition, one may also define the minimal (unfolded) surface number, and the minimal (unfolded) boundary number; these are the minimum number of 2-cells necessary to construct an unfolded lattice Seifert surface of a given knot type in the lattice, and the minimum number of edges necessary in a lattice knot to gaurantee the existence of an unfolded lattice Seifert surface. In addition, I derive some relations amongst these invariants.

Key words. Lattice Knots, Knot Complexity, Knot Invariants.

1. Introduction. Let $\mathcal{Z}^3$ be the cubic lattice with each vertex a point in $\mathcal{R}^3$ with integer coordinates. Augment the lattice by adding to it all edges between vertices with coordinates which differ in only one component. A *polygon* is a cycle (or an unlabeled vertex-avoiding closed path) in $\mathcal{Z}^3$. Define two polygons to be equivalent if they can be made identical by a translation or rotation. I abuse this terminology by calling the resulting equivalence classes *polygons*. Since the girth of $\mathcal{Z}^3$ is 4, the polygon with the least number of edges has length 4. Every polygon is a piecewise linear embedding of S^1 in $\mathcal{R}^3$, and is a tame knot. The equivalence class of polygons with fixed knot type K is called a *lattice knot* of type K. The lattice defines directions and a length scale in terms of the canonical unit vectors, and a polygon may be measured in terms of these.

A *2-cell* is a unit square with vertices in $\mathcal{Z}^3$. A collection of 2-cells in $\mathcal{R}^3$ is *vertex-avoiding* if every pair of 2-cells in the collection with a non-empty intersection are both incident on the same edge, or (together) have intersection with a third 2-cell which is a pair of incident perpendicular edges. A vertex-avoiding collection of 2-cells is a *surface* if it is connected. The boundary of a surface σ, $\partial\sigma$, is the union of a set of non-intersecting polygons. A *lattice Seifert surface* of a polygon representative of a lattice knot, P_K, is an orientable surface with P_K as its boundary. One may also define the Seifert surface of P_K as a surface in $\mathcal{R}^3$ with boundary P_K. The areas of both a lattice Seifert surface and a Seifert surface of P_K can be measured in terms of the length scale set by the lattice. It is not obvious that every polygon has a lattice Seifert surface, but by subdivision one sees that there exists a polygonal realisation of every knot type which admits a lattice Seifert surface.

The *bottom* and *top vertices* of a polygon are found by a lexicographic ordering of the vertices of the polygon. Similarly, the bottom and top

* Department of Mathematics and Statistics, York University, 4700 Keele Street, North York, Ontario, M3J 1P3, Canada.

vertices of a lattice surface are defined. The *bottom* and *top edges* of a polygon is found by a lexicographic ordering of edges by the coordinates of their midpoints. A lattice surface is said to be *unfolded* if its boundary contains its top and bottom vertices. In that case we can define a top (or bottom) edge as the lexicographic most (or least) edge in the boundary incident with the top (or bottom) vertex. By case analysis, one observes that the top and bottom edges, if they exist, are always perpendicular to the direction defined by the first component in coordinates in $\mathcal{R}^3$. In Figure 1 an unfolded surface with boundary the knot 3_1 is indicated. The knot (the boundary of the surface) has length 36, and the area of the lattice surface is 19.

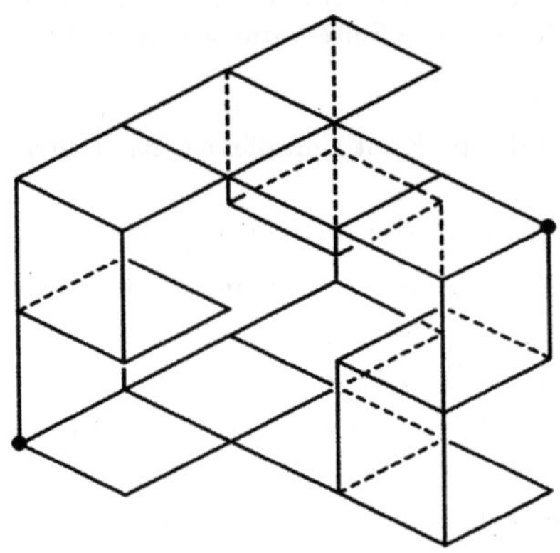

FIG. 1. An unfolded surface with boundary the trefoil knot. The top and bottom vertices are indicated.

It is possible to make the lattice knot 3_1 using only 24 edges, due to a theorem by Diao [1,2] (see Figure 2). Thus, there are two interesting questions suggested by Figure 1: (1) Which number of edges is necessary and sufficient in a realisation of a lattice knot of type K to gaurantee the existence of an (unfolded) lattice Seifert surface, and (2) what is the minimum area of an (unfolded) lattice Seifert surface. (In this manuscript I will focus only on unfolded surfaces, I can say more in that case). At this stage I have identified no less than three lattice invariants for knots: (1) The minimal edge number (the minimum number of edges necessary to realise a lattice knot of type K), indicated by $\mathcal{M}(K)$ [3]. (2) The minimal area of a unfolded lattice Seifert surface of lattice knot of type K, indicated by $\mathcal{A}(K)$. (3) The minimal number of edges necessary in the realisation of a lattice knot of type K which gaurantees the existence of an (unfolded) lattice Seifert surface, denoted $\mathcal{N}(K)$. Of course, the minimal area of a (non-lattice) Seifert surface of a lattice knot is also a geometric invariant.

The minimal edge number was considered in reference [3], and in Section 2 I briefly review the properties of this invariant. In Section 3 I consider the properties of $\mathcal{A}(K)$ and $\mathcal{N}(K)$.

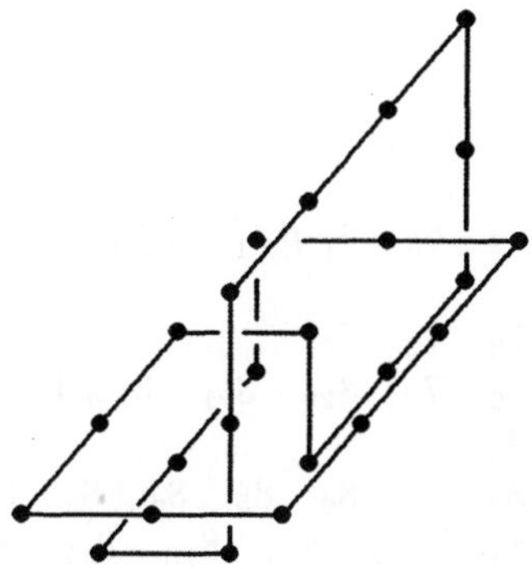

FIG. 2. A trefoil with minimal length ($\mathcal{M}(3_1) = 24$).

2. The Minimal Edge Number. Minimal edge numbers were estimated for prime knots up to 8 crossings [3] in the standard knot tables using simulated annealing overlying a Markov chain ergodic within state spaces which are polygons with fixed knot types. The Markov chain was realised using the BFACF algorithm [4], which is known to have ergodicity classes in the state space of unrooted polygons which are polygons of fixed knot type [5]. $\mathcal{M}(K)$ was also estimated for the knots 9_1, 9_2, 9_{42}, 9_{47}, 10_1, 10_{161}, $3_1\#4_1$ and the square and granny knots. In addition, it was also estimated for the compounded knots 3_1^n for $n \leq 10$; all factors with the same chirality. I show the results in Tables I and II. Observe that more extensive simulations on the knots 7_2 and 8_3 reduced the estimates of their minimal edge numbers by 2 each, improving the bounds in [3].

There seems to be a correspondence between minimal crossing numbers and minimal edge numbers in Table I; since any polygon with m edges may have at most $m(m-3)/2$ crossings in any projection, a lower bound on $\mathcal{M}(K)$ is $\sqrt{2C_K}$, where C_K is the minimal number of crossings of a knot of type K. The minimal edge numbers of various knots with $C_K = 8$ are spread over a wide range in Table I, which is probably a typical feature: There is no sharp connection between the minimal crossing number and the minimal edge number of a knot. Non-alternating knots appear to have anomolously low minimal edge numbers, as one witness by considering 8_{19}, 8_{20}, 8_{21}, 9_{42}, 9_{47} and 10_{161} in Table I. Other interesting patterns in Table I includes the minimal numbers of sequences of torus knots $T_{2,k}$, for $k = 3, 5, 7, 9$: these are 24, 34, 44 and 54, incrementing in steps of 10. This pattern probably arises because a basic pattern is repeated as k increases in $T_{2,k}$, one such possibility is illustrated in Figure 3; there are 10 edges per

TABLE I
$\mathcal{M}(K)$ Estimated by Simulated Annealing

$\mathcal{M}(K)$	K
4	0_1
24	3_1
30	4_1
34	5_1
36	5_2
40	6_1 6_2 6_3 $3_1^-\#3_1^-$ $3_1^+\#3_1^-$
42	8_{19}
44	7_1 7_2 7_3 7_7
46	7_4 7_5 7_6 8_{20} 8_{21} $3_1\#4_1$
48	8_7 9_{42}
50	8_1 8_2 8_3 8_4 8_5 8_6 8_9 8_{13} 9_{47} 10_{161}
52	8_8 8_{10} 8_{11} 8_{15} 8_{16} 8_{17} 8_{18}
54	8_{12} 8_{14} 9_1
56	9_2 $3_1^-\#3_1^-\#3_1^-$
60	10_1

twist, but with every eighth twist an additional 2 are needed. A similar sequence is observed when twist knots are considered; the knots 4_1, 6_1, 8_1, 10_1 constitutes one sequence and 5_2, 7_2, 9_2 a second, both sequences incrementing in steps of 10, presumable in the same fashion as for torus knots.

A second interesting feature of Table I is that the number of knots with $\mathcal{M}(K) = n$, where n is a fixed number, seems to increase rapidly with n. This increase is at most exponential, but I can also prove that it is at least exponential. The argument goes as follows: Let $q(n)$ be the number of distinct lattice knots with $\mathcal{M}(K) = n$, and let $Q(n) = \sum_{m \leq n} q(m)$. $q(n)$ increases at most exponentially with n, since the number of distinct polygons increases exponentially [6]. On the other hand, $Q(n)Q(m) \leq Q(n+m)$, since two knots with n and m edges, respectively, may be translated such that their top and bottom edges are identified, giving a compounded knot with $(n+m-2)$ edges. But this does not count prime knots with $\mathcal{M}(K) \leq (n+m)$. Thus, $Q(n)$ is supermultiplicative, and since $Q(n)$ is bound exponentially by the number of polygons with at most n edges, one gets that $\lim_{n\to\infty}(\log Q(n))/n = \log \mathcal{Q}$ exists and is finite. Moreover, $Q(n) \leq \mathcal{Q}^n$ [7]. The rate of exponential growth of $q(n)$ is the same as that of $Q(n)$; the data in Table I give the bound $\mathcal{Q} \geq 1.073$.

The minimal edge number (minus 2) is subadditive with respect to the connected sum of two knots K and L:

$$\mathcal{M}(K\#L) \leq \mathcal{M}(K) + \mathcal{M}(L) - 2. \tag{1}$$

This fact is easily seen: Let K and L be realisations of the lattice knots

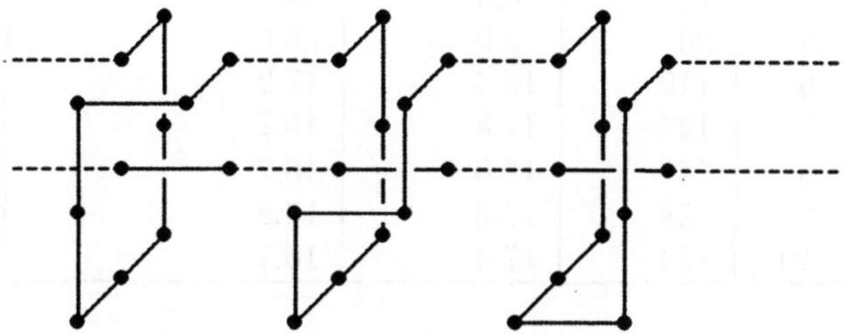

FIG. 3. Twisting two strands.

with lengths their respective minimal edge numbers. By translating and rotating L such that its bottom edge is identified with the top edge of K, one realises a lattice knot of type $K\#L$ with $\mathcal{M}(K)+\mathcal{M}(L)-2$ edges, and (1) follows.

A second interesting fact is that $\mathcal{M}(K^n)$ increases faster than $n\alpha_K$, where one may choose $\alpha_K = 4(b(K)-1)$ (and $b(K)$ is the bridge number of K), as n increases. One finds that

$$\mathcal{M}(K^n) \geq n\alpha_K, \tag{2}$$

where $\alpha_K > 0$ if and only if $K \neq \emptyset$, and where K^n is the connected sum of n copies of K. One may prove (2) by an argument involving curvature and bridge-number. Equations (1) and (2) imply the existence of the limit

$$\inf_n \frac{\mathcal{M}(K^n)-2}{n} = \lim_{n\to\infty} \frac{\mathcal{M}(K^n)-2}{n} = \mu_K \tag{3}$$

where $\mu_K \neq 0$ if and only if $K \neq \emptyset$ and μ_K is the *minimal edge index* of the knot K [4]. μ_K is a knot invariant. An additional result is that

$$2+n(\mathcal{M}(K)-2) \geq \mathcal{M}(K^n) \geq 2+n\mu_K. \tag{4}$$

Equation (4) suggests strongly that $\mathcal{M}(K)$ may be considered a measure of the complexity of knots; albeit of a subadditive nature. One observes naturally the following properties of $\mathcal{M}(K)$: $\mathcal{M}(\emptyset) = 4$ (since the girth of the lattice is 4), $\mathcal{M}(K) \geq 24$ for any knot other than the unknot [1,2], and for any pair of knots K and L, $\mathcal{M}(K^n\#L) \geq n\mu_K$ [3]. Indeed, the minimal edge number appears to count the number of factors in compounded knots.

TABLE II
$\mathcal{M}(3_1^n)$ Estimated by Simulated Annealing

n	$\mathcal{M}(3_1^n)$	$\mathcal{M}(3_1^n)/n$	$(\mathcal{M}(3_1^n) - 6.8)/n$
1	24	24.0	17.2
2	40	20.0	16.6
3	56	18.7	16.4
4	74	18.5	16.8
5	90	18.0	16.6
6	110	18.4	17.2
7	124	17.8	16.7
8	140	17.5	16.7
9	158	17.6	16.8
10	174	17.4	16.7

The minimal edge index μ_K is similarly a measure of knot complexity. It may be shown to have the following properties [3]: $\mu_\emptyset = 0$, and $\mu_K > 4$ if $K \neq \emptyset$. Moreover, $\mu_{K^n} = n\mu_K$ and $\mu_{K^n \# L} \geq n\mu_K$, for any pair of knots K and L. Thus, in exactly the same sense as above, μ_K is a subadditive measure of the complexity of a knot: $\mu_{K\#L} \leq \mu_K + \mu_L$. I conjecture this last inequality to be an *equality*. This will make μ_K an additive measure of knot complexity.

In Table II I list the number of components in 3_1^n in the first column, the minimal edge number in the second column and the bound predicted by the minimal edge number of the edge index in the third column. The best bound is obtained when $n = 10$, in which case one sees that $\mu_{3_1} \leq 17.4$. A better bound may be obtained by considering the construction in Figure 4: Each factor in the infinite string of knots takes exactly 17 edges, so $4 < \mu_{3_1} \leq 17$. The lower bound is inserted to emphasize the fact that the current bounds on the minimal indices are not very good.

There are interesting patterns in Table II. One observes that for $n = 3, 5, 7, 9$ the minimal edge numbers increment in steps of 34, and for $n = 1, 4, 7, 10$ and $2, 5, 8$ the increment is in steps of 50. Is there an infinite arithmetic subsequence of this kind? If there is, then μ_{3_1} is a rational number (if the increment is in steps of m every k factors, then μ_{3_1} equals m/k). If there is not, is there a positive density of finite progressions of this kind? Is it possible to show that μ_{3_1} is irrational?

I cannot answer any of these questions, but they do illustrate the inadequate nature of Figure 4. There, I attempt to take the limit in equation (3) as an infinite string of tight knots. However, it is unlikely that the limit is taken in that way: Instead, I expect "knots" to appear on different scales (fitting together like the pieces in a puzzle), resulting in a better packing than illustrated there. If that is the case, then infinite arithmetic progressions are presumably ruled out. An attempt to extrapolate the data in Table II is made in its last column, by a least squares linear fit. That

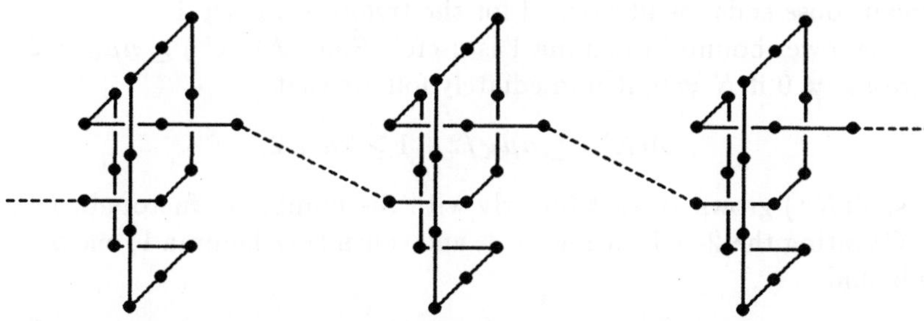

FIG. 4. An infinite string of tight trefoils.

suggests a limiting value of around 16.8, but the sequence is so short that I do not have much faith in that estimate.

3. The Minimal Unfolded Lattice Seifert Surface. In this section two lattice invariant for knots, derived from a lattice version of its Seifert surface, are explored. In Figure 1 I illustrate an unfolded lattice Seifert surface; its top and bottom vertices are in its boundary. An interesting observation is that every 2-cell in Figure 1 has one or two edges in the boundary; any 2-cell with three edges in the boundary may be removed without changing the topology of the surface. Therefore, in the best possible situation, each 2-cell will have two edges in the boundary, and the minimal possible area is $\mathcal{M}(K)/2$.

On the other hand, an unfolded Seifert surface may always be constructed by taking a Seifert surface of minimal area in $\mathcal{R}^3$, and then embedding it in $\mathcal{Z}^3$ by subdivision if necessary and pushing the surface onto the lattice. Finally, one must arrange matters such that the lattice surface is unfolded. This is easily done by subdividing if necessary, and removing a shortest linear strip of 2-cells between the top vertex and the boundary, and between the bottom vertex and the boundary. Thus, every knot has an unfolded lattice Seifert surface:

LEMMA 1. There exists a minimal unfolded lattice Seifert surface for any knot (of type K). Moreover, if the area of the minimal lattice Seifert surface is $\mathcal{A}(K)$, then $\mathcal{A}(K) \geq \mathcal{M}(K)/2$. □

One may remove 2-cells from a lattice Seifert surface until every 2-cell has at least one vertex in the boundary: Suppose that σ is a 2-cell which has at least one edge in the boundary of the knot, and suppose that σ is incident (via an edge or a vertex) with a 2-cell τ which is disjoint with the boundary. Then a case analysis shows that one may remove σ without changing the topology of the surface. Then τ will be incident with the boundary. This process may be continued until one cannot find a τ disjoint with the boundary, that is, every 2-cell is incident with the boundary. One thus expects that minimal Seifert surfaces in the lattice will consists of ribbons of 2-cells joined together at their ends into a "branched ribbon" without loose ends, as illustrated for the trefoil in Figure 1.

The lower bound in Lemma 1 is useful: Since $\mathcal{M}(K^n) \geq n\mu_K + 2 > 0$ where $\mu_K \neq 0$ if $K \neq \emptyset$, it immediately follows that

$$\mathcal{A}(K^n) \geq n\mu_K/2 + 1 \geq 2n + 2. \tag{5}$$

Thus, $\mathcal{A}(K^n)$ grows at least linearly with the number of factors in K^n.

Counting the 2-cells in Figure 1, and consulting Lemma 1, one obtains the bounds

$$12 \leq \mathcal{A}(3_1) \leq 19. \tag{6}$$

Since $\mathcal{M}(K) \geq 24$ if $K \neq \emptyset$ [1,2], I also conclude that $\mathcal{A}(K) \geq 12$ if $K \neq \emptyset$, and one notes that $\mathcal{A}(\emptyset) = 1$. Other minimal areas can be estimated by constructing them explicitly. I found for example that

$$15 \leq \mathcal{A}(4_1) \leq 23. \tag{7}$$

The example for the upper bound is given in Figure 5. 4_1 is an amphichireal knot, and the realisation of the lattice Seifert surface in Figure 5 is also amphichireal in a geometric sense: A reflection through the broken horisontal line leaves it unchanged.

Unfolded Seifert surfaces may be concatenated by translating the first surface, and rotating it if necessary, such that its bottom edge coincides with the top edge of the second surface. The new surface is also an unfolded surface, since its top and bottom edges are the top edge of the first surface, and the bottom edge of the second surface, respectively. Moreover, the boundary of the new surface is the connected sum of the boundaries of the first and second surfaces; and thus, the new surface is an unfolded lattice Seifert surface of its boundary. If $\sigma(K)$ and $\sigma(L)$ are minimal unfolded lattice Seifert surfaces of the knots K and L respectively, then concatenation produces an unfolded lattice Seifert surface of the knot $K\#L$, which is not necessarily minimal. Thus

$$\mathcal{A}(K_1\#K_2) \leq \mathcal{A}(K_1) + \mathcal{A}(K_2) \tag{8}$$

and I conclude that $\mathcal{A}(K)$ is a subadditive function with respect to the connected sum of knots. Together with equation (5) I establish the following theorem:

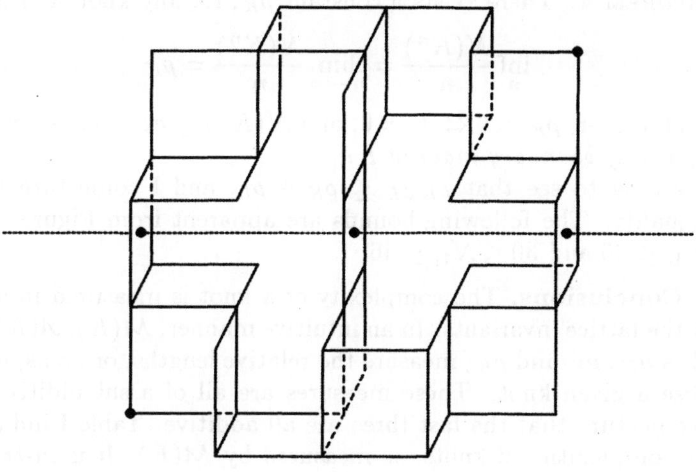

FIG. 5. $\mathcal{A}(4_1) \leq 23$.

THEOREM 2. There exists a constant ν_K, for every knot of type K, such that

$$\inf_n \frac{\mathcal{A}(K^n)}{n} = \lim_{n\to\infty} \frac{\mathcal{A}(K^n)}{n} = \nu_K$$

where $\nu_K = 0$ if and only if $K = \emptyset$, and $\mathcal{A}(K^n) \geq n\nu_K$, for all n. ν_K is the *(lattice) surface index* of K. □

Theorem 2 is an immediate consequence of the application of the theory of subadditive functions to equations (5) and (7). It is easily seen that $\nu_\emptyset = 0$, by direct construction. Since $\mu_K > 4$ if K is not trivial, I have $\nu_K > 2$ if K is not trivial. Moreover, $\nu_{K^n} = n\nu_K$, and $\nu_{K\#L} \leq \nu_K + \nu_L$, so that the surface index is subadditive. I expect it to be additive.

The minimal boundary number of a knot, $\mathcal{N}(K)$, may be treated similarly. If σ_K is a minimal unfolded lattice Seifert surface of a knot of type K, then its boundary $\partial\sigma_K$ is a polygon of knot type K, and therefore has at least $\mathcal{M}(K)$ edges. On the other hand, Lemma 1 implies that there is a finite minimal boundary number for any knot K, so we conclude that:

LEMMA 3. There exists a minimal boundary number $\mathcal{N}(K)$ for any knot K, such that $\mathcal{N}(K) \geq \mathcal{M}(K)$. □

By concatening two minimal unfolded lattice Seifert surfaces which realises the minimal boundary numbers of two knots K and L, one finds a subadditive relation:

$$\mathcal{N}(K\#L) \leq \mathcal{N}(K) + \mathcal{N}(L). \tag{9}$$

An immediate consequence of Lemma 3 and equation (9) is the following theorem:

THEOREM 4. There exists a constant ρ_K, for any knot K, such that

$$\inf_n \frac{\mathcal{N}(K^n)}{n} = \lim_{n\to\infty} \frac{\mathcal{N}(K^n)}{n} = \rho_K$$

exists. Moreover, $\rho_K = 0 \Leftrightarrow K = \emptyset$, and $\mathcal{N}(K^n) \geq n\rho_K$, for all $n > 0$. ρ_K is the *(lattice) boundary index* of K. □

It is easy to see that $\rho_{K\#L} \leq \rho_K + \rho_L$, and I conjecture that this is an equality. The following bounds are apparent from Figures 1 and 5: $24 \leq \mathcal{N}_{3_1} \leq 36$ and $30 \leq \mathcal{N}_{4_1} \leq 46$.

4. Conclusions. The complexity of a knot is measured in a natural way by the lattice invariants. In an intuitive manner, $\mathcal{M}(K)$, $\mathcal{A}(K)$, $\mathcal{N}(K)$, as well as μ_K, ν_K and ρ_K, measure the relative lengths, or areas, necessary to realise a given knot. These measures are all of a subadditive nature, and I conjecture that the last three are all additive. Table I indicates the relative complexities of knots, as measured by $\mathcal{M}(K)$. It is instructive to compare this table with rankings obtained by other measures, such as the Mobius energies of knots [8]. That spectrum generally reflects the ranking in Table I, but since energies are real numbers, the ranking contains no ties. In increasing energies, the knots are 0_1, 3_1, 4_1, 5_1, 5_2, 6_1, 6_2, 6_3, 7_1, 7_2, 7_3, 8_{19}, 7_4, ... The knot 8_{19}, which one would reasonably expect to have a higher energy than any knot with crossing number 7, appears nevertheless before some knots with crossing number 7. The ranking by $\mathcal{M}(K)$ compares well with this progression, except that 8_{19} appears even earlier in the sequence.

Acknowledgements. This research was supported by an operating grant from NSERC Canada.

REFERENCES

[1] Y. Diao, *Minimal Knotted Polygons on the Cubic Lattice*, J. of Knot Theory and its Ramifications, **2** (1993) 413-425.

[2] Y. Diao, *The Number of Smallest Knots on the Cubic Lattice*, J. Stat. Phys., **74** (1994) 1247-1254.

[3] E.J. Janse van Rensburg and S.D. Promislow, *Minimal Knots in the Cubic Lattice*, J. of Knot Theory and its Ramifications, **4** (1995) 115-130.

[4] B.Berg and D. Foester, *Random Paths and Random Surfaces on a Digital Computer*, Phys. Lett., **106B** (1981) 323-326. C. Aragao de Carvalho, S. Caracciolo and J. Fröhlich, *Polymers and $g|\phi|^4$-theory in Four Dimensions*, Nucl. Phys. B [FS7], **215** (1983) 209-248.

[5] E.J. Janse van Rensburg and S.G. Whittington, *The BFACF algorithm and Knotted Polygons*, J. Phys. A: Math. Gen., **24** (1991) 5553-5567.

[6] J.M. Hammersley, *The Number of Polygons on a Lattice*, Proc. Camb. Phil. Soc., **57** (1961) 516-523.

[7] E. Hille, *Functional Analysis and Semi-Groups*, Am. Math. Soc. Colloq. Publ. **31** (1948) (American Mathematical Society: New York).

[8] R.B. Kusner and J.M. Sullivan, *Möbius Energies for Knots and Links, Surfaces and Submanifolds*, Mathematical Sciences Research Institute Preprint No. 026-94 (1994).

TOPOLOGY AND GEOMETRY OF BIOPOLYMERS

E.J. JANSE VAN RENSBURG*, ENZO ORLANDINI†, DE WITT SUMNERS‡, M. CARLA TESI†, AND STUART G. WHITTINGTON†

Abstract. This paper is concerned with some simple lattice models of the entanglement complexity of polymers in dilute solution, with special reference to biopolymers such as DNA. We review a number of rigorous results about the asymptotic behaviour of the knot probability, the entanglement complexity and the writhe of a lattice polygon (as a model of a ring polymer) and discuss Monte Carlo results for intermediate length polygons. In addition we discuss how this model can be augmented to include the effect of solvent quality and ionic strength. We also describe a lattice ribbon model which is able to capture the main properties of an oriented ribbon-like molecule (such as duplex DNA).

1. Introduction. Biologically important polymers such as DNA and RNA share with synthetic polymers a number of essential features. They are long flexible objects which can be thought of as a set of monomers, connected together in a linear or circular arrangement. For RNA, base pairing can produce more complex spatial structures, including stem-loop structures and pseudoknots. DNA can exist both in the single-stranded and double-stranded (duplex DNA) state and single-stranded DNA is much more flexible than duplex DNA. The flexibility of these biopolymers implies that entropic effects are important in determining their conformational properties, though these can also be strongly influenced by the forces acting between the monomers. For instance, since DNA and RNA are highly charged polyelectrolytes, their properties [1,2], and particularly their entanglement complexity [3,4,5,6], will be sensitive to the ionic strength of the ambient solution in which they are contained. Being long, thread-like, periodically attached to proteins, and possessing helically intertwined backbone strands, DNA is subject to severe topological entanglement problems in the course of replication, transcription and other cellular processes. Controlling topological entanglement in DNA is therefore critically important in the functioning of the cell, and there exist enzymes such as *topoisomerases* and *recombinases* which facilitate cellular metabolism by changing the topological state of the DNA. These enzymes perform microscopic feats of topology. The enzyme topoisomerase passes DNA through itself via enzyme-bridged transient breaks in the DNA [7,8,9,10]; the enzyme recombinase breaks a pair of duplex DNA strands apart, reconnecting the DNA to different ends [11].

* Department of Mathematics, York University, Downsview, Ontario, Canada M3J 1P3.

† Department of Chemistry, University of Toronto, Toronto, Ontario, Canada M5S 1A1.

‡ Department of Mathematics, Florida State University, Tallahassee, Florida 32306-3027, USA.

Circular DNA molecules are known to occur in knotted and linked forms, both *in vitro* as an experimental assay for enzyme action [12] and as products of topoisomerase action in experiments [13], and *in vivo* as a side product of metabolic processes such as transposition [14,15] and in linked networks of kinetoplast DNA in some pathogens [16]. The *topological enzymology* experimental protocol [12] uses the topological invariance of knotted and linked DNA to study the action of enzymes which manipulate DNA topology. In such experiments, unknotted circular DNA substrate is reacted *in vitro* with an enzyme, and the products of such a reaction may include knotted and linked DNA. The reaction products form a unique enzyme signature; information on enzyme binding and enzyme mechanism can be deduced from the spectrum of product DNA knots and links [17].

In some experiments, one can regard random thermodynamic DNA knotting as "noise" which must be subtracted out to obtain an enzymatic signal. Experiments of this type have increased interest in the probability of ring polymers being knotted but, in spite of a considerable body of work, the phenomenon is still not well understood. Frisch and Wasserman [18] and Delbruck [19] conjectured that sufficiently long ring polymers would be knotted with probability one, and since then the problem has been studied numerically by several groups [20,21]. In 1988 the validity of the Frisch-Wasserman-Delbruck conjecture was established rigorously, for a lattice model of a polymer, by Sumners and Whittington [22] and, independently, by Pippenger [23]. For a self-avoiding polygon with n edges on the simple cubic lattice, the probability $P(n)$ that the polygon is knotted goes to unity exponentially rapidly as n goes to infinity. This result immediately suggests further questions. For instance, is the knot likely to be prime or composite? How do measures of the entanglement complexity, such as the average value of the crossing number or unknotting number, depend on n? These questions have been addressed by Soteros, Sumners and Whittington [24]. Section 2 gives a brief review of some pertinent results on self-avoiding walks, and Section 3 gives an account of some rigorous results on knot probabilities in polygons.

Even when a polygon is unknotted it can still be self-entangled in a geometrical sense. For instance, unknotted circular DNA can be supercoiled. Linear DNA can be highly self-entangled and entangled with other DNA molecules. An interesting measure of this type of entanglement is the *writhe* of the molecule, and this quantity has been used in modelling supercoiling in DNA [25]. In Section 4 we describe some rigorous results about the n-dependence of the writhe of lattice polygons and self-avoiding walks.

Polygons and related models are useful for describing the overall dimensions of the molecule, and for investigating some measures of the entanglement complexity. However, DNA is a double stranded molecule, with the two strands wrapping around one another in a right-handed helical fashion. In circular DNA the strands can be linked and the linking number is a topological invariant, unchanged in a physical process unless at least one of the backbone strands is broken and subsequently rejoined. To investigate these linking properties one needs an extension of the polygon model; this leads to an interest in ribbon models of circular DNA [25]. The (orientable) ribbon is a surface with two boundary curves which can be linked. There is an important conservation theorem [26] which relates the writhe of the backbone of the ribbon, the twist of one boundary curve about the other, and the linking number of the two boundary curves. In Section 5 we discuss a *lattice ribbon* model where the ribbon is made up of a set of *plaquettes*. The lattice model has the advantage that the methods which have been developed for the polygon model can be extended to prove theorems about the asymptotic behaviour of various complexity measures for the ribbon.

The results which will be discussed in Sections 3, 4 and 5 are all asymptotic in nature. They say rather little about knotting and writhing for small values of n. If one is interested in the behaviour for small n, there are at present no viable theoretical methods; the most useful current approaches are Monte Carlo simulation methods. A number of different Monte Carlo approaches have been used but perhaps the most promising is to sample along a realization of a Markov chain defined on the set of polygons. Recently Madras *et al* [27] and Janse van Rensburg *et al* [28] have developed a pivot algorithm for polygons, and this algorithm has been used [29] to investigate the n-dependence of the knot probability for polygons. Using similar methods one can investigate the n-dependence of the writhe of a polygon [30], and the algorithm can be adapted to estimate interesting properties of lattice ribbons, including the average linking number of the boundary curves. This work is described in Section 6.

Finally, in Section 7, we give a brief summary and some suggestions for future research.

2. Self-avoiding walks and polygons. In order to understand most of the rigorous results which have been obtained on knotting and writhing, one needs some background knowledge of self-avoiding walks. In this section we review some of the necessary results, and their relation to polygons. We shall be chiefly concerned with Z^3, the simple cubic lattice. The vertices of this lattice are the integer points in R^3, and the edges join pairs of vertices which are unit distance apart. A *walk* is a directed sequence of edges, such that adjacent pairs of edges in the sequence are incident on a common vertex. A walk is *self-avoiding* if no vertex is visited more than once. Two walks will be considered distinct if they cannot be superimposed (with due regard for their directed nature) by translation. If we write c_n for the number of self-avoiding walks with n edges on Z^3, it is easy to see that $c_1 = 6$, $c_2 = 30$, $c_3 = 150$, $c_4 = 5 \times c_3 - 24 = 726$, $c_5 = 3534$, etc. Perhaps the most important result about the asymptotic behaviour of c_n is the following theorem due to Hammersley [31]:

THEOREM 2.1.

$$0 < \lim_{n\to\infty} n^{-1} \log c_n = \inf_{n>0} n^{-1} \log c_n \equiv \kappa < \infty \tag{2.1}$$

where κ *is called the connective constant of the lattice.*

Proof: The proof uses the elegant idea of concatenation of two walks to derive a sub-multiplicative inequality for c_n. Consider a self-avoiding walk of m steps and a self-avoiding walk of n steps. Translate these on the lattice so that the last vertex of the first walk coincides with the first vertex of the second walk. The first walk can be chosen in c_m ways and the second in c_n ways, and the graphs resulting from the concatenation construction include all self-avoiding walks of $m + n$ steps, so that

$$c_m c_n \geq c_{m+n}. \tag{2.2}$$

Since every walk consisting only of steps in the *positive* coordinate directions is self-avoiding, $c_n \geq 3^n$, so that $\liminf_{n\to\infty} n^{-1} \log c_n \geq \log 3 > 0$. This, together with (2.2), implies (2.1).

Since we are interested in the properties of ring polymers the corresponding model is a self-avoiding polygon, i.e. an embedding of the circle graph in Z^3. We write p_n for the number of (undirected, unrooted) self-avoiding polygons embeddable in a lattice, where we count two polygons as distinct if they cannot be superimposed by translation. For instance, on the simple cubic lattice, $p_4 = 3$, $p_6 = 22$, $p_8 = 207$ and $p_{10} = 2412$. Hammersley [32] has shown that

$$\lim_{n\to\infty} n^{-1} \log p_n = \kappa \tag{2.3}$$

where κ is the connective constant defined in equation (2.1). That is, the numbers of polygons and self-avoiding walks grow at the same exponential rate.

3. Rigorous results on knot probabilities in polygons. We begin by describing a rigorous result which says that almost all sufficiently long polygons are knotted. The proof requires three separate results. The first is a pattern theorem due to Kesten [33]. A *pattern* is any finite self-avoiding walk. Given a particular pattern γ, if there exists a self-avoiding walk on which the pattern γ appears three times then we call γ a *K-pattern*. Roughly speaking, Kesten's theorem says that every K-pattern will appear with positive density on almost all sufficiently long self-avoiding walks. The precise statement is as follows:

THEOREM 3.1. *For any K-pattern* γ, *let* $c_n(\epsilon;\gamma)$ *be the number of n-step self-avoiding walks on which* γ *appears at most* ϵn *times. Then there exists a value of* $\epsilon > 0$ *for which* $\limsup_{n\to\infty} n^{-1} \log c_n(\epsilon;\gamma) < \kappa$.

We now need a corresponding result for polygons. Let γ be an undirected pattern. We now say that γ is a K-pattern if there exists a self-avoiding walk on which one of the two directed versions of the pattern appears three times.

THEOREM 3.2. *Let $p_n(\epsilon;\gamma)$ be the number of n-edge polygons on which the K-pattern γ appears at most ϵn times. Then there exists some $\epsilon > 0$ for which*

$$\limsup_{n\to\infty} n^{-1} \log p_n(\epsilon;\gamma) < \kappa.$$

Proof: Define the *bottom vertex* of a polygon by lexicographic ordering. The vertex has two incident edges, each incident on a second vertex. Choose the lower of these two vertices (lexicographically) and delete the corresponding edge. This gives an (undirected) self-avoiding walk with $n-1$ edges. Since deleting an edge cannot create a pattern, the theorem follows from Theorem (3.1) and equation (2.3).

The second ingredient in the proof is the idea of a knotted arc. We can capture the important part of a knot such as a trefoil and tie it so tightly on the lattice that the rest of the walk cannot pass through the neighbourhood of this subwalk and untie the knot. Technically, the subwalk and its associated dual 3-cells form a *knotted ball pair*, i.e. a ball pair (B^3, B^1) which is not ambient isotopic to the standard ball pair. These knotted ball pairs can be constructed so that the subwalk (B^1) is a K-pattern, leading to the following Lemma: [24]

LEMMA 3.1. *Every knot type is represented by a K-pattern on Z^3 such that the K-pattern and its associated dual 3-cells form a knotted ball pair.*

Finally we need a standard result [34] in knot theory:

LEMMA 3.2. *For a given non-trivial knot k there is no knot l such that the connected sum $k\#l$ is unknotted.*

We are now ready to state the primary result of this section as a pair of theorems:

THEOREM 3.3. *If p_n^o is the number of unknotted polygons with n edges then*

$$\lim_{n\to\infty} n^{-1} \log p_n^o = \kappa_o \tag{3.1}$$

Proof: When two unknotted polygons are concatenated the resulting polygon is unknotted. This shows that

$$p_n^o p_m^o \leq p_{n+m}^o \tag{3.2}$$

which, together with the obvious bound $p_n^o \leq 6^n$, is enough to establish the existence of the limit (3.1).

THEOREM 3.4. *The probability $P(n)$ that an n-gon is knotted is given by*

$$P(n) = 1 - \exp(-\alpha n + o(n)) \tag{3.3}$$

where $\alpha = \kappa - \kappa_o > 0$.

Proof: The set of unknotted polygons with n edges is a subset of the polygons which do not contain a trefoil, which, in turn, are a subset of the polygons which do not contain the K-pattern associated with a trefoil. Hence $\kappa_o < \kappa$. Since $P(n) = 1 - p_n^o/p_n$, this gives eqn (3.3).

Instead of using the K-pattern associated with a trefoil, we can construct a corresponding pattern for any other knot. In particular we can consider the pattern corresponding to two copies of the trefoil. Since this must appear on almost all polygons, prime knots must be exponentially rare compared to composite knots. Indeed, almost all sufficiently long polygons are very badly knotted.

This immediately allows us to say something about the numbers of polygons with *fixed* knot type. Suppose that $p_n(3_1)$ is the number of n-gons which are trefoils. What can we say about the asymptotic behaviour for large n? The above argument implies that

$$\limsup_{n\to\infty} n^{-1} \log p_n(3_1) < \kappa. \tag{3.4}$$

By concatenating a fixed trefoil with an unknotted polygon, it is easy to see that

$$\liminf_{n\to\infty} n^{-1} \log p_n(3_1) \geq \kappa_o \tag{3.5}$$

but the existence of the limit has not been established. If k is a (composite) knot which has l as a factor, similar arguments can be used to show that

$$\limsup_{n\to\infty} n^{-1} \log p_n(l) \leq \liminf_{n\to\infty} n^{-1} \log p_n(k). \tag{3.6}$$

Soteros *et al* [24] have used similar arguments to investigate how measures of the knot complexity (such as the average crossing number, average unknotting number, genus, etc.) grow for large n. Measures of knot complexity which are more or less additive with respect to knot composition can be shown to increase at least as fast as n.

Similar but slightly weaker results have been proved for Gaussian random polygons in R^3 [35] and for equilateral polygons in R^3 [36].

4. The writhe of a polygon. The work described in the previous section is concerned with knotting as a measure of the entanglement complexity of a polygon. It is clear that even unknotted polygons and linear chains can be badly embedded in the sense that they can have many crossings in every projection. One useful measure of this kind of embedding complexity is the *writhe* of the polygon. Consider the planar projection of a particular polygon in Z^3 in a given direction. In general the projection will contain crossings and, for almost all projection directions, all of the crossings will be transverse. In this case one can attach a sign to the crossing (+1 or -1) according to the right-hand rule convention as illustrated in Figure 1. The writhe of the polygon is the sum of the signed crossing numbers, averaged over all projection directions. For self-avoiding polygons in Z^3, the computation of the writhe is greatly simplified by a theorem [37]

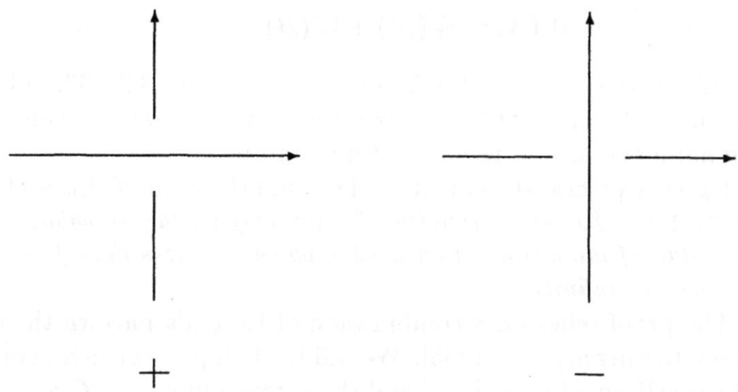

FIG. 4.1. *The sign convention for crossings*

which reduces the writhe computation to the average of linking numbers of the given curve with four selected pushoffs. This is a key result for the remainder of this section. Incidentally, since linking number is an integer, this theorem implies that four times the writhe is an integer!

The main result which we shall describe in this section concerns the asymptotic behaviour of the average writhe of a polygon with n edges, in the limit of large n. Roughly speaking, the result is that the average writhe increases at least as fast as $\sqrt{n}$ [30]. To establish this we shall require a number of steps. First we consider a particular polygon and compute its writhe. Consider the two points $P = (0,0,0)$ and $Q = (0,1,0)$, and the solid cube C of size 2x2x2 centered at $(1,0,0)$. Both P and Q are on the boundary of C. Let i, j, k denote unit vectors in a right-handed coordinate system for Z^3. Consider the self-avoiding polygon B which is contained in C and described as follows: begin at P, and take the following sequence of steps in the directions $\{i, i, -k, -j, -i, j, j, k, -i, -j\}$. Let $W(B)$ denote the writhe of curve B. It is a straight-forward computation (using the theorem of Lacher and Sumners [37] to show that

$$W(B) = +\frac{1}{2}. \tag{4.1}$$

If B^* denotes the mirror image of B (reflected in the plane $z = 0$), then B^* lies in C, and $W(B^*) = -1/2$.

Suppose now that A is a self-avoiding polygon which (given one of its two orientations) intersects cube C only in the self-avoiding walk B' which begins at P and ends at Q and traverses all the steps of B except the last one. We can truncate the polygon A by deleting the 9 steps of B' and adding in the step i which connects P to Q on the boundary of C. This gives a new polygon A'.

LEMMA 4.1. *The writhes of these polygons are related by the equation*

$$W(A) = W(A') + W(B). \tag{4.2}$$

The details of the proof of this Lemma are given in [30]. The idea is to make use of the theorem of Lacher and Sumners, so that one can work with linking numbers, and to use an isotopy argument.

We next give a precise statement of the main theorem of this section:

THEOREM 4.1. *For every function $f(n) = o(\sqrt{n})$, the probability that the absolute value of the writhe of an n-edge polygon is less than $f(n)$ goes to zero as n goes to infinity.*

Proof: The proof relies on a combination of Kesten's pattern theorem [33], and a coin tossing argument [35]. We call the ball pair consisting of any translate of the self-avoiding walk B' and the surrounding cube C a *pattern* $\mathcal{P} = \{C, B'\}$. Let the pattern $\mathcal{P}^*$ be the ball pair $\{C, B'^*\}$ where B'^* is the mirror image of B' (reflected in the plane $z = 0$). Kesten's theorem implies that there exists a positive number ϵ such that for all except exponentially few sufficiently long self-avoiding n-edge polygons, there are at least $\lfloor \epsilon n \rfloor$ pairwise disjoint translates of C each of which intersects the polygon in a translate of B' or B'^*. The distribution of the two patterns $\mathcal{P}$ and $\mathcal{P}^*$ is analogous to tossing a coin, because B' and B'^* occur independently with probability $\frac{1}{2}$ in each of the $\lfloor \epsilon n \rfloor$ translations of the cube C. Consequently the probability that B' occurs exactly k times among the $\lfloor \epsilon n \rfloor$ occurrences of either B' or B'^* is less than $1/\sqrt{\lfloor \epsilon n \rfloor}$ for every $k \leq \lfloor \epsilon n \rfloor$ provided that n is sufficiently large. (This can be shown by applying Stirling's approximation to the binomial distribution.) The fraction of polygons with at least $\lfloor \epsilon n \rfloor$ occurrences of either $\mathcal{P}$ or $\mathcal{P}^*$ is at least $1 - e^{-\gamma n}$ for some positive γ. For each of these polygons, the writhe is the sum of two terms. The first term is from the polygon formed by truncating $\lfloor \epsilon n \rfloor$ times (which is some fixed number), and the second is from the $\lfloor \epsilon n \rfloor$ copies of B or B^* formed in these truncations. If the total writhe is numerically less than $f(n)$ then the contribution to the writhe from the $\lfloor \epsilon n \rfloor$ occurrences of the patterns must be one of at most $\lceil 2f(n) + 1 \rceil$ different values. Hence

$$Prob(|W| < f(n)) \leq \frac{(1 - e^{-\gamma n})(\lceil 4f(n) + 1 \rceil}{\sqrt{\lfloor \epsilon n \rfloor}} \tag{4.3}$$

which goes to zero as $n \to \infty$ if $f(n) = o(\sqrt{n})$.

The proof of Theorem 4.1 can be used to prove some interesting corollaries [35] regarding random polygons. In particular, almost all random polygons are chiral and lie in a non-trivial cobordism class. Let $\sigma(K)$ denote the *signature* of the knot K [38]. The signature is computed from a Seifert linking matrix associated with an orientable Seifert surface spanning the knot K. The signature function takes on both positive and negative integral values, and is additive under knot composition. In fact, if K^* denotes the mirror image of K, then $\sigma(K^*) = -\sigma(K)$, so that $\sigma(K \# K^*) = 0$.

Using the fact that the signature of the (+)trefoil is +2 and the signature of its mirror image, the (-)trefoil is -2, and the fact that the occurrence of the (+) or (-) trefoil is equally likely in a long self-avoiding walk, one can prove

COROLLARY 1. *For every function $f(n) = o(\sqrt{n})$, the probability that the absolute value of the signature of an n-edge polygon is less than $f(n)$ goes to zero as n goes to infinity.*

The knot K is *chiral* if K is inequivalent to K^*. A necessary condition for chirality is that the signature vanishes. We immediately obtain the following:

COROLLARY 2. *All but exponentially few sufficiently long polygons are chiral knots.*

Two oriented knots K and L are *cobordant* if K and L^* form the ends of an annulus properly embedded in $S^3 \times I$ [38]. The set of cobordism classes of oriented knots forms a group under the operation of composition. The signature is a homomorphism from the cobordism group of knots to the additive group of the integers.

COROLLARY 3. *All but exponentially few sufficiently long polygons are in a non-trivial knot cobordism class.*

5. Lattice ribbons. In this section we shall describe a modification of the polygon models in which the polygon is replaced by a ribbon made up of a sequence of plaquettes. We shall be concerned with the simple cubic lattice Z^3 and we define a plaquette as a unit square, with vertices having integer coordinates. We define an *open ribbon* as an ordered sequence of plaquettes labelled $i = 1, 2, \ldots n$ such that

1. every two adjacent plaquettes ($|i - j| = 1$) in the sequence have a common edge
2. two plaquettes (i and j) cannot be incident on a common edge unless $|i - j| = 1$
3. two non-adjacent plaquettes cannot be incident on a common vertex unless they are also incident on a common plaquette, and
4. not more than three plaquettes can be incident on a common vertex.

We call the the number of edges which a plaquette has in common with other plaquettes the *degree* of the plaquette. In an open ribbon the first and last plaquettes have degree 1, and all other plaquettes have degree 2. We write w_n for the number of open ribbons with n plaquettes, where two ribbons are considered distinct if they can not be superimposed by translation.

A *directed rooted closed ribbon* with n plaquettes is an ordered sequence of n plaquettes, $i = 1, 2, \ldots n$, obeying conditions similar to those given above, except that conditions 1 and 2 must be interpreted modulo n. Any cyclic permutation (and the reverse permutation, and any cyclic permutation of the reverse permutation) is also a directed rooted closed ribbon, and the resulting set of ribbons can be regarded as a single geometrical object which we call an undirected unrooted closed ribbon or, simply, a *closed* ribbon. Every plaquette in a closed ribbon has degree 2.

Closed ribbons can be orientable (i.e. having two boundary curves) or non-orientable (having only one boundary curve). We write r_n for the number of closed ribbons with n plaquettes (where two ribbons are considered distinct if one can be superimposed on the other by translation). We write r_n^o and r_n^u for the numbers of closed ribbons with n plaquettes which are orientable and non-orientable, respectively. Clearly

$$r_n = r_n^o + r_n^u. \tag{5.1}$$

To obtain an upper bound on w_n we consider the set of objects obtained in the following way. The first plaquette is in any of the three coordinate planes, and the second is incident on one of the four edges of the first plaquette, but is not superimposed on the first plaquette. The k'th plaquette is added so that it has an edge incident on one of the edges of the $(k-1)$'th plaquette, other than the edge on which the $(k-1)$'th and $(k-2)$'th plaquettes are both incident. In addition the k'th and $(k-1)$'th plaquettes are not superimposed. Clearly $w_n \leq 36(3\times 3)^{n-2}$ since the first plaquette can be embedded in the lattice in 3 ways, the second can be added to this in 4×3 ways (choose an edge in 4 ways, and an orientation in 3 ways), and subsequent ones in at most 3×3 ways (choose an edge in 3 ways, since one has already been used, and an orientation in 3 ways).

To obtain a lower bound consider the number of open ribbons with n plaquettes which have the property that the barycentre of the i'th plaquette has at least one coordinate larger, and no coordinate smaller, than the corresponding coordinate of the barycentre of the $(i-1)$'th plaquette. Such objects are certainly examples of open ribbons and give the inequality $w_n \geq 3(2\times 2)^{n-1}$ since the first plaquette can be embedded in the lattice in 3 ways and, when adding subsequent plaquettes, the edge to which the next plaquette is to be added can be chosen in 2 ways, and the orientation in 2 ways.

We next state an important Lemma about the growth of the number of open ribbons.

LEMMA 5.1. *The limit*

$$\lim_{n\to\infty} n^{-1}\log w_n = \log\rho \tag{5.2}$$

exists and $4\leq\rho\leq 9$. *In addition*

$$w_n \geq \rho^{n-1}. \tag{5.3}$$

Proof: Open ribbons having say m and n plaquettes can be concatenated by translating so that the barycentre of the last plaquette of one ribbon coincides with the barycentre of the first plaquette of the other ribbon. This gives an object which may or may not be a ribbon. If it is then the ribbon has $n + m - 1$ plaquettes and every open ribbon with this number of plaquettes can be obtained by this construction. Hence $w_n w_m \geq w_{n+m-1}$. The existence of the limit $\lim_{n\to\infty} n^{-1} \log w_n = \log \rho$ and the inequality in (5.3) follow immediately [39], and the above bounds imply that $4 \leq \rho \leq 9$.

We next state a Lemma about the relative numbers of open and closed ribbons. The proof is quite technical and we give only a sketch here.

LEMMA 5.2. *The numbers of open and closed ribbons grow at the same exponential rate, so that*

$$\lim_{n\to\infty} n^{-1} \log r_n = \log \rho. \tag{5.4}$$

Sketch proof: By deleting the last plaquette in a closed ribbon we obtain an open ribbon so that it follows immediately that $2nr_n \leq w_{n-1}$. To obtain a bound in the other direction we use an adaptation of the proof of Theorem 3.2.4 in Madras and Slade [40]. The general idea is to unfold open ribbons, show that unfolded ribbons have the same growth constant as open ribbons (using an argument similar to that given by Hammersley and Welsh [41]), and construct a subset of closed ribbons by joining up sets of unfolded ribbons, so that this subset grows at the same exponential rate as open ribbons. Taking logarithms, dividing by n and letting n tend to infinity in these upper and lower bounds gives (5.4).

In a similar way we can show that the numbers of orientable and non-orientable ribbons are closely related, and we state this in the following Lemma.

LEMMA 5.3. *There exist constants A, B, l and m such that*

$$r_n^u \geq A r_{n-l}^o \tag{5.5}$$

and

$$r_n^o \geq B r_{n-m}^u. \tag{5.6}$$

Consequently orientable and non-orientable ribbons both have growth constant ρ.

Sketch proof: The proof relies on concatenation arguments. Every orientable ribbon with $n-l$ plaquettes can be concatenated with every non-orientable ribbon with $l-p$ plaquettes, using an additional p plaquettes to effect the concatenation, to give a non-orientable ribbon with n plaqettes. The ribbons produced by this process are a subset of the non-orientable ribbons so that

$$r_n^u \geq r_{n-l}^o r_{l-p}^u. \tag{5.7}$$

Similarly two non-orientable ribbons can be concatenated to form an orientable ribbon, resulting in the inequality

$$r_n^o \geq r_{n-m}^u r_{m-q}^u \tag{5.8}$$

In particular this implies that both orientable and non-orientable ribbons have the same growth constant (take logarithms, divide by n and let n go to infinity with l, m, A and B fixed). From (5.1), this is equal to ρ, since ρ is maximum of the growth constants of the orientable and non-orientable ribbons.

6. Monte Carlo results. The results described in the last three two sections are about asymptotic behaviour of polygons and ribbons. If one is interested in the numerical values of properties such as the knot probability or the writhe at modest values of n the only useful approach seems to be a Monte Carlo method. In addition this approach can be used to estimate parameters such as α in (3.3), or the exponent determining the behaviour of the absolute value of the writhe.

The idea of a Monte Carlo approach is to generate a random sample of polygons, and determine (for instance) what fraction of the sample is knotted. The difficulty is to generate a random sample in a reasonable amount of computer time. Early work [20] established that the knot probability in lattice polygons was quite small when the number of edges in the polygon was of the order of a few hundred, but it is only recently that algorithms have been developed which are adequate for generating samples of long polygons. The approach which has been most successful for lattice polygons is to sample along a realization of a Markov chain defined on the set of polygons. For most of the problems which we shall describe, the value of n is kept fixed so the states of the Markov chain are the polygons with n edges. In the simplest case, all polygons have the same probability so one wants to construct a Markov chain whose unique limit distribution is the uniform distribution. Sufficient conditions for this are (i) that the transition matrix is symmetric and (ii) that every state can be reached from every other state by a suitable sequence of moves. Appropriate algorithms have been devised for the simple cubic lattice [27] and for the face centred cubic lattice [28].

A detailed investigation of the n-dependence of the knot probability has been carried out for polygons on the face centred cubic lattice [29]. For moderate values of n the knot probability turns out to be quite small. E.g. for $n = 800$, $P(n)$ is only about 4×10^{-3}, and this value rises to about 1.2×10^{-2} for $n = 1600$. Even at this value of n almost all the knots found in the Monte Carlo calculations are trefoils (although we know from Section 3 that in the large n limit almost all knots must be composite). The data are well described by the relation

$$1 - P(n) = Ce^{-\alpha n} \tag{6.1}$$

where $\alpha = (7.6 \pm 0.9) \times 10^{-6}$ and C is about 1.

A similar calculation has been carried out for polygons on the simple cubic lattice. In this case $P(800)$ is about 2.8×10^{-3} and $P(1600)$ is about 7×10^{-3}. At $n = 1600$ more that 95% of the non-trivial knots in the sample are trefoils. The data are again well represented by the above equation and α is estimated to be $\alpha = (5.7 \pm 0.5) \times 10^{-6}$ with C being once again close to 1. Unfortunately the results do not resolve the question of the possible lattice dependence of α. All that can be said is that the 95% confidence intervals have substantial overlap.

DNA and RNA are polyelectrolytes and their conformational properties are sensitive to the ionic strength of the solution in which they are dissolved. In particular, there is direct experimental evidence [3,4] that the knot probability in circular DNA increases as the ionic strength of the solution increases. The charges are screened by the presence of the supporting electrolyte and the inter-charge repulsion is thus reduced. The polymer becomes less expanded and this leads to an increase in the knot probability. To model this effect we have considered an extension of the polygon model described above, in which an additional potential is associated with the conformations of the polygon. The total potential energy of a polygon in this model is

$$U = \sum_{i<j} \left(u(r_{ij}) + Ae^{-r_{ij}/r_o}/r_{ij} \right) \tag{6.2}$$

where r_{ij} is the distance between the i'th and j'th vertices of the polygon, measured in lattice units, and the sum is over all non- bonded pairs of vertices. The first term, $u(r)$, is equal to kTv if $r = 1$ and zero otherwise, where v is a negative constant. We chose for v the value -0.26 since it is known that, for the simple cubic lattice, it corresponds to a poor solvent regime [42], in which the knot probability is higher and easier to study. The second term in the potential energy is the Yukawa term which accounts for the screened Coulomb interaction between the charges on the monomers of the polyelectrolyte. r_o is the Debye length measured in lattice units and its value reflects the ionic strength of the solution. The parameter A is connected to the charge density along the polymer chain, and to a length scale in the polymer, such as the persistence length.

As an example we shall discuss the dependence of the knot probability on r_o for polygons of three sizes, $n = 200$, $n = 300$ and $n = 400$, with $A/kT = 0.01$ [6]. As expected, the knot probability increases as n increases, at fixed r_o, and increases as r_o decreases at fixed n. When $r_o > 10$ the knot probability is essentially zero at these values of n but increases to about 3% when $r_o \simeq 0.01$. When r_o is sufficiently small, the Coulombic repulsion is effectively negligible so that this behavior is consistent with previous findings that the knot probability is quite small at those values of n, even in a poor solvent [29]. For a 1-1 electrolyte, $1/r_o$ is proportional to the square root of the concentration of the ions, so we can compare these results with the experimental results of Shaw and Wang [3] for a solution containing $NaCl$. The behavior is very similar, and our results confirm the general shapes of the curves drawn through the experimental points in that paper. To investigate the influence of the charge density on the knot probability we repeated the calculations of knot probability for $A/kT = 0.1$ and 1, with $n = 300$. The same general behaviour occurs at each value of A, although the the sigmoidal portion of the curve moves to higher values of κ as A increases. That is, as the charge density along the polymer increases, a smaller value of the Debye length is required for the knot probability to increase markedly.

In order to obtain further information about the behaviour of the writhe, we have carried out some Monte Carlo calculations, generating polygons by a pivot algorithm [27] and computing their writhe by the pushoff technique [37]. Theorem 4.1 strongly suggests that

$$\langle |W| \rangle \sim n^{\alpha} \tag{6.3}$$

and the numerical data for modest values of n ($n \leq 1100$) can be analysed to give an estimate of the exponent, and lead to the result that $\alpha = 0.522 \pm 0.004$.

The writhe of a polyelectrolyte depends on the ionic strength of the solution, for the same physical reason that the knot probability does. We have estimated [6] the writhe of a polygon, subject to the potential given by (6.2), and the expectation of the absolute value of the writhe increases rapidly as r_o decreases, levelling off at small values of r_o. For instance, for $n = 400$ and $A/kT = 0.1$ $\langle |W| \rangle$ increases from about 1.8 at $r_o = 10$ to about 2.6 at $r_o = 0.1$ and remains at about that value for smaller values of r_o. As the ionic strength increases, the Debye length decreases, the polymer becomes less expanded, and the writhe increases.

The Monte Carlo methods which are useful for investigating the properties of polygons can be extended to handle the problem of a (closed) ribbon. In this case we have used a hybrid algorithm (related to that described in [29]) in which one set of moves (related to pivot moves) make large changes in the conformation of the ribbon, but without changing the number of plaquettes, and another set make local changes in which the number of plaquettes can change. This algorithm is *grand canonical* and samples both orientable and non-orientable ribbons with a variety of lengths, at the same time. The key observations are

1. The mean-square radius of gyration of a ribbon scales in the same way as the mean-square radius of gyration of a polygon,
2. The mean of the absolute value of the writhe of the ribbon increases as a power of n, with the power being very close to $1/2$, and
3. The mean linking number of the two boundary curves in an orientable closed ribbon also increases as a power of n, with the power being very close to $1/2$.

It appears that the writhe (a geometrical quantity) and the linking number (a topological invariant) both capture the same type of entanglement complexity.

7. Discussion. Simple models of circular polymers, such as the polygon and ribbon models described above, have contributed significantly to our understanding of the asymptotic behavior of the knot probability, knot complexity, writhe, etc. In addition, Monte Carlo methods have proved useful in filling in a number of gaps in our theoretical knowledge. They have resulted in estimates of knot probabilities and writhe for polymers of moderate length, and have been essential in the extension of these models to more realistic ones which include the effect of solvent quality [29,43,44] and the co- and counter-ions present in solution [4,6].

Most work has focussed on polymers in very dilute solution where entanglements between different polymers can be neglected. However, the problem of linking probabilities between pairs of polygons has recently been addressed [46], and there is a very nice result by Diao [47] on the corresponding problem of a very dense collection of circles.

There are still many places where our understanding is inadequate. We know little about the relative frequencies of occurrence of polygons with different knot types. Apart from a few bounds such as (3.4), (3.5) and (3.6), most of our knowledge comes from Monte Carlo studies [6,29,43,45], and it would be satisfying to have better theorems about these relative frequencies.

The lattice ribbon model described in Section 5 is a newcomer to the field and many of its properties have still to be investigated. The results for this model which we have described here only answer some of the preliminary questions. The model can be adapted in various ways. For instance one can think of adding a twist fugacity or a writhe fugacity, or a potential between plaquettes to mimic the effect of solvent quality or ionic strength. The effect of these factors on properties like the linking number of the boundary curves will certainly be of interest.

Acknowledgements. The authors acknowledge financial support from NSERC and from NSF.

REFERENCES

[1] K.V. Klenin, A.V. Vologodskii, V.V. Anshelevich, A.M. Dykhne and M.D. Frank-Kamenetskii, *Effect of excluded volume on topological properties of circular DNA*, J. Biomol. Str. Dyn. **5** (1988), 1173-1185.

[2] A.V. Vologodskii, S.D. Levene, K.V. Klenin, M. Frank-Kamenetskii and N.R. Cozzarelli, *Conformational and thermodynamic properties of supercoiled DNA*, J. Mol. Biol. **227** (1992), 1224-1243.

[3] S.Y. Shaw and J.C. Wang, *Knotting of a DNA chain during ring closure*, Science **260** (1993), 533-536.

[4] V.L. Rybenkov, N.R. Cozzarelli and A. Vologodskii, *Probability of DNA knotting and the effective diameter of the DNA double helix*, Proc. Nat. Acad. Sci. USA **90** (1993), 5307-5311.

[5] S.Y. Shaw and J.C. Wang, *DNA knot formation in aqueous solutions*, J. Knot Theory and Its Ramifications **3** (1994), 287-298.
[6] M.C. Tesi, E.J. Janse van Rensburg, E. Orlandini, D.W. Sumners and S.G. Whittington, *Knotting and supercoiling in circular DNA: A model incorporating the effect of added salt*, Phys. Rev. E **49** (1994), 868-872.
[7] F.B. Dean, A. Stasiak, T. Koller and N.R. Cozzarelli, *Duplex DNA knots produced by Escherichia coli topoisomerase I*, J. Biol. Chem. **260** (1985), 4795-4983.
[8] C.D. Lima, J.C. Wang and A. Mondragon, *Three-dimensional structure of the 67K N-terminal fragment of E. coli DNA topoisomerase I*, Nature **367** (1994), 138-146.
[9] S.A. Wasserman and N.R. Cozzarelli, *Supercoiled DNA-directed knotting by T4 topoisomerase*, J. Biol. Chem. **266** (1991), 20567-20573.
[10] J. Roca and J.C. Wang, *DNA transport by a type II DNA topoisomerase: evidence in favor of a two-gate mechanism*, Cell **77** (1994), 609-616.
[11] S.A. Wasserman, J.M. Dungan and N.R. Cozzarelli, *Discovery of a predicted DNA knot substantiates a model for site-specific recombination*, Science **229** (1985), 171-174.
[12] S.A. Wasserman and N.R. Cozzarelli, *Biochemical topology: applications to DNA recombination and replication*, Science **232** (1986), 951-960.
[13] T. Goto and J.C. Wang, *An ATP-dependent type II topoisomerase that catalyzes the catenation, decatenation, unknotting, and relaxation of double-stranded DNA rings*, J. Biol. Chem. **257** (1982), 5866-5872.
[14] K. Shishido, N. Komiyama and S. Ikawa, *Increased production of a form of plasmid pBR322 DNA in Escherichia coli DNA topoisomerase mutants*, J. Mol. Biol. **195** (1987), 215-218.
[15] K. Shishido, S. Ishii and N. Komiyama, *The presence of the region on pBR322 that encodes resistance to tetracycline is responsible for high levels of plasmid DNA knotting in Escherichia coli DNA topoisomerase I deletion mutants*, Nucleic Acids Rsh. **17** (1989), 9749-9759.
[16] P.T. Englund, S.L. Hajduk and J.C. Marini, *The molecular biology of trypanosomes*, Ann. Rev. Biochem. **51** (1982), 695-726.
[17] D.W. Sumners, *Knot theory and DNA*, in *New Scientific Applications of Geometry and Topology*, Proceedings of Symposia in Applied Mathematics **45**, D.W. Sumners, ed., AMS, Providence, RI, 1992, pp. 39-72.
[18] H.L. Frisch and E. Wasserman, *Chemical Topology*, J. Am. Chem. Soc. **83** (1968), 3789-3795.
[19] M. Delbruck, *Mathematical Problems in the Biological Sciences*, AMS, Providence, RI, 1962 p. 55.
[20] A.V. Vologodskii, A.V. Lukashin, M.D. Frank-Kamenetskii and V.V. Anshelevich, *The knot probability in statistical mechanics of polymer chains*, Sov. Phys.-JETP **39** (1974), 1059-1063.
[21] J.P.J. Michels and F.W. Wiegel, *On the topology of a polymer ring*, Proc. Roy. Soc. A **403** (1986), 269-284.
[22] D.W. Sumners and S.G. Whittington, *Knots in self-avoiding walks*, J. Phys. A: Math. Gen. **21** (1988), 1689-1694.
[23] N. Pippenger, *Knots in random walks*, Disc. Appl. Math. **25** (1989), 273-278.
[24] C.E. Soteros, D.W. Sumners and S.G. Whittington, *Entanglement complexity of graphs in Z^3*, Math. Proc. Camb. Phil. Soc. **111** (1992), 75-91.
[25] W.R. Bauer, F.H.C. Crick and J.H. White, *Supercoiled DNA*, Sci. American **243** (1980), 118.
[26] J.H. White, *Self-linking and the Gauss integral in higher dimensions*, Am. J. Math. **91** (1969), 693-728.
[27] N. Madras, A. Orlitsky and L.A. Shepp, *Monte Carlo generation of self-avoiding walks with fixed endpoints and fixed length*, J. Stat. Phys. **58** (1990), 159-183.
[28] E.J. Janse van Rensburg, S.G. Whittington and N. Madras, *The pivot algorithm and polygons*, J. Phys. A: Math. Gen. **23** (1990), 1589-1612.

[29] E.J. Janse van Rensburg and S.G. Whittington, *The knot probability in lattice polygons*, J. Phys. A: Math. Gen. **23** (1990), 3573-3590.

[30] E.J. Janse van Rensburg, E. Orlandini, D.W. Sumners, M.C. Tesi and S.G. Whittington, *The writhe of a self-avoiding polygon*, J. Phys. A: Math. Gen. **26** (1993), L981-985.

[31] J.M. Hammersley, *Percolation processes II The connective constant*, Proc. Camb. Phil. Soc. **53** (1957), 642-645.

[32] J.M. Hammersley, *The number of polygons on a lattice*, Proc. Camb. Phil. Soc. **57** (1961), 516-523.

[33] H. Kesten,*On the number of self-avoiding walks*, J. Math. Phys. **4** (1963), 960-969.

[34] R.H. Fox, *A quick trip through knot theory* in Topology of 3-manifolds and related topics ed. M.K. Fort, Jr., Prentice-Hall, New York, 1962.

[35] Y. Diao, N. Pippenger and D.W. Sumners, *On random knots*, J. Knot Theory and Its Ramifications **3** (1994), 419-429.

[36] Y. Diao, *The knotting of equilateral polygons in R^3*, J. Knot Theory and Its Ramifications **2** (1994), 413-425.

[37] R.C. Lacher and D.W. Sumners, *Data structures and algorithms for computation of topological invariants of entanglements: link, twist and writhe*, in *Computer Simulation of Polymers*, R.J. Roe, ed., Prentice-Hall, Englewood Cliffs, NJ, 1991, pp. 365-373.

[38] G. Burde and H. Zieschang, *Knots*, de Gruyter, Berlin, 1985.

[39] J.B. Wilker and S.G. Whittington, *Extension of a theorem on super-multiplicative functions*, J. Phys. A **12** (1979), L245-247.

[40] N. Madras and G. Slade, *The Self-Avoiding Walk*, Birkhäuser, Boston, 1993.

[41] J.M. Hammersley and D.J.A. Welsh, *Further results on the rate of convergence to the connective constant for a self-avoiding walk*, Q. J. Math. Oxford **13** (1962), 108.

[42] E.J. Janse van Rensburg, D.W. Sumners, E. Wasserman and S.G. Whittington, *Entanglement complexity of self-avoiding walks*, J. Phys. A **25** (1992), 6557-6566.

[43] K. Koniaris and M. Muthukumar, *Self-entanglement in ring polymers*, J. Chem. Phys. **95** (1991), 2873-2881.

[44] M.L. Mansfield, *Knots in Hamilton Cycles*, Macromolecules **27** (1994), 5924-5926.

[45] T. Deguchi and K. Tsurusaki, *Topology of closed random polygons*, J. Phys. Soc. Japan **62** (1993), 1411-1414.

[46] E. Orlandini, E.J. Janse van Rensburg, M.C. Tesi and S.G. Whittington, *Random linking of lattice polygons*, J. Phys. A: Math. Gen. **27** (1994), 335-345.

[47] Y. Diao, *Unsplittability of random links*, J. Knot Theory and Its Ramifications **3** (1994), 379-389.

ENERGY FUNCTIONS FOR KNOTS: BEGINNING TO PREDICT PHYSICAL BEHAVIOR*

JONATHAN SIMON†

Abstract. Several definitions have been proposed for the "energy" of a knot. The intuitive goal is to define a number $u(K)$ that somehow measures how "tangled" or "crumpled" a knot K is. Typically, one starts with the idea that a small piece of the knot somehow repels other pieces, and then adds up the contributions from all the pieces. From a purely mathematical standpoint, one may hope to define new knot-type invariants, e.g by considering the minimum of $u(K)$ as K ranges over all the knots of a given knot-type. We also are motivated by the desire to understand and predict how knot-type affects the behavior of physically real knots, in particular DNA loops in gel electrophoresis or random knotting experiments. Despite the physical naiveté of recently studied knot energies, there now is enough laboratory data on relative gel velocity, along with computer calculations of idealized knot energies, to justify the assertion that knot energies can predict relative knot behavior in physical systems. The relationships between random knot frequencies and either gel velocities or knot energies is less clear at this time.

"Suppose you have a knotted loop of string, and you spread an electric charge along the string and then let go; what will happen?" This question has been a common "cocktail party" topic among knot theorists for many years, but recently questions like this have been the objects of serious study. The motivation is twofold: We want to develop new knot-type invariants that describe, in intuitively satisfying ways, how complicated/tangled/interwound one knot is relative to another, and also we want to build a mathematical enviornment in which one can hope to model, understand, and predict how different types of knots behave in physical situations. If one imagines different knots tied from the same length of string, then it is at least plausible that more complicated knots would be spatially more compact, so a quantitative measure of complexity might be used as a quantitative measure of compaction.

This paper provides an expository introduction to knot energy functions, preliminary announcement of our computational results on energy minima for various knot types, and an analysis of how these energy numbers relate to other measures of knot complexity, in particular frequency of different knots in random knotting experiments and, most of all, relative velocity of differently knotted DNA loops in agarose gel electrophoresis.

To provide the most direct route to the results, we present first the main discussion, and then provide the background on knots and energy functions in the subsequent sections.

1. Gel velocity, random knot frequency, and topological ground state energy of a knot. There is clear laboratory evidence that when closed DNA loops are knotted, different types of knots behave differ-

* Partially supported by NSF DMS–9407132.

† Department of Mathematics, University of Iowa, Iowa City, IA 52242.

ently in gel electrophoresis [10][12][13][18][49]. The duplex loops are long enough and nicked, so they are conformationally flexible; the differences in gel velocity are detecting different *configurations* (or time-averaged conformations, as opposed to individual rigid conformations), what the topologists would call different *knot types*. This differential mobility rapidly evolved from being a discovery to being an analytical tool [12][39], so that in [39], the knot types are determined entirely by gel mobility, without recourse to electron microscope confirmation that was typical in earlier papers. The intuition is that if a "complicated" and a "simple" knot are made from string of the same length, then the complicated knot will be, in time averaged conformation, more tightly crumpled, hence move faster in agarose gel.

There is similarly convincing evidence that different knot types have different probabilities of being created in random DNA knotting experiments [39][40][41]. Here the intuition might be that, other factors being equal, a particular "complicated" knot is kinetically or entropically less likely to occur than a particular "simpler" one. However, this seemingly reasonable belief clashes with the equally appealing intuition (and, in fact, a theorem [33][45]) that a *very* long "random loop" is not likely to be unknotted, and in fact is likely to be a complicated knot.

One possible way to reconcile the intuitive belief that 'more complicated' implies 'less likely' with the fact that loops of long strings 'must' be complicated knots would be if, as the number of segments in a random polygonal loop increases, while the probability of finding one particular simple knot type decreases towards zero, perhaps that probability still is greater than the probability of finding *one particular* complicated knot type. However, in the computer simulations of [11], there is data saying that for certain numbers of segments, a knot of type A is more likely than a knot of type B, while for more segments, type B is more likely than type A. It might be significant that this phenomenon is observed for composite knots (see section 2 below for knot terminology), but not (yet) observed for prime knots. In [11], and in the computer simulations in [39], the ordering of prime knot types by their probabilties among random polygons of different lengths is independent of segment number.

There is a perfect (qualitative) correlation between relative gel velocity and relative frequency for the first few knot types, i.e. the unknot, trefoil, and figure-eight (see section 2 below for definitions). Knots which move faster in the gel are less likely to be produced by random knotting than knots which migrate more slowly. However, when we get to knots with five crossings (so-called knot types 5_1 and 5_2) the knot 5_2, which is slightly faster in the gel [13][12], seems to be *more* likely to occur than the slightly slower 5_1 [39].

The crossing number (i.e. node number) of a knot type also is, initially, a good predictor of relative frequency. However, according to Figure 3.5 of [11], for large numbers of segments, the *nine* node composite knot

$3_1\#3_1\#3_1$ is more likely to occur than the *seven* node knot $3_1\#4_1$.

Until we have a better understanding of the interplay between knot type and number of segments (i.e. string length), it does not seem possible to predict the relative frequencies of different knot types in random knotting experiments *in vitro* or *in virtuo* [44] based on one numerical measure of complexity. Predicting relative gel velocity seems a more attainable goal.

As a first sorting, the crossing numbers (see section 2 below) of knot-types seems to work very well as a predictor of relative gel velocity [10], though (Table I, p. 4979 of [10], esp. band 6) the separation seems to get less clear as the number of nodes increases. (For homologous series, e.g. of $(2,n)$ torus knots [42] and twist knots [18], the coincidence of gel bands and crossing numbers seems perfect.) Within the gel region associated with a given crossing number, it may be possible to distinguish bands corresponding to different knot types with the same crossing number, e.g. the 'granny knot' vs. the prime six crossing knot 6_1 [18]. Another situation where gel velocity distinguishes between different knots having the same crossing number is the observation [13][12] that the $(2,n)$ torus knots migrate slightly slower than the corresponding n-crossing twist knots. Of course it is a tautology that we would need a numerical measure of complexity that is finer than crossing number, if we want to distinguish knots of the *same* crossing number; and apparently "nature" does indeed make such finer distinctions. In Table 1.1, we list the first few knot types in order of observed relative gel velocity, as well as the behavior of some homologous families.

The *energy functions* E_2 and U_{MD} for knots, described in sections 3.2 and 3.3 below, seem to make the distinctions we want. In Table 1.2, we list the computationally estimated "topological ground state energies" for various knots. The numbers $\tilde{U}^0_{MD}$ are our own calculations. The values of $\min E_{\cos}$ (here $E_{\cos}$ is an approximation of $E_2 - 4$) are taken from Table 1.2 of [20], except the value for the square knot and granny, which are taken from elsewhere in that paper. The two energy functions order the knots identically through 7_1. The more tightly crumpled a knot is, the higher its "energy" and, for all knots through six crossings and most through seven, the energy ranking respects crossing number. The sudden appearance of the $(3,4)$ torus knot 8_{19} early among the seven crossing knots is consistent with its anomalous physical behavior [9] (and also its anomalously low minimal edge number on the cubic lattice [37]). The minimum energy for knot type 5_2 is slightly higher than the minimum for 5_1, so the energy functions would predict that 5_2 moves faster in the gel than 5_1. Likewise, the energy functions would predict that the composite six crossing knots, i.e. the granny knot or square knot, would move slower in the gel than the twist knot 6_1. When we get to 7 and 8 crossings, the two energy functions differ slightly in their rankings. Even through all the eight and nine crossing prime knots, the two lists differ only occasionally, and then only by 1–3 places, in how knots are ranked. We plan in the future to

TABLE 1.1
Ranking of DNA knots by relative gel velocity.

knot types (slow to fast)	general observations
unknot	All knots are generally ordered by crossing number, especially through six nodes, but less distinctly as number grows [10].
trefoil knot 3_1	
figure-8 knot 4_1	
(2,5) torus knot 5_1	
twist knot 5_2	Twist knots are slightly faster than $(2, n)$ torus knots with the same crossing number [13].
granny knot $3_1\#3_1$ vs. 6_1 [18]	
prime knots 6_1, 6_2, 6_3 (no reports that these were separated by gel)	Homologous family of twist knots are ordered by crossing number [18]
higher crossing knot types	
	Homologous family of $(2, n)$ torus knots are ordered by crossing number [42]

obtain better estimates of the various minima (in particular to make more systematic the search for possibly different local minima) and will then be better able to reconcile, or contrast, the two rankings. For now, we accept the overall similarity of the lists, along with conceptual similarities in the definitions of these functions, as a strong statement that the two notions of energy are measuring essentially the same quality of relative knot complexity. Regarding gel velocity of DNA loops, **the qualitative ranking of knots by estimated minimum energy is consistent with all the observed velocity differences noted in Table 1.1.**

HERE ARE SOME ADDITIONAL NOTES ON TABLE 1.2.

- The (so far estimated) energy $\tilde{U}^0_{MD}$ for the granny knot is less than for the square knot, but the difference is less than the likely error in the data, so no difference is reported here.
- The numbers $\tilde{U}^0_{MD}$ for knots of six or more crossings are rounded to the nearest integer, to remind us that these are rougher approx-

TABLE 1.2
Ranking of knot types by estimated "topological ground state" energy.

knot type	$\tilde{U}^0_{MD}$	min $E_{\cos}$
unknot	0	0
trefoil 3_1	37.2	70.4
figure-8 4_1	56.9	104.9
(2,5) torus knot 5_1	74.9	126.8
twist knot 5_2	81.6	134.6
granny knot $3_1\#3_1$	110	140.8
square knot $3_1\#3_1^*$	110	140.8
twist knot 6_1	120	162.8
6_2	128	168.5
6_3	134	172.9
(2,7) torus knot 7_1	139	181.0
(3,4) torus knot 8_{19}	144	197
twist knot 7_2	153	190.3
7_3	155	192.7
8_{20}	157	203.9
7_4	160	197.7
7_5	167	199.7
7_6	170	203.7
7_7	175	207.1

imations than for the simpler knots.

- We have not yet computed a value for the composite $3_1\#4_1$, so that knot should be inserted, perhaps first among the seven crossing knots (the conjecture on additivity of minimum conformal energy [20] would place it just after 6_3).
- According to both $\tilde{U}^0_{MD}$ and min $E_{\cos}$, the prime seven crosing knots are ordered 7_1, 7_2, 7_3, 7_4, 7_5, 7_6, 7_7. That is, the same as their numbering in the classical knot tables. This is a remarkable coincidence, inviting speculation on what aesthetic sense, or intuition about complexity, led the early tablers to list the knots in that order.

It is appropriate to say "qualitative" in the last sentence of the paragraph above, because of the pseudo-physical nature of the knot energy functions: these are purely geometric quantities, albeit physically motivated. However, there is a provocative parallel between the energies discussed here, and the calculated free energies of formation in [40][41]. While comparing the absolute numbers is meaningless without some agreement on units, if we compare the ratios, we get surprisingly close agreement. In [40], the calculated energy of formation of the figure-8 knot at 20 [resp. 30] Kuhn lengths and maximum $MgCl_2$ concentration (maximum salt $\Longrightarrow$

polymer is most able to knot) is 1.69 [resp. 1.64] times the energy for a trefoil knot. On the other hand, the ratio of minimum energies $\tilde{U}^0_{MD}$ of the two knot types is 1.66. Obviously, more data is needed to test whether the energy U_{MD} really can predict so closely the relative energies of formation of different knots *in vitro*.

If the prime six-crossing knots can be distinguished by gel velocity, that would provide a good further test of our assertion that energy functions predict relative velocity. Another source of verification (or challenge, of course) would be to carry out the detailed gel electrophoresis simulations of [21] and [22] for knots beyond the trefoil. Another project that would seem very interesting would be to try to relate our energy functions to other direct geometric measures of compaction, such as mean square radius of gyration or mean span [36]. The Monte Carlo study of random polygons in [36] can be interpreted to say that if polygonal loops with many segments are viewed from far away, so that only gross features such as overall diameter are discernable, then one won't be able to see any difference between unknots and (even complicated) knots. This seems vaguely similar to the observation one can make from the table of energy minima, especially [20], that as the crossing number gets large, there is not much difference between the energy minima for different knot types.

While we are focusing on knots, it should also be noted that DNA catenanes exhibit interesting and useful gel behavior. In [1], and the survey [12], it is noted that catenanes that differ only in their linking pattern (e.g. n-form where four loops are linked in a linear chain vs. iso-form where there are three in a row, with the fourth linked to the one in the middle—see [Figure 1.1]) are examples of DNA stereoisomers that have the same crossing numbers but different gel velocities. Can knot energies capture this difference? Here it will be necessary to modify the existing minimization algorithms so as to maintain the string length of each loop. For the catenanes in [Figure 1.1A], we do not (yet) know whether forcing the four loops to maintain identical lengths will produce the same energy ranking as if we allow the loops to change relative lengths. In [Figure 1.1B], the *only* difference between the two catenanes is the location of the long loop relative to the short ones.

Finally, we may note that the ranking of knots by gel velocity, energy, and (for the simplest knots, or for all knots studied if we don't make very fine distinctions) frequency, even is generally consistent with results in [37][34][26] on the minimum number of "sticks" needed to represent a given knot type (the studies differ in what restrictions are placed on the geometry of the sticks).

In view of the many correspondences, it seems reasonable to assert that our naively defined "knot energies", numerical measures of the tightness of a knot (from which one obtains a measure of the complexity of a knot-type by taking the minimum over all knots representing a given type) are succeeding in capturing whatever it is that causes the different knot types to

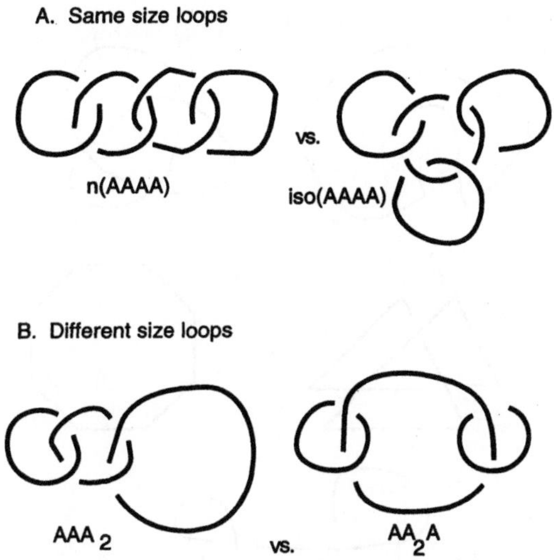

FIG. 1.1. *DNA catenanes with different gel velocities.*

have different gel velocities or different behavior in other physical situations.

In the sections below, we provide background for the preceeding discussion and, in particular, for Table 1.2. In Section 2, we give a micro-introduction to knot theory, including terminology we have been using. In Section 3, we present several definitions of *energy* of a knot and describe how the data in Table 1.2 was obtained.

2. Knots. A *knot* is a simple closed curve in 3-space [Figure 2.1]. One can talk about smooth knots or about polygonal ones; in the former case, ideas of differential geometry such as total curvature [27] can be used to begin measuring how complicated a knot is. For polygons, one might compute total curvature or just count how many sticks are needed to represent the knot; also one might restrict to situations where all the sticks are the same length, or one might allow varying lengths. (For the polygonal energy U_{MD}, we *do* allow lengths to very.)

Typically, knots are represented and communicated by their projections into a plane, with the graphical convention that an apparent gap in a curve indicates a place where one part of the knot is passing over/under the other relative to the direction of projection [Figures 2.2, 2.3]. For a given smooth or polygonal knot, almost all directions yield projections with only finitely many points of singularity, all of which are all double points. The number of double points in a particular projection of a knot is called the *crossing number* of the projection.

Two knots are equivalent if one can be deformed to the other without

FIG. 2.1. *Smooth and polygonal knots.*

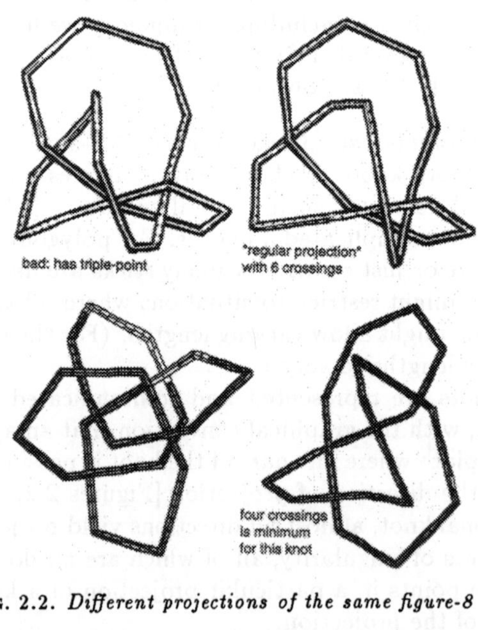

FIG. 2.2. *Different projections of the same figure-8 knot.*

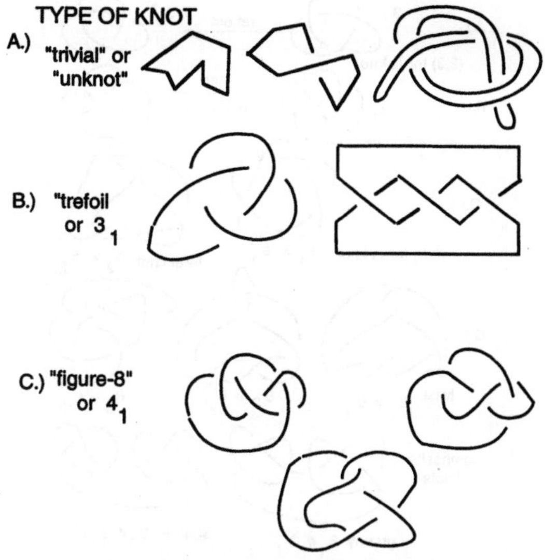

FIG. 2.3. *Several types of knots.*

passing the knot through itself; formally, if one is ambient isotopic to the other; in this case, we say the two knots are of the same *knot type*. So a *knot type* is an equivalence class of knots [Figure 2.3]. We shall use $\langle K \rangle$ to denote the knot type of a particular knot K. The *crossing number* of a knot type is the minimum, over all knots realizing that type, of the crossing numbers of the individual knots. In DNA literature, the term *node* is used instead of crossing.

The only knot that can be drawn with zero, one, or two crossings is the *unknot* or *trivial* knot. (All the knots listed in this and the next paragraph are illustrated in [Figure 2.4] or [Figure 2.3].) There are two knot types that can be represented with a diagram having three crossings; one is the *trefoil* knot, the other just the reflection of a trefoil in the plane of projection. The trefoil type also is denoted 3_1, while its mirror image is 3_1^*. Sometimes this distinction is ignored; we might say that 3_1 and 3_1^* are the same *knot type* but are different *ambient isotopy types.* With four crossings, we can realize knots that are trivial or trefoils, and one new kind, called the *figure-8*, and often denoted 4_1. It is an interesting exercise to show that a (hence every) knot of the figure-8 type is equivalent to its mirror image, so there's no knot type denoted 4_1^*. At five crossings, there are four knot types, two and their respective mirror images. The knot type denoted 5_1 is one of an infinite class called *torus knots* because they can be drawn on the surface of a standard torus in space. The other five crossing knot type, denoted 5_2, is also one of an infinite class, called *twist knots*, where the "clasp" at the top remains the same, but the number of "twists" at the bottom varies.

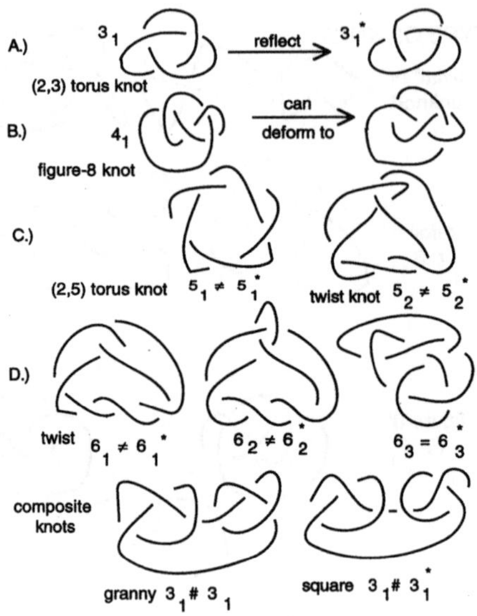

FIG. 2.4. *Knots through six crossings.*

At six crossings, the standard knot tables list three types, 6_1, 6_2, and 6_3 (two of these do not interconvert with their mirror images, so we also need to record 6_1^* and 6_2^*). However, a new phenomenon appears at 6 crossings, that is *composition* of knots. The final two knots in [Figure 2.4] are obtained by tying, in succession, in the same string, first a knot 3_1 and then either a knot 3_1 or else a knot 3_1^*. So some knots are *composite*, others *prime*. The classical knot tables only tried to list prime knots, and just one representative for each chiral pair; even with these restrictions, the number of knot types grows rapidly with crossing number [14].

3. Energy functions

3.1. Motivation and vertex-energy. It seems most plausible to try to define a self-repelling "energy" of a knot by trying to extend the standard notion of electrostatic repelling from a finite set of charged points to a continuous charge distribution along a curve or polygon. However, this straightforward process *fails* (see e.g. [4]). If $p_1, \ldots, p_n$ are n points in 3-space, each carrying charge $= \frac{1}{n}$, so the total charge $= 1$, then the electrostatic potential of the ensemble $\{p_1, \ldots, p_n\}$ is

$$\sum_{i=1}^{n-1} \sum_{j=p+1}^{n} \frac{\frac{1}{n}\frac{1}{n}}{\operatorname{dist}(p_i, p_j)} . \tag{3.1}$$

If we imagine a single unit charge distributed evenly along a curve or line segment X, we might then try to define the potential energy to be half (because each pair (x, y) appears twice) of

$$\int_X \int_X \frac{dx\, dy}{|x-y|} \,. \tag{3.2}$$

Unfortunately, this integral is *infinite*; equivalently, if the points $\{p_i\}$ are spaced equally along a line segment or a circle, then the limit, as $n \to \infty$, of the sums (3.1) is infinite. The divergence of the integral (3.2) and the limit of the finite sums are on the order of $\int_0^1 \frac{1}{t}\, dt \approx \sum_1^\infty \frac{1}{n} \approx -\log(0)$.

To define a usable energy function along these lines, one must somehow eliminate the "near-neighbor" effects, which are what force the above integral to diverge. In the case of a polygon, we accomplish this (see below) by ignoring the interactions between a segment and itself or adjacent segments of the polygon. In the case of smooth curves, the divergent integral is "regularized" by subtracting off the (infinite) energy of a standard round circle. In fact, it turns out useful to do a similar reduction of the polygonal energy.

There are several mathematical properties that have generally been accepted as characterizing a "good" energy for knots. Here is part of the list in [4]:

1. The energy $u(K)$ should be a real number, varying continuously with the position of the polygon or curve K.
2. The function u should be well-behaved under change of scale. (The energies U_{MD} and E_2 described below are both scale-invariant; that is, energy is a property of the *shape* of a knot rather than its absolute *size*.)
3. The function u should be defined in a way that is consistent with the idea that repelling energy varies inversely with distance, or some power of it, between repelling objects.
4. The energy u should respect the topology, in the sense that in order to deform a knot in a way that changes the knot type, that is by passing part of the string through another part, the energy has to become infinite. Alternatively put, the different knot types are separated by infinitely high potential walls. (A different, perhaps physically more realistic, approach has been suggested by Lomonaco [23], where the energy barrier to changing knot type is finite.)

If we allow knots to flow in their configuration space, we can imagine a knot gradually deforming to reduce energy, heading in the limit to some conformation that is a (local) energy minimum. One implication of Property 4 above is that knot type cannot change during such an energy reducing flow.

Before going on to describe energies for polygons and smooth curves that have the desired mathematical properties, we should cite the seminal paper [17], which analyzed vertex-only potentials for polygonal knots. Let K be a polygonal knot with vertices $p_1, \ldots, p_n$. Motivated by equation

(3.1), Fukuhara defined an infinite family of energies,

$$U_d(K) = \sum_{i=1}^{n-1} \sum_{j=p+1}^{n} \frac{1}{\text{dist}(p_i, p_j)^d} \quad . \tag{3.3}$$

(This formula is a little simpler than Fukuhara's, which includes a constant of proportionality that depends on the number, n, of edges of the polygon.) Fukuhara obtained the following result:

THEOREM 3.1. *Consider the space K of polygonal knots of n edges, in which each edge has fixed length = 1. Let K_0 be a polygon in K with vertices $p_1, \dots, p_n$ such that for each $i \neq j$, $\text{dist}(p_i, p_j) > 1$. Then there exists an exponent d (which depends on n) such that the knot type of K_0 remains unchanged as K_0 flows via the gradient of U_d.*

This result is surprising because the energy function only "sees" vertices, so it does not "know" if two edges of a polygon pass through each other at interior points.

Before discussing the energies reflected in Table 1.2, we might also note that a quite different approach is developed in [28], based on idealized fluid flow in knotted tubes.

3.2. Energies for smooth curves. The first knot energy with Properties 1–4 above, E_2, was proposed by J. O'Hara [29]. It turned out that this energy has an invariance property that is stronger than being invariant under change of scale; it is, in fact, invariant under conformal transformations of 3-space [3] (so E_2 is often called the *conformal energy*). Freedman and his colleagues also established a strong relationship between the energy E_2 of a knot K and the *average crossing number* of K [15]. Here is their formulation of the energy E_2.

Let K be a smooth knot in 3-space $\Re^3$. So we can view K as the image of the standard unit circle S in the plane under a smooth map $\gamma : S \longrightarrow \Re^3$. Then

$$E_2(K) = \int_S \int_S \left(\frac{1}{\text{dist}(\gamma(s) - \gamma(t))^2} - \frac{1}{\text{arcdist}(\gamma(s), \gamma(t))^2} \right) |\gamma'(s)|\, |\gamma'(t)|\, ds\, dt \tag{3.4}$$

Here arcdist (s, t) denotes the minimum distance long the knot between points s and t.

If we assume the parametrization γ is by arclength, that is the speed $|\gamma'(s)|$ is always 1, then the definition simplifies to

$$E_2(K) = \int_S \int_S \left(\frac{1}{\text{dist}(\gamma(s) - \gamma(t))^2} - \frac{1}{\text{arcdist}(\gamma(s), \gamma(t))^2} \right) ds\, dt \tag{3.5}$$

When we compare the above energy with the polygonal energy U_{MD} defined below, we should note that E_2 counts each interaction of $\gamma(s)$ with

$\gamma(t)$ twice, whereas U_{MD} counts it just once, so in comparing theorems or computational results, we should compare E_2 with $2 \cdot U_{MD}$.

The conformal energy has been studied theoretically [3][15][16] [29][30] [32] and computationally [2][19][20] [31] .

The data presented in [20] (called $E_{\cos}$ in Table 1.2 above) is based on a somewhat different approach to regularizing the divergent integral. So, as stated in [20], the numbers $E_{\cos}$ are the minima so far observed for $E_2(K) - 4$.

Instead of "subtracting the infinity" as is done to regularize the integral in (3.4), another approach [5][6] is to view the energy represented by two infinitesimal segments of the curve K as having a vector attribute, and consider only the component normal to the curve. This intuition yields another energy for smooth knots, satisfying Properties 1–4, called the *projection energy*, where the term *projection* refers to the projection of the vector, not a special invariance property of the energy function. To avoid confusing the projection energy with others discussed here, we shall use the notation U_{proj}, which is different from [6]. As before, let K be a smooth knot in 3-space $\Re^3$. At each point x of K there is (since K is smooth) a well-defined tangent line and, hence, a well-defined normal plane N_x. For each point $y \neq x$ of K, let θ_{xy} be the angle between the vector $(x - y)$ and the plane N_x. Then we define

$$U_{\text{proj}}(K) = \int_{y \in K} \int_{x \in K} \left(\frac{\cos(\theta_{xy})}{|x - y|} \right)^2 dy\, dx \quad . \tag{3.6}$$

The projection energy also is implemented in the program [2]. The relationship between U_{proj} and the average crossing number of a knot is similar to that of the conformal energy, though the proof is simpler [7].

3.3. Energies for polygons. For polygonal knots, the first energy satisfying Properties (1)–(4) was defined in [4]. A simpler polygonal energy, and the one we shall discuss here, is presented in [43].

Let X, Y be disjoint line segments in $\Re^3$. Then the minimum distance between X and Y is a positive number which we denote $MD(X, Y)$. Letting $\ell(X)$ denote the length of a segment X, define

$$U_{MD}(X, Y) = \frac{\ell(X) \cdot \ell(Y)}{MD(X, Y)^2} \quad .$$

For a polygon K, we define

$$U_{MD}(K) = \sum_{\substack{X,Y \text{ non-consecutive} \\ \text{segments of } K}} U_{MD}(X, Y) \quad .$$

While we are interested in knots, in particular nonplanar polygons with six or more sides, it may be helpful to work out the numbers for some simpler cases.

Example. *For a triangle, there are no pairs of nonconsecutive edges, so we define U_{MD}(any triangle) $= 0$. For a plane quadrilateral, there are two pairs of edges that contribute, and the energy is minimum when the figure is a square; U_{MD}(square) $= 2 \cdot (\frac{a \cdot a}{a^2}) = 2$. For a regular pentagon, we have five pairs of edges, each pair contributing 1, so U_{MD}(regular pentagon) $= 5$. For a regular hexagon, there are six pairs of edges that contribute 1 each and three pairs of edges that contribute $\frac{a \cdot a}{(a\sqrt{3})^2} = \frac{1}{3}$ each, so U_{MD}(regular hexagon) $= 7$. For regular n-gons with $n \geq 7$, the numbers become irrational.*

These examples begin to illustrate the general phenomenon that if K is any polygonal knot with six or more edges, then $U_{MD}(K) > n$.

The energy function U_{MD} satisfies Properties 1–4; in particular, it is invariant under change of scale. The function is not differentiable at some points, so we cannot talk about standard gradient flow. However, it does make sense to do a simulated annealing ("at zero degrees") type minimization:

- Start with a polygonal knot K_0,
- randomly perturb the polygon to a new conformation K_1 (making sure that the perturbation is small enough that the knot type of K_0 is not changed),
- compare the energies $U_{MD}(K_0)$ and $U_{MD}(K_1)$,
- if $U_{MD}(K_0) > U_{MD}(K_1)$ then replace K_0 with K_1.

In this way, we generate a sequence of knots converging to a knot whose energy is a (local) minimum within the knot type of K_0. We illustrate in [Figure 3.1] how an 18 segment figure-8 knot evolves from an initially compressed conformation towards a (local) energy minimizing conformation under U_{MD}.

If a knot type $\langle K \rangle$ can by realized among polygons with, say, 18 edges, then there exists [43] a nonsingular polygon K with minimum energy among all 18 segment polygons representing knots of that type. (This is true whether the knot K is prime or composite, in contrast to the conformal energy E_2, where minimum energy conformations are believed to exist only for prime knots.) We may hope to discover (estimate) the minimum energy value for a particular knot type within a given number of segments by starting with any representative knot and letting it flow, as in the preceeding paragraphs, to a local minimum. Of course one needs to worry about local vs. global minima. However, in experiments run so far, it appears that for each knot type and given number of segments, there are relatively few distinct local minima, so we have some basis for hoping that the best local minima we have found so far are indeed global minima. There's a danger of circular reasoning here, and we are beginning a more systematic analysis of local vs. global minima for U_{MD}.

In order to obtain *ground state energy* numbers for each knot type, we need to consider what happens when the same knot is made from different

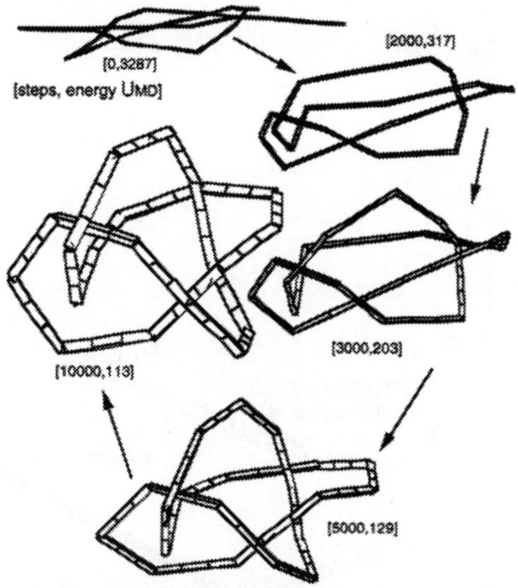

FIG. 3.1. *Evolution of a 20-segment figure-8 knot.*

numbers of segments. We have two approaches for this.

It appears that as the number of segments increases, the minimum energy first decreases until an "ideal" number of segments is reached, then increases from there on. We know from [43] that the energy satisfies the relation $U_{MD}(K) >$(number of segments), so the energies must eventually grow; but the theorem does not predict the unmodality so far observed. In [Figure 3.2] we show the observed U_{MD} energy minima for trefoil knots made from 6-to-49 segments. One possible way to define the topological ground state energy of a particular knot type would be to identify the optimum number of segments and use the energy minimum for that number: So for the trefoil knot, we would say that the minimum energy is (approximately) 74.5, attained at 14 segments. This approach may prove to be a useful one for relating energy to knotting probability, since the relative probabilities are sensitive to segment numbers [11][24][25][35][39][46][47][48].

Another way we can define the minimum energy for a particular knot type is to somehow remove the phenomenon that causes U_{MD} to grow with the number of segments; this is analogous to the regularization of the divergent integral 3.4. In [Figure 3.2], we have plotted, along with the U_{MD} minima for the trefoil knot, the energies of regular planar n-gons, and then the differences between the trefoil energies and the regular n-gon energies. The first observation that seems graphically obvious is that the energies of regular planar n-gons lie on a straight line; actually, there is a bit of wiggle, with odd/even oscillation, but the numbers are asymptotic to

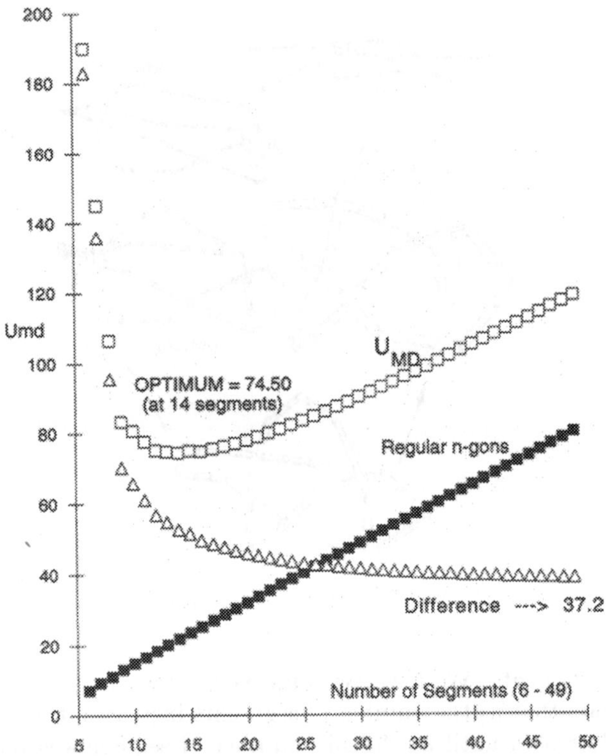

FIG. 3.2. *Observed minimun energies for trefoil knots.*

a straight line whose slope turns out to be $\frac{\pi^2}{6}$ [38]. The second observation is that the graph of U_{MD} minima is asymptotically parallel to that straight line. (While it is "obvious" from the graphs for the trefoil knot in [Figure 3.2] and the figure-8 knot in [Figure 3.3], the fact that energy minima grow asymptotically linearly in n (= number of segments) is a computational observation, not yet a theorem.) Assuming the graphs of U_{MD} minima and energies of regular n-gons are asymptotically parallel, their difference approaches a constant, and that constant is the value we can use as the topological ground state energy of the knot, $\tilde{U}^0_{MD}$. This is how (referring to [Figure 3.2] and [Figure 3.3] we obtain the values $\tilde{U}^0_{MD}$(trefoil) ≈ 37.2 and $\tilde{U}^0_{MD}$(figure-8 knot) ≈ 70.4. (Rather than do curve fitting to a model we have yet to postulate and justify, the estimates in Table 1.2 are simply the values obtained from the largest segment number we have calculated for the given knot type; these segment numbers vary from knot to knot, which is one reason we expect to sharpen the table in the future.)

We need a great deal more computational data, or a theorem, to solidify the belief that all knot types exhibit the kind of convergence described

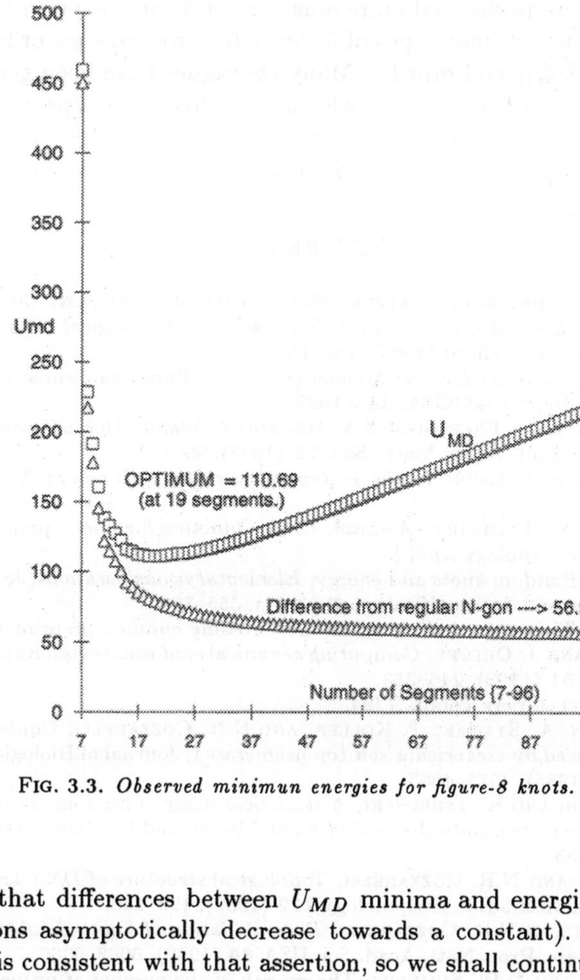

FIG. 3.3. *Observed minimun energies for figure-8 knots.*

above (i.e. that differences between U_{MD} minima and energies of regular planar n-gons asymptotically decrease towards a constant). But all our data so far is consistent with that assertion, so we shall continue to assume it as a working hypothesis. This approach is very appealing because it allows us to use any one computation of a minimum to estimate the limiting value, along with the belief that if one computes a minimum for a still larger number of segments, the estimate will be more accurate.

In comparing the energy minima for U_{MD} and for $\min E_{\cos}$ in Table 1.2, it seemed initially plausible to conjecture that $\min E_{\cos}\langle K\rangle = 2\tilde{U}^0_{MD}\langle K\rangle - 4$, because one might think of approximating E_2 for some smooth knot K by computing U_{MD} for an inscribed polygon and then doubling. There is excellent agreement for the trefoil, but the data so far do not support the conjecture for other knots. The provocative similarity in how the two energies rank knots remains a challenge to further study.

Acknowledgements. Kenneth Hunt's programs have been essential for computations; also he was the first to observe the unimodality (with segment number) of energy minima of U_{MD} for a given knot type.

Aaron Trautwein performed numerous computational experiments and determined that prime knot types of 6 through 9 crossings are ordered almost identically by $\tilde{U}^0_{MD}$ and min E_2. Many colleagues have been generous with their enthusiasm, ideas, and knowledge: for this paper, special thanks are due to C. Benham, G. Buck, N. Cozzarelli, E.J.J. VanRensburg, D. Roseman, S. Spengler, A. Stasiak, N. Stellwagen, and J. Sullivan.

REFERENCES

[1] H.W. Benjamin, M.M. Matzuk, M.A. Krasnow, and N.R. Cozzarelli, *Recombination site selection by Tn3 resolvase: topological tests of a tracking mechanism*, Cell **40** (1985), 147–158.

[2] K. Brakke, *Surface Evolver Manual* (v. 1.92), Minnesota Geometry Center, Research Report **GCG55**, July 1993.

[3] S. Bryson, M.H. Freedman, Z.X. He, and Z. Wang, *Mobius invariance of knot energy*, Bull. Amer. Math. Soc. **28** (1993), 99–103.

[4] G. Buck and J. Simon, *Knots as dynamical systems*, Topology Appl. **51** (1993), 229–246.

[5] G. Buck and J. Orloff, *A simple energy function for knots*, preprint 12/92 (to appear, Topology Appl.).

[6] G. Buck, *Random knots and energy: Elementary considerations*, Journal of Knot Theory and its Ramifications **3** (1994), 355–364.

[7] G. Buck, *The projection energy bounds crossing number*, preprint 10/93.

[8] G. Buck and J. Orloff, *Computing canonical conformations for knots*, Topology Appl. **51** (1993), 246–253.

[9] N. Cozarelli, pers. comm. 7/94.

[10] F.B. Dean, A. Stasiak, T. Koller, and N.R. Cozzarelli, *Duplex DNA knots produced by escherichia coli topoisomerase* I, Journal of Biological Chemistry **260** (1985), 4975–4983.

[11] T. Deguchi and K. Tsurusaki, *A statistical study of random knotting using the Vassiliev invariants*, Journal of Knot Theory and its Ramifications **3** (1994), 321–353.

[12] P. Dröge and N.R. Cozzarelli, *Topological structure of DNA knots and catenanes*, Methods in Enzymology **212** (1992), 120–130.

[13] P. Dröge and N.R. Cozzarelli, *Recombination of knotted substrates by Tn3 resolvase*, Proc. Natl. Acad. Sci. USA **86** (1989), 6062–6066.

[14] C. Ernst and D.W. Sumners, *The growth of the number of prime knots*, Math. Proc. Camb. Phil. Soc. **102** (1987), 303–315.

[15] M. Freedman and Z.-X. He, *Divergence free fields: Energy and asymptotic crossing number*, Ann. Math. **133** (1991), 189–229.

[16] M. Freedman, X. He, [and Z. Wang], *On the "energy" of knots and unknots*, preprint 12/91 [12/92].

[17] S. Fukuhara, *Energy of a knot*, Fete of Topology (Matsumoto et al., Eds.), Academic Press, New York, 1988, 443–451.

[18] R. Kanaar, A. Klippel, E. Shekhtman, J.M. Dungan, R. Kahmann, and N.R. Cozzarelli, *Processive recombination by the phage Mu Gin system: Implications fo rthe mechanisms of DNA strand exchange, DNA site alignment, and enhancer action*, Cell **62** (1990), 353–366.

[19] D. Kim and R. Kusner, *Torus knots extremizing the Möbius energy*, Experimental Math. **2** (1993), 1–9.

[20] R. Kusner and J. Sullivan, *Möbius energies for knots and links, surfaces and submanifolds*, preprint 4/94.

[21] H.A. Lim and E.J. Janse van Rensburg, *A numerical simulation of electrophoresis of knotted DNA*, Supercomputer Computations Research Institute (Florida

State Univ.), report **#FSU-SCRI-91-163**, to appear J. Modelling Sci. Comput., Oxford.

[22] H.A. LIM, M.T. CARROLL, AND E.J. VAN RENSBURG, *Electrophoresis of knotted DNA in a regular and random electrophoretic medium*, Biomedical Modeling and Simulation (J. Eisenfeld et al., Eds.), Elsevier Science Pub., New York, 1992, 213–223.

[23] S. LOMONACO, *The modern legacies of Thompson's atomic vortex theory in classical electrodynamics*, Amer. Math. Soc. Proc. in Appl. Math., to appear; also pers. comm., preprint 1994, and talks at several meetings.

[24] J.P.J. MICHELS AND F.W. WIEGEL, *On the topology of a polymer ring*, Proc. Roy. Soc. A **403** (1986), 269–284.

[25] J.P.J. MICHELS AND F.W. WIEGEL, Phys. Let. **90A** (1982), 381–384.

[26] K. MILLETT, *Knotting of regular polygons in 3-space*, Journal of Knot Theory and its Ramifications, **3** (1994), 263–278.

[27] J. MILNOR, *On the total curvature of knots*, Ann. Math. **52** (1950), 248–257.

[28] K. MOFFATT, *The energy spectrum of knots and links*, Nature **347** (Sept. 1990), 367–369.

[29] J. O'HARA, *Energy of a knot*, Topology **30** (1991), 241–247.

[30] J. O'HARA, *Family of energy functionals of knots*, Topology Appl. **48** (1992), 147–161.

[31] J. O'HARA, *Energy functionals of knots*, Topology-Hawaii (K.H. Doverman, Ed.), (Proc. of 1991 conference), World Scientific, Singapore, 1992, 201–214.

[32] J. O'HARA, *Energy functionals of knots II*, Topology Appl. **56** (1994), 45–61.

[33] N. PIPPENGER, *Knots in random walks*, Disc. Appl. Math. **25** (1989), 273-278.

[34] R. RANDELL, *An elementary invariant of knots*, Journal of Knot Theory and its Ramifications, **3** (1994), 279–286.

[35] E.J. JANSE VAN RENSBURG AND S.G. WHITTINGTON, *The knot probability in lattice polygons*, J. Phys. A: Math. Gen. **23** (1990), 3573–3590.

[36] E.J. JANSE VAN RENSBURG AND S.G. WHITTINGTON, *The dimensions of knotted polygons*, J. Phys. A: Math. Gen **24** (1991), 3935-3948.

[37] E.J.J. VAN RENSBURG AND S.D. PROMISLAW, *Minimal knots in the cubic lattice*, preprint 12/93.

[38] At the July 1994 IMA conference, after the talk which included the data and computationally estimated slope of 1.647, several colleagues (including E.J.J. Van Rensburg, A. Stasiak, and J. Sullivan) suggested that the limiting slope was, in fact, $\frac{\pi^2}{6}$, and this has been verified.

[39] V.Y. RYBEKOV, N.R. COZZARELLI, AND A.V. VOLOGODSKII, *Probability of DNA knotting and the effective diameter of the DNA double helix*, Proc. Natl. Acad. Sci. USA **90** (1993), 5307-5311.

[40] S.Y. SHAW AND J.C. WANG, Science **260** (1993), 533.

[41] S.Y. SHAW AND J.C. WANG, *DNA knot formation is aqueous solutions*, Journal of Knot Theory and its Ramifications, **3** (1994), 287–298.

[42] S. SPENGLER, A. STASIAK, AND N.R. COZZARELLI, *The stereostructure of knots and catenanes produced by phage λ integrative recombination: implications for mechanism and DNA structure*, Cell **42** (1985), 325–334.

[43] J.K. SIMON, *Energy functions for polygonal knots, Journal of Knot Theory and its Ramifications*, **3** (1994), 299–320.

[44] In a conversation during the July 1994 IMA conference, Andrzej Stasiak suggested the appealing term *in virtuo* to describe computer experiments.

[45] DE WITT SUMNERS AND S.G. WHITTINGTON, *Knots in self-avoiding walks*, J. Phys. A: Math. Gen. **21** (1988), 1689–1694.

[46] M.C. TESI, E.J. JANSE VAN RENSBURG, E. ORLANDINI, AND S.G. WHITTINGTON, *Knot probability for lattice polygons in confined geometries*, J. Phys. A: Math. Gen. **27** (1994), 347–360.

[47] M.C. TESI, E.J. JANSE VAN RENSBURG, E. ORLANDINI, D.W. SUMNERS, AND S.G. WHITTINGTON, *Knotting and supercoiling in circular DNA: A model incorpo-*

rating the effect of added salt, Phys. Rev. E: **49** (1994), 868–872.

[48] A.V. Vologodski, A.V. Lukashin, M.D. Frank-Kaminetskii, and A.V. Ahshelevich, *The knot probability in statistical mechanics of polymer chains*, Sov. Phys. JETP **39** (1974), 1059–1063.

[49] S. Wasserman and N.R. Cozzarelli, *Supercoiled DNA-directed knotting by T4 topoisomerase*, Journal of Biological Chemistry **266** (1991), 20567–20573.

[50] S. Whittington, *Topology of polymers*, in New Scientific Applications of Geometry and Topology (D.W. Sumners, Ed.), Amer. Math. Soc. PSAM **45** (1992), 73–95.

THE ELASTIC ROD PROVIDES A MODEL FOR DNA AND ITS FUNCTIONS

JOHN E. HEARST* AND YAOMING SHI*

1. Introduction. The processes of transcription and replication are catalysed by processive enzyme complexes which move translationally along the DNA helix, unwinding the DNA helix ahead of the complex and reforming a duplex helix behind the complex (Gamper and Hearst 1982; Cook et al. 1994). These processes are known to torsionally stress DNA.

The biological implications of the torsional tension associated with transcription are potentially significant (Lilley and Higgins 1991; Pruss and Drlica 1989), and a fundamental question is the extent to which transcription determines the level of DNA supercoiling *in vivo*. Transcription can induce supercoiling of the template by virtue of the topological relationship between DNA and elongating RNA polymerase (Liu and Wang 1987). Some models of transcription elongation require that polymerase follow the helical screw of the DNA such that there is one 360° rotation between the enzyme and DNA for each 10.5 bp transcribed (Gamper and Hearst 1982). Since RNA polymerase elongates at the rate of about 40 nucleotides per sec, an efficiently anchored transcription complex should introduce approximately four negative superturns upstream and four positive superturns downstream from an actively expressed gene each second. This would suggest extraordinarily fast rates of localized supercoiling after the onset of transcription.

The idea that transcription can introduce localized supercoiling in the DNA template was originally based on several studies of the topology of plasmid pBR322 in *E. coli*. Inhibition of DNA gyrase by novobiocin results in the formation of positively supercoiled pBR322 topoisomers (Lockshon and Morris 1983). Conversely, transcription of the *tet* gene on pBR322 in $\Delta topA$ strains leads to the accumulation of a heterogeneous population of topoisomers, some of which are hyper-negatively supercoiled (Pruss and Drlica 1986). These experiments led Liu and Wang (1987) to propose that the relative rotation of RNA polymerase about DNA (Gamper and Hearst 1982) can lead to driven rotation of the double helix. When the DNA template is topologically anchored, this introduces positive supercoils downstream from and negative supercoils upstream from the elongating polymerase. The absence of gyrase (Lockshon and Morris 1983) or topoisomerase I (Pruss and Drlica 1986) in these early studies enabled the detection of this supercoiling due to transcription as a change in the linking number of pBR322 DNA. Many studies have verified the essential features of this model both *in vitro* and in bacteria and yeast (Brill and Sternglanz

* Department of Chemistry, University of California, Berkeley, California 94720-1460.

1988; Giaever and Wang 1988; Kim and Wang 1989; Ostrander et al. 1990; Rahmouni and Wells 1989; Tsao et al. 1989; Wu et al. 1988). Cook et al. (1992) have recently demonstrated that under special conditions DNA does indeed rotate one turn about the helix axis for every 10.5 bases of RNA synthesized.

The ultimate goal of this theoretical work is a mathematical description of transcription and replication. The necessary conclusion from experimental results with prokaryotes is that changes in DNA torsional tension associated with transcription, and by inference those associated with other template utilizing processes such as replication and DNA repair, do occur, but are restricted to local regions in the vicinity of the template utilizing events.

In eukaryotes these issues are still more complex: DNA in the eukaryotic cell is not generally under torsional tension, the way it is in the prokaryote, yet local regions of torsional tension have been reported (Sinden and Ussery 1992).

Other structural features of the eukaryotic genome include 11 and 30 nanometer fibers. The 11 nm fiber, 11 nanometers in diameter, is made up of a linear chain of nucleosomes, or protein core units, around which DNA molecules are helically wrapped. This 11 nm fiber itself wraps into a helical structure, consisting of six nucleosomes per turn, with an overall diameter of 30 nanometers. This is the 30 nanometer fiber. The 30 nm fiber is the most prevalent structure observed by electron microscopy in the interphase nucleus. From the point of view of the DNA, the 11 nm fiber is a linear single helix which describes the position of the duplex cylindrical axis of the DNA. Such a linear helix can be represented by periodic boundary conditions applied to the ends of the helix. The DNA molecule in the 30 nm fiber has the form of a helix-on-a-linear-helix, again with periodic boundary conditions.

The stationary solutions for the elastic rod which are presented here are the first step in an effort which must include the mathematical descriptions of all of the phases mentioned above, as well as the dynamics of phase changes occurring between them.

The use of a rod which is isotropic with respect to its bending properties as a model for DNA dates back to Hearst and Stockmayer (1962), when it was demonstrated that, at the least, the hydrodynamic properties of DNA solutions could be modeled using such an approximation. Harris and Hearst (1966) wrote the first statistical mechanical theory for this model. While one would expect that bending the DNA towards one groove or the other might result in different bending force constants in different directions, DNA is rather stiff relative to the periodicity of the duplex helix. The persistence length of DNA is approximately 155 basepairs while there are only 10.4 basepairs per turn of duplex helix. Therefore, there are approximately 15 full twist turns of duplex for each persistence length. Although anisotropy of bending associated with the helix might slightly modulate the shape of the backbone space curve at the periodicity of the helix, the larger scale properties of the rod will not be markedly affected by such a fine scale modulation. On the other hand, if there are major structural features which locally alter the elastic properties of the rod, such as a protein induced kink, or even a naturally denatured region, the boundary conditions required to describe the elastic rod must be altered to take such features into account.

2. The Kirchhoff rod model of DNA. The mathematics of elastic rods has been studied for over two centuries since the days of Daniel Bernoulli and Euler in the 1730's (Truesdell 1960, 1987). In 1859 Kirchhoff discovered that the equations that describe the thin elastic rod in equilibrium are mathematically identical to those used to describe the dynamics of the heavy top (Kirchhoff 1859, 1876; Clebsch 1862, 1883; Love 1927). The problem of the heavy symmetric top (now known as the Lagrange top) was solved exactly by Lagrange in 1788 (Lagrange 1788; Whittaker 1965; MacMillan 1936). Although it was very successful, Kirchhoff's analogy solved only the initial value problem of the thin symmetric rod in equilibrium. It did not solve the boundary value problem where the Cartesian coordinates are specified at both ends and the direction of the force in the rod is unknown. Besides the boundary value problem, stability and bifurcation problems have been and continue to be of considerable interest to many people. The literature of elasticity theory has also moved on to more sophisticated integrable (or nonintegrable) models that describe effects of shear and compression as well as bend and twist, etc. (See *e.g.* Ericksen and Truesdell 1958, Cohen 1966, Antman 1972, Reissner 1973, Maddocks 1984, Simo, Marsden and Krishnaprasad 1988, and the references therein.)

Because the thin elastic rod is often used as a model for the DNA molecule, there has been a recent resurgence of interest in applying the theory for the Kirchhoff elastic rod to the phenomenon of DNA supercoiling (Benham 1977, 1979, 1983, 1987; LeBret 1979, 1984; Tanaka and Takahashi 1985; Wadati and Tsuru 1986; Tsuru and Wadati 1986; Hunt and Hearst 1991; Hearst and Hunt 1991; Schlick and Olson 1992; Yang et al. 1993; Coleman et al. 1993; Vologodskii and Cozzarelli 1993). In particular, the mathematics of closed DNA molecules with a linking number deficit or excess, ΔLk, has become important. According to the rules of topology, such a linking number deficit or excess must manifest itself as a combination of writhe of the axis of the DNA duplex helix as a space curve, Wr, and twist excess or deficit of the DNA duplex helix, ΔTw. *In vivo*, DNA is always found with an average linking number deficit of -0.05 turns per turn of DNA duplex helix. This linking number deficit is very important in reducing the number of configurational states allowed to DNA in nature, resulting in a substantial reduction of the volume occupied by the genetic carrier *in vivo* (Hearst and Hunt 1991). The low entropic state is nevertheless capable of conformational transitions which are suspected of being important in gene regulation and cell cycle dynamics. It is thus essential that an appropriate mathematics be developed to model the high linking number deficit states of DNA.

A variety of techniques has been used to model these systems, including Monte Carlo calculations (Vologodskii and Cozzarelli 1993), finite element analysis (Schlick and Olson 1992; Yang et al. 1993), and Euler angle method (Benham 1977, 1979, 1983, 1987; LeBret 1979, 1984; Tanaka and Takahashi 1985; Wadati and Tsuru 1986; Tsuru and Wadati 1986). The former two, while providing numerical solutions, lack parametric closed solutions for most of the interesting physical properties. Little physical insight and expandability are provided by these techniques.

The Euler angle method provides a way for writing down differential equations describing such systems. These equations are usually highly nonlinear, in terms of the Euler angles $\{\theta(s), \varphi(s), \psi(s)\}$, and thus closed-form solutions to these equations have been elusive. We have recently demonstrated that focusing on curvature $\kappa(s)$ and geometric torsion $\tau(s)$, as opposed to focusing on the Euler angles $\{\theta(s), \varphi(s), \psi(s)\}$, leads to well known differential equations with known solutions for the simplest cases. It is likely that this insight will provide a path to the solutions of the more difficult problems as well.

3. Strategy for solving the equilibrium equations. The DNA duplex is usually treated as an inextensible and unshearable thin elastic rod with circular cross section, characterized by a twist constant, C, a bending constant, A, and a radius of cross section, r_{DNA}. Here and elsewhere in this manuscript "DNA" and "rod" have the same meaning, as do "centerline of the rod" and "axis of the DNA."

At each point s on the centerline, $\boldsymbol{R}(s)$, of the thin rod, a localized Cartesian coordinate frame (rod frame) $\{\boldsymbol{e}_1(s), \boldsymbol{e}_2(s), \boldsymbol{e}_3(s)\}$ is affixed with $\boldsymbol{e}_3(s) = \boldsymbol{t}(s)$ (the unit tangent vector) and with $\boldsymbol{e}_1(s)$ and $\boldsymbol{e}_2(s)$ in the directions of the principal axes of inertia of its cross section. The variable, s, is a contour parameter analogous to time for the Lagrange (heavy symmetric) top. The localized coordinate frame at $s + \Delta s$ is obtained by an infinitesimal rotation $\Delta\Theta$ of the coordinate frame at s . The deformed state of the axis of the thin rod is determined by the Darboux (curvature) vector $\boldsymbol{\omega} = (\omega_1, \omega_2, \omega_3) = \lim_{\Delta s \to 0}(\Delta\Theta/\Delta s)$, which is analogous to the angular velocity of the Lagrange top.

At a given position (say $s = s_0$) along the centerline, there is a cross section upon which internal forces are exerted. One side of the cross section ($s < s_0$) acts on the other side ($s > s_0$) and *vice versa*. The internal forces are resolvable into a force $\boldsymbol{F}(s_0)$ and a torque $\boldsymbol{M}(s_0)$. At each cross section such a force and torque may be found, giving rise to functions $\boldsymbol{F}(s)$ and $\boldsymbol{M}(s)$ describing a system of stresses on the rod. The force, $\boldsymbol{F}$, is analogous to gravity in the Lagrange top problem, whereas the torque, $\boldsymbol{M}$, is analogous to the angular momentum. For a recent complete comparison of the thin elastic rod and the Lagrange top, see Table 1 of Benham (1977).

If $\boldsymbol{F}^{(e)}(s)$ and $\boldsymbol{M}^{(e)}(s)$ are the externally applied force and torque per unit length, then the stationary state conditions, in the *body fixed frame* (or rod frame), are (Love 1927, Landau and Lifshitz 1959):

$$\frac{d}{ds}\boldsymbol{F}(s) + \boldsymbol{\omega}(s) \times \boldsymbol{F}(s) + \boldsymbol{F}^{(e)}(s) = 0 \tag{3.1a}$$

$$\frac{d}{ds}\boldsymbol{M}(s) + \boldsymbol{\omega}(s) \times \boldsymbol{M}(s) + \boldsymbol{t}(s) \times \boldsymbol{F}(s) + \boldsymbol{M}^{(e)}(s) = 0 \tag{3.1b}$$

where $\mathbf{t}(s)$ is a unit tangent vector along the centerline. The case $\{\boldsymbol{F}^{(e)}, \boldsymbol{M}^{(e)}\} = \{0, 0\}$ is analogous to that of a freely spinning top.

For small local deformations of the thin rod, the torque $\boldsymbol{M}(s)$ is related to the curvature vector $\boldsymbol{\omega}(s)$ by the linear constitutive relation (Hooke's law):

$$\boldsymbol{M}(s) = \boldsymbol{I} \cdot (\boldsymbol{\omega}(s) - \boldsymbol{\omega}^{(0)}) \tag{3.2}$$

where $\boldsymbol{I}$ denotes the stiffness tensor of the thin rod and is diagonal in the rod frame; namely, $\boldsymbol{I} = A\boldsymbol{e}_1\boldsymbol{e}_1 + A\boldsymbol{e}_2\boldsymbol{e}_2 + C\boldsymbol{e}_3\boldsymbol{e}_3$. The $\boldsymbol{\omega}^{(0)} = \omega_3^{(0)}\boldsymbol{e}_3$ is the intrinsic constant curvature vector expressed in the rod frame. In other words, the relaxed state of DNA is assumed to be straight but is twisted by $\omega_3^{(0)}$ radian per unit length. (The DNA duplex helix contains 10.4 basepairs per turn for which $\omega_3^{(0)} = 1.78$ radian/nm.)

In the absence of intrinsic constant curvature and external force and torque, i.e., $(\boldsymbol{\omega}^{(0)}, \boldsymbol{F}^{(e)}(s), \boldsymbol{M}^{(e)}(s)) = (0,0,0)$, Eq.(3.1) is dynamically equivalent, by the Kirchhoff kinematic analog, to that describing the motion of a Lagrange top. The conventional method for solving these equations is to express the curvature vector $(\omega_1, \omega_2, \omega_3)$ in terms of three Euler angles (θ, φ, ψ) with respect to the laboratory frame; and then solve that equation for $u = \cos\theta$ as a function of the contour length parameter, s, in terms of elliptical functions and then solve for angles (φ, ψ) in terms of $u = \cos\theta$; and then determine the laboratory position of the DNA axis by quadratures; namely,

$$x(s) = \int_0^s \sin\theta\cos\varphi ds,\; y(s) = \int_0^s \sin\theta\sin\phi ds, \text{ and } z(s) = \int_0^s \cos\theta ds.$$

This Euler angle procedure has been extensively studied by Benham, LeBret, and Wadati and Tsuru. The disadvantage of this approach is that it is very hard in general to get analytical expressions for the Cartesian coordinates $\{x(s), y(s), z(s)\}$. Thus it is even harder to relate the parameters, which determine the shape of the centerline of the rod, to known physical quantities, such as the total length L, the bending constant A, the twist constant C, and the linking number difference ΔLk if the rod is closed. In addition, this treatment cannot be applied to another class of DNA supercoiled structures which involve self-contact and which are typified by the plectonemic (interwound) conformation. Although they have been analyzed by numerical simulations, these conformations are very difficult to treat analytically.

Recently we have approached this problem from a different direction (Shi and Hearst 1994). We show that if $\omega(s) = (\omega_1(s), \omega_2(s), \omega_3(s))$ as functions of the contour parameter, s, is known, then the centerline of the rod is determined. The advantage of our approach is that, in the absence of external force and torque, we can obtain mathematical expressions for the cylindrical coordinates of the centerline of the rod and for its total elastic energy in closed form.

4. The nonlinear Schrödinger equation and closed-form solutions for the axis of DNA. We now consider that the thin elastic rod may have self- contact, and that some frictionless forces are exerted which help to maintain the stationary configurations. The self-contact force will be in the direction perpendicular to the tangent direction $\boldsymbol{t}(s)$.

Let $\boldsymbol{b}(s)$ and $\boldsymbol{n}(s)$ be the unit binormal and normal vectors at s, then we may express the most general self-contact force as

$$\boldsymbol{F}^{(e)} = A(N_b^{(e)}\boldsymbol{b} + N_n^{(e)}\boldsymbol{n}) \tag{4.1}$$

where the force components $N_b^{(e)}(s)$ and $N_n^{(e)}(s)$ are real. Since the forces are frictionless, the resulting self-contact torques are, of course, taken to be zero, $M_1^{(e)} = M_2^{(e)} = M_3^{(e)} = 0$.

A strategy of Shi and Hearst (1994) is to define a complex curvature $\eta(s)$ as $\eta(s) = e^{ias/2}[\omega_1(s) + i\omega_2(s)]$ and a complex force $\mathcal{N}(s)$ as $\mathcal{N}(s) = N_1(s) + iN_2(s)$. Substituting (4.1) into (3.1) and writing the results in component form, after much calculation (see Shi and Hearst 1994 for details), we finally obtain

$$\frac{d^2}{ds^2}\eta - (\mathcal{E} - \frac{1}{4}\mathcal{Q}^2)\eta + \frac{1}{2}|\eta|^2\eta - (N_n^{(e)} + iN_b^{(e)})\frac{\eta}{|\eta|} = 0 \tag{4.2a}$$

$$\frac{d}{ds}(\eta e^{-ias/2}) + i\omega_3\eta e^{-ias/2} - i\mathcal{Q}\eta e^{-ias/2} + i\mathcal{N} = 0 \tag{4.2b}$$

$$\frac{1}{2}|\eta|^2 + N_3 = \mathcal{E}, \qquad \lambda(\omega_3 - \omega_e^{(0)}) = \mathcal{Q} \tag{4.2c}$$

where $\mathcal{E}$ and $\mathcal{Q}$ are two constants of motion and the quantity λ is defined as $\lambda = \frac{C}{A} = \frac{1}{1+\sigma'}$, and where σ' is the "Poisson's ratio" for the DNA molecule. The parameter a in (4.2b) is determined by $(\omega_3^{(0)}, \lambda, \mathcal{Q})$.

Thus one may start with two constants of motion, $\mathcal{E}$ and $\mathcal{Q}$, as well as certain known functions for $N_b^{(e)}$ and $N_n^{(e)}$, then solve the complex curvature $\eta(s)$ from (4.2a), then solve the complex force $\mathcal{N}(s)$ from (4.2b), and finally solve N_3 from (4.2c). Thus the six unknowns of Eqs.(3.1) and (3.2), $(\omega_1, \omega_2, \omega_3, N_1, N_2, N_3)$, are solved in principle.

Eq.(4.2a) is a time-independent cubic nonlinear Schrödinger equation with an extra potential term. The complex curvature $\eta(s)$ is related to the curvature $\kappa(s)$ and geometric torsion $\tau(s)$ of the centerline of the rod, via

(4.3) $$\eta(s) = \kappa(s)\exp\{-\frac{i}{2}\mathcal{Q}s + i\int_0^s \tau(s)ds\}$$

When $N_b^{(e)} = N_n^{(e)} = 0$, Eq.(4.2a) becomes the following Euler-Lagrange Eqs:

(4.4a) $$\frac{d^2}{ds^2}\kappa - \kappa(\tau - \frac{1}{2}\mathcal{Q})^2 - (\mathcal{E} - \frac{1}{4}\mathcal{Q}^2)\kappa + \frac{1}{2}\kappa^3 = 0$$

(4.4b) $$\kappa^2(\tau - \frac{1}{2}\mathcal{Q}) = J$$

Eqs.(4.4a,b) are known to have a solution of the form

(4.5a) $$\kappa^2(s) = \alpha[1 - \frac{\mu}{v}sn^2(ws|\mu)]$$

(4.5b) $$\tau(s) = \frac{1}{2}\mathcal{Q} + \frac{J}{\kappa^2(s)}$$

where $sn(ws|\mu)$ is the Jacobi sinus amplitudinus elliptic function (Abramowitz and Stegun 1970), and where μ is the modulus of the elliptical function and α, J, v, and w are real parameter.

The $\mathcal{Q} = 0$ version of Eqs.(4.4a,b) has been extensively studied by Langer and Singer (1984). They have obtained three first order differential equations for $\{\rho(s), \phi(s), z(s)\}$. The explicit expressions for $\{\rho(s), \phi(s), z(s)\}$ can be obtained by quadratures but are not presented in their paper.

The true knowledge of the cylindrical coordinates $\{\rho(s), \phi(s), z(s)\}$ is essential to the study of DNA supercoiling. Therefore, in Shi and Hearst (1994) we generalized the integration procedure of Langer and Singer, obtaining the explicit formulas for the cylindrical coordinates $\{\rho(s), \phi(s), z(s)\}$. The results are

(4.6a) $$\rho^2(s) = a_1 + \frac{\alpha\mu}{2vN^2}[1 - 2sn^2(ws|\mu)]$$

(4.6b) $$\phi(s) = a_2 s + \frac{J'}{w\alpha'}[\Pi(\frac{\mu'}{v}; ws|\mu)]$$

(4.6c) $$z(s) = a_3 s + \frac{2w}{N} Z(ws|\mu)$$

where various parameters are defined in Table 4.1.

TABLE 4.1

We list here the formulas that one needs to calculate 12 dependent parameters $\{\alpha, \mathcal{E}, J, B, N, P, \alpha', \mu', J', a_1, a_2, a_3\}$ from the four independent parameters $\{w, \mu, v, Q\}$. The first nine were defined in Shi and Hearst (1994), and the last three are just introduced in this paper.

$$\alpha = 4w^2 v \tag{T1}$$

$$\mathcal{E} = \tfrac{1}{4}Q^2 + w^2(-1 - \mu + 3v) \tag{T2}$$

$$J^2 = 16w^6 v(1 - v)(v - \mu) \tag{T3}$$

$$B = 2w^4[\mu(2v - 1) + v(2 - 3v)] \tag{T4}$$

$$N^2 = 2B - JQ + \mathcal{E}^2 \tag{T5}$$

$$J = Q\mathcal{E} - PN \tag{T6}$$

$$\alpha' = \alpha + Q^2 - P^2 \tag{T7}$$

$$\mu' = \tfrac{\alpha}{\alpha'}\mu \tag{T8}$$

$$J' = P\mathcal{E} - QN + \tfrac{P}{2}(Q^2 - P^2) \tag{T9}$$

$$a_1 = \tfrac{\alpha'}{N^2}[1 - \tfrac{\mu'}{2v}] \tag{T10}$$

$$a_2 = \tfrac{J'}{\alpha'}\Pi(\tfrac{\mu'}{v}|\mu)K^{-1}(\mu) - \tfrac{P}{2} \tag{T11}$$

$$a_3 = \tfrac{1}{N}[-\mathcal{E} + 2w^2(v - 1) + 2w^2(v - 1) + 2w^2 E(\mu)K^{-1}(\mu)] \tag{T12}$$

In (T10)-(T12) the functions $K(\mu), E(\mu)$, and $\Pi(x|\mu)$ are the complete elliptic integral of the first kind, second kind, and third kind, respectively.

In (4.6) the function $Z(x|\mu)$, known as the Jacobi's zeta function, defined as (Abramowitz and Stegun 1970)

(4.7) $$Z(x|\mu) \equiv E(x|\mu) - xE(\mu)K^{-1}(\mu),$$

is the periodic version of the elliptic integral of the second kind, $E(x|\mu)$ (Abramowitz and Stegun 1970). The function $Y(p; x|\mu)$ is defined (in Appendix D of Shi and Hearst 1994) as the periodic version of the elliptic integral of the third kind, $\Pi(p; x|\mu)$ (Abramowitz and Stegun 1970); namely,

$$Y(p; x|\mu) \equiv \Pi(p; x|\mu) - x\Pi(p|\mu)K^{-1}(\mu) \tag{4.8}$$

In (4.7) and (4.8) the functions $K(\mu)$, $E(\mu)$, and $\Pi(p|\mu)$ are the complete elliptical integrals of first kind, second kind, and third kind, respectively (Abramowitz and Stegun 1970).

In an effort to visualize the geometric forms represented by Eqs.(4.6a-c) we may rewrite them as

$$\rho^2(s) = a_1 + f_1(s) \tag{4.6d}$$

$$\phi(s) = a_2 s + f_2(s) \tag{4.6e}$$

$$z(s) = a_3 s + f_3(s) \tag{4.6f}$$

where $f_i(s)$ $(i = 1, 2, 3)$ are periodic functions of s [with period $2K(\mu)/w$, the same as that of $sn^2(ws|\mu)$]. Thus we say that the general solution (4.6) can be viewed as a helix-on-a-linear-helix, i.e., DNA wraps around a rod whose centerline is itself a linear helix. The first terms of the cylindrical coordinates, $\{\sqrt{a_1}, a_2 s, a_3 s\}$, describe the linear helix and the second terms, $\{f_1(s), f_2(s), f_3(s)\}$, describe the wrapping of DNA around the rod whose centerline is described by the first terms.

In Table 4.1 we show the formulas that one needs to calculate 12 dependent parameters $\{\alpha, \mathcal{E}, J, B, N, P, \alpha', \mu', J', a_1, a_2, a_3\}$ from the four independent parameters $\{w, \mu, v, \mathcal{Q}\}$. Thus the elliptic function solutions (4.5a,b), and (4.6a-c) for the $\{\kappa(s), \tau(s), \rho(s), \phi(s), z(s)\}$ are uniquely determined by six parameters $\{w, \mu, v, \mathcal{Q}, \lambda, L\}$.

5. Conclusions. The general solutions to the time-independent nonlinear Schrödinger equation (4.2a) represent the stationary states of the elastic rod with torsion. Closed end boundary conditions without intersegmental contact of the DNA have led to toroidal stationary states, examples of which have been presented in Shi and Hearst (1994). While these solutions are of considerable historical importance, they are not of immediate import to the representation of DNA, since plectonemic superhelical structures of the elastic rod are known to have lower energies than the corresponding toroidal forms (Hunt and Hearst 1991). The plectonemic structures can also be determined from (4.2a) and will be the topic of a subsequent manuscript.

The solutions to the nonlinear Schrödinger equation presentedhere are in a form which allows for the representation of the structural phases of chromatin. Both the linear helix and the helix-on-a-linear-helix are solutions to Eq.(4.2a). In fact, with appropriate boundary conditions, Eqs.(4.5a,b) readily represent the curvature and the torsion of such structures. While these structures are not the most stable structures accessible to free DNA, it is clear that, in the presence of external forces such as those exerted by the proteins in the 11 nm fibers of chromatin, they can become stable. Thus, the mathematical modeling of such structures depends upon the judicious choices of $N_n^{(e)}$ and $N_b^{(e)}$ of Eq.(4.2a), followed by the subsequent solution of the differential equation.

We thus consider Eq.(4.2a) to be the route to the mathematical representation of all interesting DNA structures, subject to the constraint of the in vivo linking number deficit. In this respect, the fundamental generality provided by Eq.(4.2a), the nonlinear Schrödinger equation including external forces, has great potential.

Upon determination of the static solutions of Eq.(4.2a) with certain choices of $N_n^{(e)}$ and $N_b^{(e)}$, the natural next step is the inclusion of time dependence. The solutions to the time-dependent equation will provide representation of local transitions among the chromatin 30 nm structure, the 11 nm structure, and free DNA. Such transitions may be at the heart of gene regulation. We expect that this mathematical model will play an important role in understanding the energetics and dynamics of such transitions.

Recently, we (Shi, Borovik and Hearst 1995) have found 2 sets of closed-form solutions for the centerline of a shearable, extensible, bendable, twistable rod. One set of solutions resembles the solution (4.6) of this paper, which can be viewed as a helix-on-a-linear-helix (HH). The other set is new and can be viewed as overtone-modulated-helix-on-a-linear-helix (OMHH).

REFERENCES

[1] Abramowitz, M., and Stegun, A., eds. *Handbook of mathematical functions*, 9th Printing, Dover, New York (1970). See Chapter 16 and 17.

[2] Antman, S.S. *The theory of rods*, Handbuch der Physik, Vol. VIa/2, Springer, Berlin (1972).

[3] Benham, C.J. *Elastic model of supercoiling*, Proc. Nat. Acad. Sci. USA 74: 2397–2401 (1977).

[4] Benham, C.J. *An elastic model of the large-scale structure of duplex DNA*, Biopolymers 18: 609–623 (1979).

[5] Benham, C.J. *Geometry and mechanics of DNA superhelicity*, Biopolymers 22: 2477–2495 (1983).

[6] Benham, C.J. *The role of the stress resultant in determining mechanical equilibria of superhelical DNA*, Biopolymers 26: 9–15 (1987).

[7] Brill, S.J. and Sternglanz, R. *Transcription-dependent DNA supercoiling in yeast DNA topoisomerase mutants*, Cell 54: 403–411 (1988).

[8] Clebsch, A. *Theorie der elasticität Fester Körper* , B. G. Teubner, Leipzig (1862).

[9] Clebsch, A. *Théorie de l'elasticité des corps solides*, Translation of Clebsch (1862) by Saint-Venant & Flamant, Dunod, Paris (1883).

[10] Cohen, H. *A nonlinear theory of elastic directed curves*, Int. J. Engng Sci. 4: 511–524 (1966).

[11] Coleman, B.D., Dill, E.H., Lembo, M., Lu, Z. and Tobias, I. *On the dynamics of rods in the theory of Kirchhoff and Clebsch*, Arch. Rational Mech. Anal. 121: 339–359 (1993).

[12] Cook, D.N., Ma, D. and Hearst, J.E. *Supercoiling induced by transcription*, Nucleic Acids and Molecular Biology 8: 133–146 (1994).

[13] Cook, D.N., Ma, D., Pon, N.G. and Hearst, J.E. *Dynamics of DNA supercoiling by transcription in E. coli*, Proc. Natl. Acad. Sci., USA 89: 10603–10607 (1992).

[14] Ericksen, J.L. and Truesdell, C. *Exact theory of stress and strain in rods and shells*, Arch. Rational Mech. Anal. 1: 293–323 (1958).

[15] Gamper, H.B. and Hearst, J.E. *A topological model for transcription based on unwinding angle analysis of E. coli RNA polymerase binary, initiation, and ternary complexes*, Cell 29: 81–90 (1982).

[16] Giaever, G.N. and Wang, J.C. *Supercoiling of intracellular DNA can occur in eukaryotic cells*, Cell 55: 849–856 (1988).

[17] Harris, R.A. and Hearst, J.E. *On polymer dynamics*, Journal of Chemical Physics 44: 2595–2602 (1966).

[18] Hearst, J.E. and Hunt, N.G. *Statistical mechanical theory for the plectonemic DNA supercoil*, J. Chem. Phys. 95: 9322–9327 (1991).

[19] Hearst, J.E. and Stockmayer, W.E. *Sedimentation constants of broken chains and wormlike coils*, Journal of Chemical Physics 37: 1425–1433 (1962).

[20] Hunt, N.G. and Hearst, J.E. *Elastic model of DNA supercoiling in the infinite-length limit*, J. Chem. Phys. 95: 9329–9336 (1991).

[21] Kim, R.A. and Wang, J.C. *A subthreshold level of DNA topoisomerases leads to the excision of yeast rDNA as extrachromosomal rings*, Cell 57: 975–985 (1989).

[22] Kirchhoff, G. *Über das Gleichgewicht und die Bewegung eines unendlich dünnen elastischen Stabes*, J. f. reine. angew. Math. (Crelle) 56: 285–313 (1859).

[23] Kirchhoff, G. *Vorlesungen über mathematische physik, mechanik*, Vol. 28, B. G. Teubner, Leipzig (1876).

[24] Lagrange, J.L. *Mecanique analytique*, Paris, Chez la Veuve Desaint (1788).

[25] Landau, L.D. and Lifshitz, E.M. *Theory of Elasticity*, 3rd ed., Pergamon, New York, Sec. 19 (1959).

[26] Langer, J. and Singer, D.A. *Knotted Elastic Curves in R^3*, J. London Math. Soc. 30: 512–520 (1984).

[27] Le Bret, M. *Catastrophic variation of twist and writhing of circular DNAs with constraint?*, Biopolymers 18: 1709–1725 (1979).

[28] Le Bret, M. *Twist and writhing in short circular DNAs According to first-order elasticity*, Biopolymers 23: 1835–1867 (1984).

[29] Lilley, D.M.J. and Higgins, C.F. *Local DNA topology and gene expression: the case of the leu-500 promoter*, Mol. Microbiol. 5: 779–783 (1991).

[30] Liu, L.F. and Wang, J.C. *Supercoiling of the DNA template during transcription*, Proc. Natl. Acad. Sci., USA 84: 7024–7027 (1987).

[31] Lockshon, D. and Morris, D.R. *Positively supercoiled plasmid DNA is produced by treatment of Escherichia coli with DNA gyrase inhibitors*, Nucleic Acids Res. 11: 2999–3017 (1983).

[32] Love, A.E.H. *A treatise on the mathematical theory of elasticity*, 4th ed. Dover, New York, Sec. 260 (1927).

[33] MacMillan, W.D. *Dynamics of rigid bodies*, McGraw-Hill, New York, Secs. 102–112 (1936).

[34] Maddocks, J.H., *Stability of nonlinearly elastic rods*, Arch. Ration. Mech. Anal. 85: 311–354 (1984).

[35] Ostrander, E.A., Benedetti, P. and Wang, J.C. *Template supercoiling by a chimera of yeast GAL4 protein and phage T7 RNA polymerase*, Science 249: 1261–1265 (1990).

[36] Pruss, G.J. and Drlica, K. *Topoisomerase I mutants: the gene of pBR322 that encodes resistance to tetracycline affects plasmid DNA supercoiling*, Proc. Natl. Acad. Sci. USA 83: 8952–8956 (1986).

[37] Pruss, G.J. and Drlica, K. *DNA supercoiling and prokaryotic transcription*, Cell 56: 521–523 (1989).

[38] Rahmouni, A.R. and Wells, R.D. *Stabilization of Z DNA in vivo by localized supercoiling*, Science 246: 358–363 (1989).

[39] Reissner, R. *On a one-dimensional, large-displacement, finite-strain beam-theory*, Stud. Appl. Math. 52: 87–95 (1973).

[40] Schlick, T. and Olson, W.K. *Supercoiled DNA energetics and dynamics by computer simulation*, J. Mol. Biol. 223: 1089–1119 (1992).

[41] Shi, Y. and Hearst, J.E. *The Kirchhoff elastic rod, the nonlinear Schrödinger equation, and DNA supercoiling*, J. Chem. Phys. 101(6): 5186–5200 (1994).

[42] Shi, Y., Borovik, A.E. and Hearst, J.E. *Elastic rod model incorporating shear and extension, generalized nonlinear Schrödinger equations, and novel closed-form solutions for supercoiled DNA*, J. Chem. Phys. 103: 3166–3183 (1995).

[43] Simo, J.C., Marsden, J.E. and Krishnaprasad, P.S. *The Hamiltonian structure of nonlinear elasticity: the material and convective representations of solids, rods, and plates*, Arch. Ration. Mech. Anal. 104: 125–183 (1988).

[44] Sinden, R.R. and Ussery, D.W. *Analysis of DNA structure in vivo using psoralen photobinding: measurement of supercoiling, topological domains, and DNA-protein interactions*, Methods in Enzymology 212: 319–335 (1992).

[45] Tanaka, F. and Takahashi, H. *Elastic theory of supercoiled DNA*, J. Chem. Phys. 83: 6017–6026 (1985).

[46] Truesdell, C.A. *The rational mechanics of elastic or flexible bodies*, 1638–1788. L. Euleri Opera Omnia, Series II, Volume 11, Part 2, Füssli, Zürich (1960).

[47] Truesdell, C.A. *Origin of the theory of vibrating systems*, Res Mechanica 21: 291–311 (1987).

[48] Tsao, Y-P., Wu, H-Y., and Liu, L.F. *Transcription-driven supercoiling of DNA: direct biochemical evidence from in vitro studies*, Cell 56: 111–118 (1989).

[49] Tsuru, H. and Wadati, M. *Elastic model of highly supercoiled DNA*, Biopolymers 25: 2083–2096 (1986).

[50] Vologodskii, A.V. and Cozzarelli, N.R. *Monte Carlo analysis of the conformation of DNA catenanes*, J. Mol. Biol. 232: 1130–1140 (1993).

[51] Wadati, M. and Tsuru, H. *Elastic model of looped DNA*, Physica D 21: 213–226 (1986).

[52] Whittaker, E.T. *A treatise on the analytical dynamics of particles and rigid bodies*, 4th ed., Cambridge, London (1965).

[53] Wu, H-Y., Shyy, S., Wang, J.C. and Liu, L.F. *Transcription generates positively and negatively supercoiled domains in the template*, Cell 53: 433–440 (1988).

[54] Yang, Y., Tobias, I. and Olson, W.K. *Finite element analysis of DNA supercoiling*, J. Chem. Phys. 98: 1673–1686 (1993).

HAMILTONIAN FORMULATIONS AND SYMMETRIES IN ROD MECHANICS

DONALD J. DICHMANN*, YIWEI LI*, AND JOHN H. MADDOCKS*

1. Introduction. This article provides a survey of contemporary rod mechanics, including both dynamic and static theories. Much of what we discuss is regarded as classic material within the mechanics community, but the objective here is to provide a self-contained account accessible to workers interested in modelling DNA. We also describe a number of recent results and computations involving rod mechanics that have been obtained by our group at the University of Maryland. This work was largely motivated by applications to modelling DNA, but our approach reflects a background of research in continuum mechanics. In particular, we emphasize the role that Hamiltonian formulations and symmetries play in the effective computation of special solutions, conservation laws of dynamics and integrals of statics.

We begin in chapter 2 with an overview of the kinematics, balance laws and constitutive relations that collectively make up the director theory of rods, and which together determine the equations of motion. We then discuss some of the conservation laws which are implied by the evolution equations. In chapter 3 we describe various Hamiltonian formulations of the partial differential equations governing rod dynamics. In the case of a rod which is constrained to be inextensible and unshearable, the derivation of the Hamiltonian formulation is nonstandard and involves an approach we call the impetus-striction method. We conclude the chapter by delineating the Hamiltonian form of the invariants of motion. In chapter 4 we describe some Hamiltonian formulations of the ordinary differential equations governing the statics of rods. Following the same pattern as in chapter 3, we explicitly present Hamiltonians appropriate for both an extensible, shearable rod and for an inextensible, unshearable rod, and we then list the associated integrals. Finally in chapter 5 we discuss numerical continuation methods and describe the set of equilibria associated with what is now perhaps the prototypical problem for continuum models of DNA, namely a straight uniform rod, with circular cross-section, that is bent into a loop and then subjected to a prescribed twist (or linking difference).

In this article we have limited ourselves to a discussion of rod mechanics, and will not address the very interesting questions of how accurately a rod model actually describes DNA. In particular we do not discuss effects of solvents, nor the question of the statistical mechanics of rods. Nevertheless, the generality of the formulations we describe is somewhat suggestive

* Institute for Physical Science and Technology, and Department of Mathematics, University of Maryland, College Park, MD 20742. Research supported by AFOSR, NSF and ONR.

as a precursor to such analyses. In particular, the rod models we describe do account for natural curvature and twist, nonuniformity (as might arise from sequence dependence), anisotropy (e.g. unequal bending stiffnesses) and effects of self-contact. The Hamiltonian formulations lay bare the connections between conservation laws, integrals, special solutions, and the symmetries inherent to a straight, uniform rod, with circular cross-section. Moreover the numerical continuation methods we employ can be adapted so that the most symmetric, and by now comparatively standard case, can be exploited as an organizing center for computation of solutions in models where each of the symmetries is broken.

2. Classical Rod Mechanics

2.1. Introduction. We adopt a relatively standard model based on the special Cosserat rod theory (cf. e.g. [1]). In a Cosserat rod theory, the concept of a rod is reduced to two components: a curve and a field of directors. The curve can be interpreted as describing the center line of the rod. The director field assigns to each point along the curve a set of three vectors (called directors), which provide a basis from which to describe the local bending, twisting, stretching and shearing of the rod. In the special Cosserat rod theory, the set of directors at each point is taken to be orthonormal, connoting the idea that the shape of each cross-section of the rod does not change significantly. In this general setting, the constitutive laws which relate the forces and moments within the rod to its bending, twisting, stretching and shearing need not be regarded as being determined by any assumed properties of a three-dimensional solid material such as linear elasticity. In the case of a model for DNA, the constitutive laws might instead be derived from experiments, or might be calculated from an atomistic simulation over a short segment.

The kinematics of a rod are first described in section 2.2. Next the force and moment balance laws are presented in section 2.3. The constitutive relations are discussed in section 2.4, with particular attention given to the modelling of an inextensible, unshearable elastic rod, i.e. an elastica. In section 2.5 we delineate the known conservation laws associated with rod mechanics, including a conservation law connected with translational symmetry which was found only recently. In section 2.6 we represent the kinematics of an elastica using Euler parameters, which, for computational purposes, have proved to be a very useful parametrization of the rotation group. The material presented in sections 2.2–2.4 is largely standard [1], some of the conservation laws described in section 2.5 were first presented in [29], and the material of section 2.6 is taken from [8].

2.2. Kinematics of rods. We begin by reviewing the director description of rod kinematics used in the special Cosserat theory. The dependent variables are a vector function $\boldsymbol{r}(s,t) \in \boldsymbol{R}^3$ and a right-handed orthonormal frame of directors $\boldsymbol{d}_k(s,t) \in \boldsymbol{R}^3$, $k = 1, 2, 3$. (Hereafter subscripts will be taken to run from 1 to 3, and repeated indices should be summed unless stated to the contrary.) The independent variables are two scalars, namely undeformed arc-length s, and time t. At each time t, the curve $\boldsymbol{r}(\cdot, t)$ can be interpreted as the configuration of the center line in a long, slender elastic body. The triad $\{\boldsymbol{d}_k\}$ can be interpreted as providing information concerning the orientation of the material cross-section of the rod. (See Figure 2.1.) The kinematics of the rod are encapsulated in the relations

$$\boldsymbol{r}' = \boldsymbol{v} \ , \tag{2.2.1}$$

$$\boldsymbol{d}_k' = \boldsymbol{u} \times \boldsymbol{d}_k \ , \tag{2.2.2}$$

$$\dot{\boldsymbol{d}}_k = \boldsymbol{\omega} \times \boldsymbol{d}_k \ . \tag{2.2.3}$$

Here $\boldsymbol{d}_k'$ denotes the partial derivative with respect to arc-length s, while $\dot{\boldsymbol{d}}_k$ denotes the partial derivative with respect to t, etc. The components $v_k \equiv \boldsymbol{v} \cdot \boldsymbol{d}_k$ of the vector $\boldsymbol{v}(s,t)$ with respect to the triad $\{\boldsymbol{d}_k\}$ are the strains associated with stretching and shear, the components $u_k \equiv \boldsymbol{u} \cdot \boldsymbol{d}_k$ of the (Darboux) vector $\boldsymbol{u}$ are the strains associated with bending and twist, and the components $\omega_k \equiv \boldsymbol{\omega} \cdot \boldsymbol{d}_k$ of the vector $\boldsymbol{\omega}(s,t)$ can be regarded as the director components of the angular velocity of the material cross-section at s when viewed as a rigid body. We will use u to denote the set of components (u_1, u_2, u_3), and similarly use v to denote (v_1, v_2, v_3).

Because v_k, u_k and ω_k are components with respect to a variable frame we obtain relations such as

$$\boldsymbol{\omega}' = \omega_k' \boldsymbol{d}_k + \boldsymbol{u} \times \boldsymbol{\omega} \ , \tag{2.2.4}$$

and

$$\boldsymbol{r}'' = \boldsymbol{v}' = v_k' \boldsymbol{d}_k + \boldsymbol{u} \times \boldsymbol{v} \ . \tag{2.2.5}$$

(By convention expressions such as v_k' involve the derivative of the kth component of $\boldsymbol{v}$ with respect to the triad $\{\boldsymbol{d}_j\}$, which is of course not the same as the kth component of $\boldsymbol{v}'$.) Equations (2.2.2), (2.2.3) and (2.2.4) in the identities

$$\frac{\partial^2}{\partial s \partial t} \boldsymbol{d}_k = \frac{\partial^2}{\partial t \partial s} \boldsymbol{d}_k \tag{2.2.6}$$

imply compatibility conditions such as

$$\dot{\boldsymbol{u}} = \omega_k' \boldsymbol{d}_k \ , \tag{2.2.7}$$

which place restrictions upon initial data in rod dynamics.

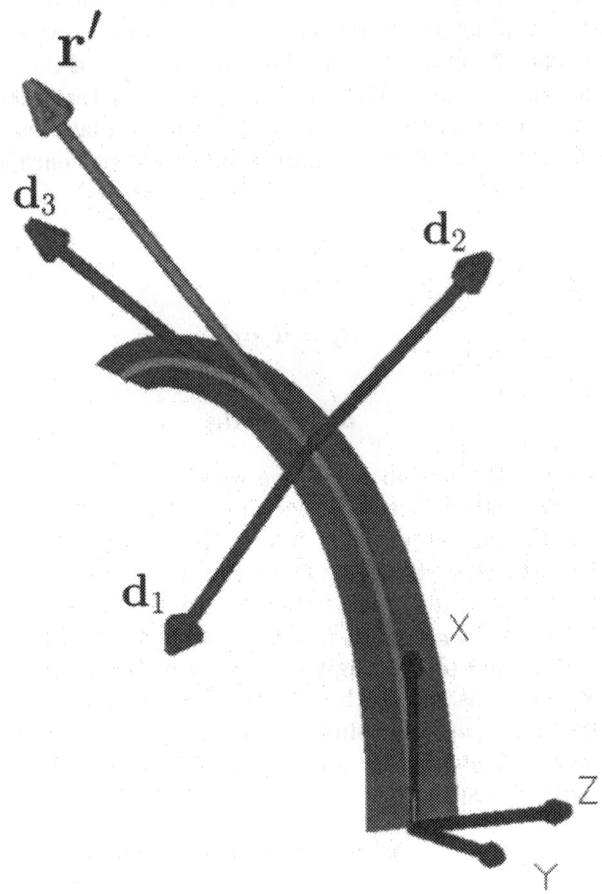

FIG. 2.1 *A ribbon representation of the configuration of a rod. The center curve is parameterized by undeformed arc-length s. The location of the material point at s is given by its position $\boldsymbol{r}(s)$ with respect to the origin. The directors $\boldsymbol{d}_1$, $\boldsymbol{d}_2$ and $\boldsymbol{d}_3$ describe the orientation of the cross-section of the rod at s, with $\boldsymbol{d}_1$ determining the ribbon. The components of $\mathbf{r}'$ with respect to the $\boldsymbol{d}_i$ basis determine the strains of shear and extension.*

2.3. Balance laws. The densities of linear and angular momentum are

$$\boldsymbol{p}(s,t) \equiv \rho(s)\dot{\boldsymbol{r}}, \tag{2.3.1}$$

(2.3.2) $$\boldsymbol{\pi}(s,t) \equiv I_{ij}(s)\omega_i \boldsymbol{d}_j.$$

Here $\rho(s)$ is the mass per unit arc-length associated with the material cross-section at the point s, which is determined entirely by the reference configuration. The quantities $I_{ij}(s)$ are the components of the positive-definite inertia tensor of the material cross-section at s expressed with respect to the triad $\{\boldsymbol{d}_k\}$. In the special Cosserat rod model, it is assumed that these inertia coefficients are independent of the strains u_k and v_k. Thus the quantities $I_{ij}(s)$ are also completely determined by the reference configuration and will therefore be regarded as known. In particular the symmetry conditions

(2.3.3) $$I_{ij} = I_{ji}$$

are valid. It is consistent to restrict to the case in which the cross-section is modelled as a lamina with principal axes of inertia $\boldsymbol{d}_1$ and $\boldsymbol{d}_2$, so that

(2.3.4) $$I_{ij} = 0, \quad i \neq j, \qquad I_{33} = I_{11} + I_{22}.$$

Henceforth we make this assumption, and we write I_{jj} simply as I_j.

In a rod model, the stresses acting across each material cross-section are reduced to a net force $\boldsymbol{n}(s,t)$ and moment $\boldsymbol{m}(s,t)$ (of the material on the side s^+ acting on the material on the side s^- at the point $\boldsymbol{r}(s,t)$). In addition, the rod may be subject to an external distributed force $\boldsymbol{f}(s,t)$ and an external distributed moment $\boldsymbol{l}(s,t)$. Then balance of linear and angular momentum yield the equations

(2.3.5) $$\dot{\boldsymbol{p}} = \boldsymbol{n}' + \boldsymbol{f},$$

and

(2.3.6) $$\dot{\boldsymbol{\pi}} = \boldsymbol{m}' + \boldsymbol{r}' \times \boldsymbol{n} + \boldsymbol{l}.$$

In most rod models the effects of external loads are neglected. In the modelling of DNA, two natural external loads to consider are self-interaction and the effects of a viscous solvent. We will not discuss viscous effects here. Self-interaction forces arise when the rod deforms in such a manner that points separated by large differences in arc-length become close to one another in space. The force in this case can be modelled in the general form

(2.3.7) $$\boldsymbol{f}(s) = -\int M(s,\sigma)\frac{U'(|\boldsymbol{r}(s)-\boldsymbol{r}(\sigma)|)}{|\boldsymbol{r}(s)-\boldsymbol{r}(\sigma)|}\,[\boldsymbol{r}(s)-\boldsymbol{r}(\sigma)]\,d\sigma\,,$$

where U is a self-potential generating a central force between pairs of points along the rod, and M is a mollifier leading to finite total energy and corresponding, for example, to a nonzero radius of the rod. Here and henceforth, an integral is assumed to be taken over the length of the rod, unless otherwise specified. For the electrostatic Coulomb force, the potential U is proportional to $|\boldsymbol{r}(s) - \boldsymbol{r}(\sigma)|^{-1}$, but screening effects could well lead to more rapidly decaying potentials being appropriate. Notice, however, that in this context even the Coulomb potential is not long range, because (2.3.7) involves only a single integral, so that the total force does decay rapidly with distance. We take the corresponding external moment $\boldsymbol{l}(s)$ to be zero.

We shall be particularly interested in equilibrium solutions, for which the time derivatives in equations (2.3.5) and (2.3.6) must vanish, and a system of ordinary differential equations remains. The governing equations for an equilibrium are therefore

$$\boldsymbol{n}'(s) + \boldsymbol{f}(s) = \boldsymbol{0}, \tag{2.3.8}$$

and

$$\boldsymbol{m}'(s) + \boldsymbol{r}'(s) \times \boldsymbol{n}(s) + \boldsymbol{l}(s) = \boldsymbol{0}. \tag{2.3.9}$$

Note that in our derivation of a Hamiltonian form of equations (2.3.8) and (2.3.9), the arc-length s will play the role of 'time' in the evolution of an equilibrium solution. However it should be remembered that the physically appropriate supplementary conditions to (2.3.8) and (2.3.9) include boundary conditions at two values of s.

2.4. Constitutive relations. To complete the general formulation of rod dynamics, we must discuss the constitutive relations for both constrained and unconstrained motions. The nine equations contained in (2.2.3) and the three equations in each of (2.3.1), (2.3.5) and (2.3.6) together govern the time evolution of the motion of a rod. However these eighteen equations are not sufficient to determine the twenty-four dynamic variables, namely the components of the configuration $\boldsymbol{r}$, the directors $\{\boldsymbol{d}_k\}$ (i.e. an element of $SO(3)$), the strains $\boldsymbol{u}$ and $\boldsymbol{v}$, the linear momentum $\boldsymbol{p}$, the angular momentum $\boldsymbol{\pi}$, the force $\boldsymbol{n}$ and the moment $\boldsymbol{m}$. In the theory of elasticity it is postulated that the force $\boldsymbol{n}$ and moment $\boldsymbol{m}$ can be determined as functions of the current configuration of the rod. These constitutive relations are unlike the balance laws in that they are specific to the material of the rod. The manner in which $\boldsymbol{n}$ and $\boldsymbol{m}$ are determined depends upon whether the rod is materially unconstrained or constrained. Roughly speaking, the motion of a rod is said to be materially unconstrained if there are no restrictions upon the variables $\{v_k\}$ and $\{u_k\}$; otherwise, the motion is called constrained. This distinction must however be qualified, because most models of rod dynamics specify some conditions upon the strains, independent of the material properties, which are required for physically meaningful solutions. For example, one generally requires that

$$v_3 = \boldsymbol{v} \cdot \boldsymbol{d}_3 > 0 \tag{2.4.1}$$

be satisfied for all motions, so that a section of the rod of positive length cannot be compressed to zero length.

We consider only simple materials, for which $\boldsymbol{n}(s,t)$ and $\boldsymbol{m}(s,t)$ are functions of s, $\partial_s \boldsymbol{r}(s,t)$, and $\partial_s \boldsymbol{d}_k(s,t)$ only. It is known by the principle of frame indifference (cf. e.g. [1]) that $n_j \equiv \boldsymbol{n} \cdot \boldsymbol{d}_j$ and $m_j \equiv \boldsymbol{m} \cdot \boldsymbol{d}_j$ can be expressed as functions of the arc-length s and the six strains, $\{u_i\}$ and $\{v_i\}$. A case of special interest is the hyperelastic rod. A rod is said to be hyperelastic if there exists a scalar valued strain energy density function $W(v_i, u_i, s)$, dependent upon the six strains and arc-length, with the property that the components of stress satisfy the relations

$$n_j = W_{v_j}, \tag{2.4.2}$$

$$m_j = W_{u_j}. \tag{2.4.3}$$

In this article we will develop various Hamiltonian formulations assuming a general strain energy density function W. For concreteness, we will then focus upon the important special case, often called linearly elastic, where the strain energy is quadratic in the strains. In the case of an extensible shearable rod, the strain energy density then takes the form

$$W(\mathsf{u}-\hat{\mathsf{u}}, \mathsf{v}-\hat{\mathsf{v}}) = \tfrac{1}{2}K_j(u_j-\hat{u}_j)^2 + \tfrac{1}{2}A_j(v_j-\hat{v}_j)^2. \tag{2.4.4}$$

Here A_j, K_j, $\hat{v}_j$ and $\hat{u}_j$ are prescribed functions of s, possibly constant. In equation (2.4.4), the terms $\hat{\mathsf{u}} \equiv (\hat{u}_1, \hat{u}_2, \hat{u}_3)$ and $\hat{\mathsf{v}} \equiv (\hat{v}_1, \hat{v}_2, \hat{v}_3)$ describe the configuration of the unstressed rod. We will always assume that the unstressed state is unstretched and unsheared, so that $\hat{v}_1 = 0 = \hat{v}_2$ and $\hat{v}_3 = 1$. If in addition the unstressed state is also straight then $\hat{u}_j = 0$ for $j = 1, 2$, and if the unstressed state is untwisted $\hat{u}_3 = 0$. It follows from equations (2.4.2), (2.4.3) and (2.4.4) that the forces and moments in the linearly elastic case are given by

$$n_j = W_{v_j}(\mathsf{u}-\hat{\mathsf{u}}, \mathsf{v}-\hat{\mathsf{v}}) = A_j(v_j-\hat{v}_j), \tag{2.4.5}$$

$$m_j = W_{u_j}(\mathsf{u}-\hat{\mathsf{u}}, \mathsf{v}-\hat{\mathsf{v}}) = K_j(u_j-\hat{u}_j). \tag{2.4.6}$$

Models of rod mechanics often begin with the assumption that the rod is a solid object whose diameter is small compared to its length. This is the approach taken in the theory of Kirchhoff and Clebsch, which is described in [28], for example. If we adopt this point of view, and also assume that the material composing the rod is homogeneous and isotropic, then we can further interpret the elastic coefficients as $A_1 = A_2 = GA$, $A_3 = EA$, $K_j = EJ_j$ for $j = 1, 2$, $K_3 = GJ_3$, and $I_j = \rho J_j$, where A is the cross-sectional area of the rod, J_j is the geometric moment about the $\boldsymbol{d}_j$ axis (with $J_3 = J_1 + J_2$), E is Young's modulus, $G = E/[2(1+\nu)]$ is the shear modulus, and ν is Poisson's ratio. However, when modelling DNA as a rod, such an interpretation of the coefficients may not be appropriate.

One can also consider strain energies more general than that given by equation (2.4.4). In that case the quadratic function (2.4.4) may be viewed as the first terms in a Taylor series expansion of the strain energy, and so serve as an approximation to the exact strain energy. One can even consider a strain energy which is not convex as a function of the strains (cf. [18]), which is a model that might be appropriate for describing Z-form DNA.

We are particularly interested in the material constraints of inextensibility and unshearability, which are described by

$$\boldsymbol{r}_s = \boldsymbol{d}_3, \tag{2.4.7}$$

or equivalently

$$v_1 = 0 = v_2 \qquad \text{and} \qquad v_3 = 1. \tag{2.4.8}$$

Note that equation (2.4.8) automatically implies the inequality (2.4.1). For brevity equation (2.4.7) will henceforth be referred to as the inextensibility condition. It is stipulated that the constraint (2.4.7) holds for all possible forces and moments. In the model under consideration a constitutive equation is assumed only for the moments, in the form of equation (2.4.3). The force must instead be determined by the requirement that the constraint (2.4.7) be maintained. The rod will be assumed to be hyperelastic, and we will again begin with a general formulation of each Hamiltonian system, and then treat the specific example where the strain energy density is quadratic in the strains, namely

$$W(\mathsf{u} - \hat{\mathsf{u}}) = \tfrac{1}{2} K_j (u_j - \hat{u}_j)^2 \; . \tag{2.4.9}$$

The corresponding moment is given by equation (2.4.6).

2.5. Conservation laws. This section summarizes the conservation laws that are implied by the presence of continuous symmetry groups in the partial differential equations describing the dynamics of a rod. The material in this section is more fully discussed in [29].

For our purposes a conservation law means an expression of the form

$$D_t(s,t) = F_s(s,t), \tag{2.5.1}$$

where D, the density, and F, the flux, are each functions of the (scalar) independent variables s and t, and the (vector) dependent variable $\boldsymbol{y}$ and its derivatives:

$$D(s,t) = \tilde{D}(\boldsymbol{y}(s,t), \boldsymbol{y}'(s,t), \dot{\boldsymbol{y}}(s,t), s, t) \tag{2.5.2}$$

$$F(s,t) = \tilde{F}(\boldsymbol{y}(s,t), \boldsymbol{y}'(s,t), \dot{\boldsymbol{y}}(s,t), s, t) \tag{2.5.3}$$

One reason for our interest in conservation laws is that for boundary conditions such that the flux vanishes at the boundaries, each conservation law (2.5.1) can be integrated over the space variable s to yield an associated integral of motion

$$\frac{d}{dt}\int D(s)\,ds = 0. \tag{2.5.4}$$

On the other hand, in the context of statics the time derivative in equation (2.5.1) vanishes. Consequently, the flux is constant as a function of s along an equilibrium solution, and if the flux is nontrivial on static configurations, an integral of the ordinary differential equations governing equilibria is implied.

Conservation laws arise naturally in rod dynamics. It may be observed immediately that the balance of linear momentum (2.3.5) is in the form of a conservation law

$$\boldsymbol{p}_t = (\boldsymbol{n} + \boldsymbol{F})_s\,, \tag{2.5.5}$$

with density $\boldsymbol{p}$ and flux $\boldsymbol{n} + \boldsymbol{F}$, where

$$\boldsymbol{F} \equiv \int^s \boldsymbol{f}(\sigma)\,d\sigma \tag{2.5.6}$$

is the net external force on the rod up to the material point s. Moreover, we can use (2.3.1) and (2.3.5) to rewrite equation (2.3.6) in the conservation form

$$(\boldsymbol{\pi} + \boldsymbol{r} \times \boldsymbol{p})_t = (\boldsymbol{m} + \boldsymbol{r} \times \boldsymbol{n} + \boldsymbol{L})_s\,, \tag{2.5.7}$$

where

$$\boldsymbol{L} \equiv \int^s \{\boldsymbol{r}(\sigma) \times \boldsymbol{f}(\sigma) + \boldsymbol{l}(\sigma)\}\,d\sigma \tag{2.5.8}$$

is the net external moment on the rod up to material point s. Suppose that the net external force and moment on the rod vanish, as would occur in the case where external loads are due entirely to self-contact. Then for appropriate boundary conditions, such as the periodic ones pertinent for a closed loop, the total linear momentum $\int \boldsymbol{p}\,ds$ and total angular momentum $\int (\boldsymbol{\pi} + \boldsymbol{r} \times \boldsymbol{p})\,ds$ about a fixed origin are integrals of the motion, as could have been surmised at the outset. Conversely, equation (2.5.5) and (2.5.7) show that along an equilibrium solution the fluxes $\boldsymbol{n}(s) + \boldsymbol{F}(s)$ and $\boldsymbol{m}(s) + \boldsymbol{r}(s) \times \boldsymbol{n}(s) + \boldsymbol{L}(s)$ are constant.

Equations (2.5.5) and (2.5.7) are conservation laws independent of the constitutive relations. By contrast, the additional conservation laws that we discuss are associated with special properties of the constitutive relations, especially symmetries. In the important case where the rod is hyperelastic, one obtains the conservation law

$$(\tfrac{1}{2}\boldsymbol{p} \cdot \dot{\boldsymbol{r}} + \tfrac{1}{2}\boldsymbol{\pi} \cdot \boldsymbol{\omega} + W)_t = (\boldsymbol{m} \cdot \boldsymbol{\omega} + \boldsymbol{n} \cdot \dot{\boldsymbol{r}})_s. \tag{2.5.9}$$

In equation (2.5.9), W denotes $W(\mathsf{u}-\hat{\mathsf{u}},\mathsf{v}-\hat{\mathsf{v}})$, and $\frac{1}{2}\boldsymbol{p}\cdot\dot{\boldsymbol{r}}+\frac{1}{2}\boldsymbol{\pi}\cdot\boldsymbol{\omega}+W$ is the energy density. Therefore if the flux $\boldsymbol{m}\cdot\boldsymbol{\omega}+\boldsymbol{n}\cdot\dot{\boldsymbol{r}}$ vanishes at the boundaries, as would occur if the ends are stationary or if periodic boundary conditions are imposed, then the total energy of the rod is conserved by the dynamics. In the case of statics, both $\dot{\boldsymbol{r}}$ and $\boldsymbol{\omega}$ vanish, so the resulting integral is trivial.

There is an additional conservation law for a rod that is both hyperelastic and uniform. The rod is said to be uniform if the coefficients ρ and I_{ij} are actually constants,

$$\rho' \equiv I'_{ij} \equiv 0, \tag{2.5.10}$$

and, in addition, the constitutive law (2.4.2) and (2.4.3) has no explicit dependence on the arc-length s. The associated symmetry is translation in arc-length. The following conservation law is then a consequence of the governing equations (2.3.5) and (2.3.6) that express balance of linear and angular momentum:

$$(\boldsymbol{\pi}\cdot\boldsymbol{u}+\boldsymbol{p}\cdot\boldsymbol{v})_t = \left(\boldsymbol{m}\cdot\boldsymbol{u}+\boldsymbol{n}\cdot\boldsymbol{v}-W+\tfrac{1}{2}\boldsymbol{\omega}\cdot\boldsymbol{\pi}+\tfrac{1}{2}\dot{\boldsymbol{r}}\cdot\boldsymbol{p}\right)_s . \tag{2.5.11}$$

It should be remarked that it is not assumed that $\hat{u}_j$ and $\hat{v}_j$ vanish, only that they are constant. Consequently the analysis presented here is valid for the dynamics of uniform hyperelastic rods whose unstressed state is helical.

The quantity

$$\int (\boldsymbol{\pi}\cdot\boldsymbol{u}+\boldsymbol{p}\cdot\boldsymbol{v})\, ds \tag{2.5.12}$$

is called the impulse, where it is again understood that the range of integration is the entire length of the rod. The impulse is actually a time-invariant quantity, i.e., a conserved quantity or integral of the dynamics, for systems with appropriate boundary conditions. For example in a finite closed loop of a rod, all the dependent variables must be periodic functions of s, and the impulse is therefore a conserved quantity. The impulse is not simply related to total linear momentum, which, for rod dynamics, is the integral generated by translational symmetry in space.

Notice that the flux appearing in the conservation law (2.5.11) is the Legendre transform of the density of total energy of the rod. In the special case of the statics of rods, the conservation law (2.5.11) implies that the Legendre transform of the strain energy density W, which is also the Hamiltonian for statics, is an integral.

We now drop the hypotheses that the rod is uniform and hyperelastic and instead assume that the rod is transversely isotropic. By transversely isotropic it is meant that the constitutive relations for the stresses and momenta are symmetric with respect to rotations about the director $\boldsymbol{d}_3$. In the special case of linear elasticity, transverse isotropy is equivalent to the conditions that $K_2 = K_1$, $I_2 = I_1$ and the unstressed configuration in not curved, so $\hat{u}_1 = \hat{u}_2 = 0$. As full isotropy, i.e., independence of the constitutive relation to arbitrary rotations of the basis $\{\boldsymbol{d}_i\}$, is not a pertinent concept in the context of rods, we shall henceforth allow the adjective transversely to stand by implication. Antman ([1], Theorem XIII.9.16) demonstrates that the hypothesis of isotropic elastic response is equivalent to the specialized constitutive relations

$$\boldsymbol{n} = N_1\boldsymbol{v} + N_2\boldsymbol{u} + N_3\boldsymbol{d}_3, \tag{2.5.13}$$

$$\boldsymbol{m} = M_1\boldsymbol{v} + M_2\boldsymbol{u} + M_3\boldsymbol{d}_3, \tag{2.5.14}$$

where the N_i, and M_i are each scalar functions of the six scalar arguments $|\boldsymbol{u}|$, $|\boldsymbol{v}|$, $\boldsymbol{u}\cdot\boldsymbol{v}$, u_3, v_3 and s. Isotropy of the dynamic properties of the rod follows when the expression (2.3.2) for the angular momentum is specialized to

$$\boldsymbol{\pi} \equiv I_1\boldsymbol{\omega} + (I_3 - I_1)\omega_3\boldsymbol{d}_3. \tag{2.5.15}$$

Here $\boldsymbol{d}_3$ is a principal axis of inertia with associated principal inertia $I_3(s)$, and the other two principal inertias both equal $I_1(s)$ (cf. (2.3.4) with $I_1 = I_2$). The general expression (2.3.1) for the linear momentum is already isotropic.

With the constitutive relations (2.5.13), (2.5.14) and (2.5.15), and the additional restriction

$$M_1(|\boldsymbol{u}|, \boldsymbol{u}\cdot\boldsymbol{v}, |\boldsymbol{v}|, u_3, v_3, s) \equiv N_2(|\boldsymbol{u}|, \boldsymbol{u}\cdot\boldsymbol{v}, |\boldsymbol{v}|, u_3, v_3, s), \tag{2.5.16}$$

the conservation law

$$(\boldsymbol{\pi}\cdot\boldsymbol{d}_3)_t = (\boldsymbol{m}\cdot\boldsymbol{d}_3)_s, \tag{2.5.17}$$

is a consequence of the angular momentum balance (2.3.6). In particular the derivation of (2.5.17) does not exploit the linear momentum balance (2.3.5) so the conservation law remains valid even for systems with external forces. The delimitation (2.5.16) is actually implied in the case of an isotropic hyperelastic rod where there is a strain energy function of the form $W(|\boldsymbol{u}|, \boldsymbol{u}\cdot\boldsymbol{v}, |\boldsymbol{v}|, u_3, v_3, s)$. Another plausible special case is $M_1 \equiv N_2 \equiv 0$.

The quantity

$$\int \boldsymbol{\pi}\cdot\boldsymbol{d}_3\, ds \tag{2.5.18}$$

will be called the (component of) isotropic angular momentum. The isotropic angular momentum is an integral of the motion for appropriate boundary conditions, for example the periodic conditions associated with a closed loop, as has already been observed by Simo *et al.* in [36]. On the other hand, in the case of statics, equation (2.5.17) implies that the component of moment $\boldsymbol{m}(s)\cdot\boldsymbol{d}_3(s)$ is an integral.

2.6. Representation of directors by Euler parameters. Thus far we have not specified any particular parametrization of the directors $\{\boldsymbol{d}_k\}$. For three-dimensional motion, a parametrization of the directors $\{\boldsymbol{d}_k\}$ is equivalent to selecting a representation for the group of orthogonal transformations $SO(3)$. In mechanics there are several common choices for such a representation. (Cf. e.g. [16].) Euler angles are often used because this représentation uses the the minimum number of parameters, namely three. There are two significant drawbacks in using Euler angles. First, due to the topology of $SO(3)$ a set of three parameters cannot provide a global representation. In the case of Euler angles there is a direction in space at which the representation becomes singular. This singularity makes the study of dynamical systems a delicate matter, since the particular set of Euler angles must then be selected to avoid the singularity, or different charts must be used as the dynamics evolve. A second drawback in the use of Euler angles is the appearance of trigonometric functions, which can be computationally expensive.

Euler parameters provide a four-dimensional, global parametrization of $SO(3)$. In addition, only polynomial and rational functions arise in this representation. The difficulty with the Euler parameter representation is that the squares of the parameters must sum to one, which imposes a constraint upon the dynamics. A discussion of Euler parameters can be found, for example, in [5]. An elementary but self-contained description of the kinematics of Euler parameters, which appears in [8], is given below.

A set of Euler parameters is a quadruple of real numbers $\boldsymbol{q} = (q_1, q_2, q_3, q_4)^T$ (where T denotes the transpose) that satisfies the identity

$$\boldsymbol{q} \cdot \boldsymbol{q} = {q_1}^2 + {q_2}^2 + {q_3}^2 + q_4^2 = 1 \,. \tag{2.6.1}$$

The set of all quadruples satisfying (2.6.1) forms the unit sphere $\boldsymbol{S}^3$ in $\boldsymbol{R}^4$. A set of Euler parameters $\boldsymbol{q}$ is often called a unit quaternion, in connection with the algebra of quaternions [35]. In fact one reason that Euler parameters are so useful for the representation of $SO(3)$ is because the composition of two rotations may be expressed in an elegant manner which is closely related to the multiplication of quaternions.

By Euler's Theorem, the orientation of a frame of directors $\{\boldsymbol{d}_k\}$ with respect to a fixed frame $\{\boldsymbol{e}_k\}$ is described by a rotation through an angle Φ about an axis determined by a unit vector $\boldsymbol{k} = k_j \boldsymbol{e}_j$. The Euler parameters (q_1, q_2, q_3, q_4) are defined in terms of $\boldsymbol{k}$ and Φ by

$$q_j = k_j \sin(\Phi/2) \,, \qquad \text{for } j = 1, 2, 3 \tag{2.6.2}$$

and

$$q_4 = \cos(\Phi/2) \,. \tag{2.6.3}$$

Euler parameters have the feature that if $\boldsymbol{q}$ corresponds to the pair $(\boldsymbol{k}, \Phi)$, then $-\boldsymbol{q}$ corresponds to the pair $(\boldsymbol{k}, \Phi + 2\pi)$ which represents a coterminal rotation. Thus there is a two-to-one correspondence between Euler parameters and rotations. To obtain a Hamiltonian system describing rod dynamics, we express the directors in terms of Euler parameters as

$$(2.6.4)\qquad \boldsymbol{d}_1 = \frac{1}{|\boldsymbol{q}|^2}\begin{pmatrix} q_1^2 - q_2^2 - q_3^2 + q_4^2 \\ 2(q_1q_2 + q_3q_4) \\ 2(q_1q_3 - q_2q_4) \end{pmatrix},$$

$$(2.6.5)\qquad \boldsymbol{d}_2 = \frac{1}{|\boldsymbol{q}|^2}\begin{pmatrix} 2(q_1q_2 - q_3q_4) \\ -q_1^2 + q_2^2 - q_3^2 + q_4^2 \\ 2(q_2q_3 + q_1q_4) \end{pmatrix},$$

$$(2.6.6)\qquad \boldsymbol{d}_3 = \frac{1}{|\boldsymbol{q}|^2}\begin{pmatrix} 2(q_1q_3 + q_2q_4) \\ 2(q_2q_3 - q_1q_4) \\ -q_1^2 - q_2^2 + q_3^2 + q_4^2 \end{pmatrix}.$$

The factor of $1/|\boldsymbol{q}|^2$ appearing in equations (2.6.4) – (2.6.6) is not generally used in this representation. We introduce the term in order to make the directors explicitly orthonormal. By doing so, we introduce a scaling symmetry into the Hamiltonian system whereby $\boldsymbol{q}$ may be replaced by $\epsilon\boldsymbol{q}$ without altering the dynamics. The advantage of this scaling is discussed further in section 3.3.

The strains u_j and the angular velocities ω_j appearing in equations (2.2.2) and (2.2.3) are given in terms of the Euler parameters by

$$(2.6.7)\qquad u_j = \frac{2}{|\boldsymbol{q}|^2}\boldsymbol{B}_j\boldsymbol{q}\cdot\boldsymbol{q}_s$$

and

$$(2.6.8)\qquad \omega_j = \frac{2}{|\boldsymbol{q}|^2}\boldsymbol{B}_j\boldsymbol{q}\cdot\boldsymbol{q}_t\,,$$

where $\boldsymbol{B}_1\boldsymbol{q}$, $\boldsymbol{B}_2\boldsymbol{q}$ and $\boldsymbol{B}_3\boldsymbol{q}$ are vectors given by

$$(2.6.9)\qquad \boldsymbol{B}_1\boldsymbol{q} = (\ q_4,\ \ q_3,\ -q_2,\ -q_1\ )^T\,,$$

$$(2.6.10)\qquad \boldsymbol{B}_2\boldsymbol{q} = (\ -q_3,\ \ q_4,\ \ q_1,\ -q_2\ )^T\,,$$

$$(2.6.11)\qquad \boldsymbol{B}_3\boldsymbol{q} = (\ q_2,\ -q_1,\ \ q_4,\ -q_3\ )^T\,.$$

The vectors $\{\boldsymbol{B}_j\boldsymbol{q}\}$, together with $\boldsymbol{q}$, form an orthonormal basis for $\boldsymbol{R}^4$. Moreover $\boldsymbol{B}_j\boldsymbol{q}$ is linear in the Euler parameters, which simplifies computations. The vector $\boldsymbol{B}_k\boldsymbol{q}$ can be used to describe the rotation about the $\boldsymbol{d}_k$ axis in terms of Euler parameters. There is a related set of vectors $\{\boldsymbol{B}^o_k\boldsymbol{q}\}$ which describe rotations about the $\boldsymbol{e}_k$ axes,

$$\boldsymbol{B}^o_1\boldsymbol{q} = (\ q_4,\ -q_3,\ q_2,\ -q_1\ )^T , \tag{2.6.12}$$

$$\boldsymbol{B}^o_2\boldsymbol{q} = (\ q_3,\ q_4,\ -q_1,\ -q_2\ )^T , \tag{2.6.13}$$

$$\boldsymbol{B}^o_3\boldsymbol{q} = (\ -q_2,\ q_1,\ q_4,\ -q_3\ )^T . \tag{2.6.14}$$

To facilitate the writing of the Hamiltonian equations of motion, we define the matrices

$$\boldsymbol{D}_k[\boldsymbol{q}] \equiv \left(\frac{\partial \boldsymbol{d}_k}{\partial \boldsymbol{q}}\right)^T , \tag{2.6.15}$$

and note that these matrices can be expressed in terms of the bases $\{\boldsymbol{B}_j\boldsymbol{q}\}$ and $\{\boldsymbol{d}_j(\boldsymbol{q})\}$ as

$$\boldsymbol{D}_k[\boldsymbol{q}] = \frac{2}{|\boldsymbol{q}|^2}\epsilon_{ijk}\boldsymbol{B}_j\boldsymbol{q}\otimes\boldsymbol{d}_i = \frac{2}{|\boldsymbol{q}|^2}\boldsymbol{B}_j\boldsymbol{q}\otimes(\boldsymbol{d}_j\times\boldsymbol{d}_k) , \tag{2.6.16}$$

where ϵ_{ijk} is the permutation tensor and $\otimes$ denotes the outer product. Geometrically, a change in $\boldsymbol{q}$ by $\epsilon\boldsymbol{B}_k\boldsymbol{q}$ can be interpreted as the instantaneous rotation about the $\boldsymbol{d}_k$ axis through an angle 2ϵ, which does not change $\boldsymbol{d}_k$ itself.

The kinetic energy can be expressed in terms of the Euler parameters as

$$T(\boldsymbol{r},\boldsymbol{q}) = \int \tfrac{1}{2}\left\{\rho|\boldsymbol{r}_t|^2 + I_j\omega_j^2\right\} ds , \tag{2.6.17}$$

where ω_j is given by equation (2.6.8). Similarly the potential energy becomes

$$V[\boldsymbol{q}] = \int W(\mathsf{u}-\hat{\mathsf{u}},\mathsf{v}-\hat{\mathsf{v}})\, ds , \tag{2.6.18}$$

where u is given by equation (2.6.7).

In statics, it transpires that there is more freedom in the scaling of Euler parameters and the factor $1/|\boldsymbol{q}|^2$ appearing in equations (2.6.4) – (2.6.6) can be replaced with $1/|\boldsymbol{q}|^\beta$ for any β and a symmetry in the Hamiltonian system is still obtained. It is therefore convenient to employ the standard formulation with $\beta = 0$, namely

$$\boldsymbol{d}_1 = \begin{pmatrix} q_1^2 - q_2^2 - q_3^2 + q_4^2 \\ 2(q_1q_2 + q_3q_4) \\ 2(q_1q_3 - q_2q_4) \end{pmatrix} , \tag{2.6.19}$$

$$d_2 = \begin{pmatrix} 2(q_1q_2 - q_3q_4) \\ -q_1^2 + q_2^2 - q_3^2 + q_4^2 \\ 2(q_2q_3 + q_1q_4) \end{pmatrix}, \tag{2.6.20}$$

$$d_3 = \begin{pmatrix} 2(q_1q_3 + q_2q_4) \\ 2(q_2q_3 - q_1q_4) \\ -q_1^2 - q_2^2 + q_3^2 + q_4^2 \end{pmatrix}. \tag{2.6.21}$$

The strains u_j are still given in terms of the Euler parameters by

$$u_j = \frac{2}{|q|^2} B_j q \cdot q_s \,, \tag{2.6.22}$$

but this scaling, the matrices $D_k[q]$ take a slightly different form, namely

$$D_k[q] = \frac{2}{|q|^2} \epsilon_{ijk} B_j q \otimes d_i + \frac{2}{|q|^2} q \otimes d_k \,. \tag{2.6.23}$$

In statics it is the potential energy density, rather than the total potential energy, which arises in a variational formulation. In this case the potential energy density of the rod is

$$W(\mathsf{u} - \hat{\mathsf{u}}, \mathsf{v} - \hat{\mathsf{v}}) \,, \tag{2.6.24}$$

where u is given by equation (2.6.22).

3. Hamiltonian Formulations of Dynamics

3.1. Introduction. In this chapter we describe various Hamiltonian formulations of rod dynamics, for both unconstrained and constrained motion. In each instance, the Hamiltonian system is a set of partial differential equations which can be expressed compactly as

$$z_t = J\,\delta H(z) \,. \tag{3.1.1}$$

In equation (3.1.1), $z(s,t)$ is a vector with fourteen components consisting of $r(s,t)$, $q(s,t)$ and their conjugate variables, and J is the 14-by-14 skew symmetric matrix given by

$$J = \begin{pmatrix} \mathbf{0} & \mathbf{1} \\ -\mathbf{1} & \mathbf{0} \end{pmatrix}, \tag{3.1.2}$$

where $\mathbf{0}$ is the 7-by-7 zero matrix, and $\mathbf{1}$ is the 7-by-7 identity matrix. The Hamiltonian H is defined as an integral of the energy density over the length of the rod, and δH denotes its variational derivative.

Hamiltonian formulations are valuable for both analytical and computational reasons. The dynamics are first-order in time, which can simplify both analysis and numerics. There is a very simple relationship, via Noether's Theorem, between symmetries and integrals of motion. The existence of integrals strongly affects the qualitative behavior of Hamiltonian systems. In addition, through a careful choice of numerical integration scheme one can preserve a finite number of integrals of motion in order to better reflect the exact solution.

In the study of rod dynamics, unconstrained constitutive relations are much more tractable than constrained constitutive relations. The first Hamiltonian formulation of unconstrained rods of which we are aware is by Simo *et al.* [36]. Simo and Vu-Quoc [37] have also employed Euler parameters in a numerical model of rod dynamics.

The dynamics of constrained rods are more complicated. In the limit of unshearability and inextensibility, the speed of shear and compression waves becomes infinite. In this limit the shear and extensive forces also become basic unknowns, replacing the respective strains. Moreover the equations governing the forces are elliptic boundary value problems, so one obtains hyperbolic equations coupled to elliptic ones. (This is familiar in the context of incompressible fluid flow where there is an elliptic boundary value problem for the pressure field.) The first place we are aware that this observation was applied within the context of rods is Caflisch & Maddocks [6], which proved analytic results regarding existence of solutions for the planar case. Falk and Xu [15] then built a related planar numerical scheme, and Coleman and Xu [7] studied planar travelling waves both analytically and numerically. All of these works involved constrained Lagrangian dynamics. In his thesis Dichmann [8] showed how to obtain unconstrained Hamiltonian formulations of constrained Lagrangian systems, employing a novel approach we call the impetus-striction method, with the main example being constrained rods. The planar version of this Hamiltonian formulation for an inextensible, unshearable rod was used to produce an analytic proof of the stability of travelling waves [9]. The full three-dimensional equations are derived in [8]. The planar case has been implemented numerically, and the three-dimensional version of the code is in preparation [10].

There are two related works of which we are aware that consider rod dynamics in the context of DNA modelling. Schlick *et al.* [34] (and references therein) have developed an approach in which the centerline of the rod is described by B-splines, and the effects of twist are captured using a prescribed linking difference and the relation Lk = Tw + Wr. Time evolution is introduced by allowing the spline points to evolve according to a certain molecular dynamics model. While this model is perfectly viable when used as part of a Monte-Carlo procedure for the minimization of static energy, there are two drawbacks in its interpretation as a true dynamic evolution. First, the relation between the dynamics assigned to the spline points and the actual dynamics of the rod are unclear. Second, the use of the Lk = Tw + Wr formula to eliminate the local twist from the total energy assumes that the local twist is constant, which is the case only if the twist is equilibriated at all times, and the rod is isotropic. In particular there can be no curvature in the unstressed state, and the two bending stiffnesses must be equal. Even in the case of an isotropic uniform rod, the ensuing dynamics are quasi-static in the twist evolution. In another work Klapper and Tabor [21] make an assumption that is in some sense complementary to that of [34]. Klapper and Tabor examine full twist dynamics, but they make a different simplifying assumption that in a certain sense equilibriates the two bending moments, and gives quasi-static evolution in bending while having dynamic effects in twist. A similar approximation was made by Beliaev and Illichev [3] in the case of planar dynamics where the approximation is certainly consistent, but a justification of the approximation in the case of three dimensional dynamics is less clear. The parameter regimes and circumstances in which the approximations made in [34] and [21] would yield reasonably accurate models of rod dynamics is an interesting, but open area of investigation.

For simplicity we assume in this chapter that there are no external loads. In particular we neglect effects of self-contact. In section 3.2 we present a Hamiltonian formulation for the three-dimensional dynamics of an extensible, shearable rod. Then in section 3.3 we describe a Hamiltonian formulation for the dynamics of an elastica. In each case we begin with the formulation for a general strain energy density, then specialize to the case of linear elasticity. Finally in section 3.4 we delineate the associated invariants of motion for both Hamiltonian systems, including those listed in section 2.5 as well as two families of invariants which are specific to the Hamiltonian formulations.

3.2. Hamiltonian formulation of the dynamics of an extensible, shearable rod. Let us first consider a general strain energy of the form $W(\mathsf{u}-\hat{\mathsf{u}}, \mathsf{v}-\hat{\mathsf{v}})$, where $\mathsf{u} = (u_1, u_2, u_3)$ is given by (2.6.7). The Hamiltonian, which is the total energy, is given by

$$\begin{aligned} H(\boldsymbol{r},\boldsymbol{q},\boldsymbol{p},\boldsymbol{\zeta}) &= \int \left\{ \tfrac{1}{2}\rho^{-1}|\boldsymbol{p}|^2 + \tfrac{1}{2}I_j^{-1}\left(\tfrac{1}{2}\boldsymbol{B}_j\boldsymbol{q}\cdot\boldsymbol{\zeta}\right)^2 \right\} ds \\ &\quad + \int W(\mathsf{u}-\hat{\mathsf{u}}, \mathsf{v}-\hat{\mathsf{v}})\, ds\,, \end{aligned} \tag{3.2.1}$$

where $\boldsymbol{p}$ is the linear momentum (2.3.1) which is conjugate to the position $\boldsymbol{r}$, and $\boldsymbol{\zeta}$ is the momentum variable conjugate to $\boldsymbol{q}$, given by

$$\boldsymbol{\zeta} = I_j \left(\frac{2\boldsymbol{B}_j\boldsymbol{q}\cdot\boldsymbol{q}_t}{|\boldsymbol{q}|^2} \right) \frac{2\boldsymbol{B}_j\boldsymbol{q}}{|\boldsymbol{q}|^2} + \nu\boldsymbol{q}\,. \tag{3.2.2}$$

In equation (3.2.2), ν is a Lagrange multiplier associated with the constraint (2.6.1). Here, and in the subsequent Hamiltonian formulations, we use

Greek letters to represent conjugate variables that have no classic physical interpretation.

To solve equation (3.2.2) for $\boldsymbol{q}_t$ and ν, one must use the constraint (2.6.1) together with the orthogonality of the basis $\{\boldsymbol{q}, \boldsymbol{B}_j\boldsymbol{q}\}$ in order to obtain

$$\boldsymbol{q}_t = I_j^{-1}\left(\tfrac{1}{2}\boldsymbol{B}_j\boldsymbol{q}\cdot\boldsymbol{\zeta}\right)\left(\tfrac{1}{2}\boldsymbol{B}_j\boldsymbol{q}\right) \tag{3.2.3}$$

and

$$\nu = \frac{\boldsymbol{q}\cdot\boldsymbol{\zeta}}{|\boldsymbol{q}|^2}\,. \tag{3.2.4}$$

It is necessary to invoke the constraint (2.6.1) in order to solve (3.3.3) for $\boldsymbol{q}_t$, for otherwise the component $\boldsymbol{q}\cdot\boldsymbol{q}_t$ is indeterminate. This is a special case of a general theory developed by Dirac [11], which constructs Hamiltonian formulations of Lagrangian systems that are not strictly convex in the velocities.

It follows from equation (3.2.1) that the Hamiltonian dynamics are given by

$$\boldsymbol{r}_t = \frac{\delta H}{\delta \boldsymbol{\chi}} = \rho^{-1}\boldsymbol{p}\,, \tag{3.2.5}$$

$$\boldsymbol{q}_t = \frac{\delta H}{\delta \boldsymbol{\zeta}} = I_j^{-1}\left(\tfrac{1}{2}\boldsymbol{B}_j\boldsymbol{q}\cdot\boldsymbol{\zeta}\right)\tfrac{1}{2}\boldsymbol{B}_j\boldsymbol{q}\,, \tag{3.2.6}$$

$$\boldsymbol{p}_t = -\frac{\delta H}{\delta \boldsymbol{r}} = \left(W_{v_j}\boldsymbol{d}_j\right)_s\,, \tag{3.2.7}$$

$$\begin{aligned}\boldsymbol{\zeta}_t = -\frac{\delta H}{\delta \boldsymbol{q}} &= I_j^{-1}\left(\tfrac{1}{2}\boldsymbol{B}_j\boldsymbol{q}\cdot\boldsymbol{\zeta}\right)\tfrac{1}{2}\boldsymbol{B}_j\boldsymbol{\zeta} \\ &\quad + W_{u_j}\left(\frac{2\boldsymbol{B}_j\boldsymbol{q}_s}{|\boldsymbol{q}|^2} + \frac{2\boldsymbol{B}_j\boldsymbol{q}\cdot\boldsymbol{q}_s}{|\boldsymbol{q}|^2}\frac{2\boldsymbol{q}}{|\boldsymbol{q}|^2}\right) \\ &\quad + \left(W_{u_j}\frac{2\boldsymbol{B}_j\boldsymbol{q}}{|\boldsymbol{q}|^2}\right)_s - W_{v_j}\boldsymbol{D}_j\boldsymbol{r}_s\,,\end{aligned} \tag{3.2.8}$$

where W is a function of $\mathsf{u} - \hat{\mathsf{u}}$ and $\mathsf{v} - \hat{\mathsf{v}}$, the matrices $\boldsymbol{B}_j$ and $\boldsymbol{D}_j$ are defined by equations (2.6.9) – (2.6.11) and (2.6.16), and there is an implied summation on the j index. In the special case of linear elasticity, the Hamiltonian dynamics can be simplified using equations (2.4.5) and (2.4.6).

3.3. Hamiltonian formulation of the dynamics of an elastica

We now turn our attention to the three-dimensional motion of an inextensible and unshearable rod (i.e., an elastica) in terms of Euler parameters. The details of the derivation of this Hamiltonian system of equations can be found in [8], Chapter 6.

For a general strain energy of the form $W(\mathsf{u}-\hat{\mathsf{u}})$, where $\mathsf{u}=(u_1,u_2,u_3)$ is given by (2.6.7), the Hamiltonian is

$$H(\boldsymbol{r},\boldsymbol{q},\boldsymbol{\chi},\boldsymbol{\gamma}) = \int \left\{ \tfrac{1}{2}\rho^{-1}|\boldsymbol{\chi}+\tilde{\boldsymbol{\Lambda}}_s|^2 + \tfrac{1}{2}I_j^{-1}\left(\tfrac{1}{2}\boldsymbol{B}_j\boldsymbol{q}\cdot\boldsymbol{\gamma}+\epsilon_{ij3}\boldsymbol{d}_i\cdot\tilde{\boldsymbol{\Lambda}}\right)^2\right\} ds$$

$$+\int W(\mathsf{u}-\hat{\mathsf{u}})\, ds\ . \tag{3.3.1}$$

In equation (3.3.1), $\boldsymbol{\chi}$ is the momentum variable conjugate to the position $\boldsymbol{r}$ defined by

$$\boldsymbol{\chi} = \rho\boldsymbol{r}_t - \boldsymbol{\Lambda}_s\ , \tag{3.3.2}$$

while $\boldsymbol{\gamma}$ is the momentum variable conjugate to $\boldsymbol{q}$ and is given by

$$\boldsymbol{\gamma} = I_j\left(\frac{2\boldsymbol{B}_j\boldsymbol{q}\cdot\boldsymbol{q}_t}{|\boldsymbol{q}|^2}\right)\frac{2\boldsymbol{B}_j\boldsymbol{q}}{|\boldsymbol{q}|^2} - \boldsymbol{D}_3[q]\boldsymbol{\Lambda}+\nu\boldsymbol{q}\ . \tag{3.3.3}$$

In equations (3.3.2) and (3.3.3), ν has the same meaning as in section 3.2, and $\boldsymbol{\Lambda}$ is a vector of three multipliers, called the striction, which are associated with the inextensibility constraint (2.4.7). For a given state $\boldsymbol{z}=(\boldsymbol{r},\boldsymbol{q},\boldsymbol{\chi},\boldsymbol{\gamma})$, the striction $\boldsymbol{\Lambda}=\tilde{\boldsymbol{\Lambda}}(\boldsymbol{z})$ is required to satisfy the system of three elliptic ordinary differential equations

$$\begin{aligned}\boldsymbol{0} &= -\rho^{-1}(\boldsymbol{\chi}+\boldsymbol{\Lambda}_s)_s + I_j^{-1}\left(\tfrac{1}{2}\boldsymbol{B}_j\boldsymbol{q}\cdot\boldsymbol{\gamma}+\epsilon_{ij3}\boldsymbol{d}_i\cdot\boldsymbol{\Lambda}\right)\epsilon_{ij3}\boldsymbol{d}_i\\ &= -\rho^{-1}(\boldsymbol{\chi}+\boldsymbol{\Lambda}_s)_s - I_1^{-1}\left(\tfrac{1}{2}\boldsymbol{B}_1\boldsymbol{q}\cdot\boldsymbol{\gamma}-\boldsymbol{d}_2\cdot\boldsymbol{\Lambda}\right)\boldsymbol{d}_2\\ &\quad + I_2^{-1}\left(\tfrac{1}{2}\boldsymbol{B}_2\boldsymbol{q}\cdot\boldsymbol{\gamma}+\boldsymbol{d}_1\cdot\boldsymbol{\Lambda}\right)\boldsymbol{d}_1\ ,\end{aligned} \tag{3.3.4}$$

subject to an appropriate set of boundary conditions. For example, periodic boundary conditions arise for a closed loop. It can be shown that the condition (3.3.4) on the striction is simply an enforcement of the time derivative of the constraint (2.4.7).

It follows from equation (3.3.1) that the Hamiltonian dynamics are given by

$$\boldsymbol{r}_t = \frac{\delta H}{\delta\boldsymbol{\chi}} = \rho^{-1}(\boldsymbol{\chi}+\tilde{\boldsymbol{\Lambda}}_s)\ , \tag{3.3.5}$$

$$\boldsymbol{q}_t = \frac{\delta H}{\delta\boldsymbol{\gamma}} = I_j^{-1}\left(\tfrac{1}{2}\boldsymbol{B}_j\boldsymbol{q}\cdot\boldsymbol{\gamma}+\epsilon_{ij3}\boldsymbol{d}_i\cdot\tilde{\boldsymbol{\Lambda}}\right)\tfrac{1}{2}\boldsymbol{B}_j\boldsymbol{q}\ , \tag{3.3.6}$$

$$\boldsymbol{\chi}_t = -\frac{\delta H}{\delta\boldsymbol{r}} = \boldsymbol{0}\ , \tag{3.3.7}$$

$$\begin{aligned}\boldsymbol{\gamma}_t = -\frac{\delta H}{\delta\boldsymbol{q}} &= I_j^{-1}\left(\tfrac{1}{2}\boldsymbol{B}_j\boldsymbol{q}\cdot\boldsymbol{\gamma}+\epsilon_{ij3}\boldsymbol{d}_i\cdot\tilde{\boldsymbol{\Lambda}}\right)\left(\tfrac{1}{2}\boldsymbol{B}_i\boldsymbol{\gamma}-\epsilon_{ij3}\boldsymbol{D}_i[q]\,\tilde{\boldsymbol{\Lambda}}\right)\\ &\quad + W_{u_j}\left(\frac{2\boldsymbol{B}_j\boldsymbol{q}_s}{|\boldsymbol{q}|^2}+\frac{2\boldsymbol{B}_j\boldsymbol{q}\cdot\boldsymbol{q}_s}{|\boldsymbol{q}|^2}\frac{2\boldsymbol{q}}{|\boldsymbol{q}|^2}\right)\\ &\quad + \left(W_{u_j}\frac{2\boldsymbol{B}_j\boldsymbol{q}}{|\boldsymbol{q}|^2}\right)_s\ ,\end{aligned} \tag{3.3.8}$$

where W is a function of $\mathsf{u} - \hat{\mathsf{u}}$, the matrices $\boldsymbol{B}_j$ and $\boldsymbol{D}_j$ are defined by equations (2.6.9) – (2.6.11) and (2.6.16), and there is an implied summation in the j index. In the special case of linear elasticity, the Hamiltonian dynamics can be simplified using equation (2.4.6).

The Hamiltonian dynamics for an elastica are also simplified by the fact that $\boldsymbol{r}$ is an ignorable coordinate, i.e., $\boldsymbol{r}$ does appear explicitly in the Hamiltonian (3.3.1). Consequently we see from equation (3.3.7) that the conjugate momentum $\boldsymbol{\chi}$ is constant. Therefore instead of analyzing fourteen equations, it suffices to fix a value for $\boldsymbol{\chi}$, and then consider the reduced Hamiltonian system consisting of the eight equations (3.3.6) and (3.3.8).

Finally, we remark that one could consider other materially constrained rod models, such as a rod which is extensible but not shearable. In that case the system of three equations (3.3.4) would be replaced by only two elliptic equations for the strictions, together with an additional constitutive law relating the stress n_3 to the strain v_3.

3.4. Hamiltonian form of invariants of motion. For each of the Hamiltonian formulations above, we now list the known conservation laws in terms of the Hamiltonian state variables, together with their associated symmetries. For details of the derivations, we refer to [8]. Of course, in order that these quantities are actually invariants, the boundary terms must be such that the appropriate flux terms vanish.

One of the elegant features of a Hamiltonian formulation is that according to Noether's Theorem (cf., e.g., [30]), for each invariant of motion there exists a continuous symmetry. Perhaps the best known example is the invariance of the Hamiltonian itself for autonomous systems, in which t can be replaced by $t+\epsilon$. The knowledge of the invariants of motion can be important for both analytical and computational purposes. In numerical simulations, the invariants can be used to check the accuracy of a numerical approximation to a solution. Moreover, in section 5.2 we describe how the invariants can be exploited in order to formulate a well-posed boundary value problem governing the equilibria of a loop.

We begin by discussing the invariants of motion for the Hamiltonian system given in section 3.2 which arise without any assumptions about the material properties of the rod. For example, it is apparent from equation (3.2.7) that the three components of the total linear momentum

$$P_k \equiv \int \rho \boldsymbol{r}_t \cdot \boldsymbol{e}_k \, ds \, , \qquad k = 1,2,3 \tag{3.4.1}$$

are conserved. The conservation of linear momentum is associated with the translational symmetry in space in which $\boldsymbol{r}$ is replaced by $\boldsymbol{r} + \epsilon \boldsymbol{e}_k$. We denote the total angular momentum about the origin by

$$\boldsymbol{\Pi} \equiv \int \{\boldsymbol{\pi} + \boldsymbol{r} \times \boldsymbol{p}\} \, ds \tag{3.4.2}$$

and let $\Pi_k \equiv \boldsymbol{\Pi} \cdot \boldsymbol{e}_k$. In terms of the Hamiltonian variables,

$$\Pi_k = \int \left\{ \tfrac{1}{2} \boldsymbol{B}_k^o \boldsymbol{q} \cdot \boldsymbol{\zeta} + (\boldsymbol{r} \times \boldsymbol{p}) \cdot \boldsymbol{e}_k \right\} ds. \tag{3.4.3}$$

which is associated with the symmetry of rotation about the inertial axis $\boldsymbol{e}_k$.

The total linear and angular momentum are both integrals over the length of the rod. In contrast, there are two sets of pointwise invariants which arise due to our Euler parameter formulation, which have no clear physical interpretation. First, at each point s there is the invariant $\boldsymbol{q}(s) \cdot \boldsymbol{q}(s)$ for the Euler parameters which was discussed in section 2.6. This is associated with the invariance of the Hamiltonian under the transformation in which $\boldsymbol{\gamma}$ is replaced by $\boldsymbol{\gamma} + \epsilon \boldsymbol{q}$. In addition, $\boldsymbol{q}(s) \cdot \boldsymbol{\gamma}(s)$ is an invariant at each point s. The symmetry associated with $\boldsymbol{q}(s) \cdot \boldsymbol{\gamma}(s)$ is given by the group of transformations in which $\boldsymbol{q}$ is replaced by $(1+\epsilon)\boldsymbol{q}$, and $\boldsymbol{\gamma}$ is simultaneously replaced by $(1+\epsilon)^{-1}\boldsymbol{\gamma}$. In fact, one of the primary reasons for scaling the directors as we did for this model of rod dynamics was to obtain this second family of invariants.

The remaining two invariants were already discussed in section 2.5, and are associated with assumptions on the material properties of the rod. Translational symmetry with respect to arc-length s implies invariance of the linear impulse I, which can be written in Hamiltonian variables as

$$I \equiv \int \{ \boldsymbol{p} \cdot \mathbf{r}_s + \boldsymbol{\zeta} \cdot \boldsymbol{q}_s \} \, ds. \tag{3.4.4}$$

Suppose instead that the rod is isotropic, that is, it is symmetric with respect to rotation about the tangent vector $\boldsymbol{d}_3$. Then we can conclude the invariance of the isotropic angular momentum A which takes a very simple form in terms of the Hamiltonian state variables, namely

$$A = \int \tfrac{1}{2} \boldsymbol{B}_3 \boldsymbol{q} \cdot \boldsymbol{\zeta} \, ds. \tag{3.4.5}$$

For each of the invariants listed above for the extensible, shearable rod there is a corresponding invariant for the Hamiltonian system in section 3.3 governing the elastica, in which the variables $\boldsymbol{p}$ and $\boldsymbol{\zeta}$ are replaced by $\boldsymbol{\chi}$ and $\boldsymbol{\gamma}$, respectively. In particular, it follows immediately from equation (3.3.7) that the three components of the total linear impetus

$$X_k \equiv \int \boldsymbol{\chi} \cdot \boldsymbol{e}_k \, ds, \qquad k = 1, 2, 3 \tag{3.4.6}$$

are invariants of motion. Moreover, assuming periodic boundary conditions, it follows that the components of linear momentum (3.4.1) are also conserved. In addition, for the elastica there is a family of pointwise invariants $\boldsymbol{r}_s(s) - \boldsymbol{d}_3(s)$ which describe the inextensibility condition. Associated with this family of invariants, the Hamiltonian formulation possesses a symmetry linked to the strictions $\boldsymbol{\Lambda}$, in which $\boldsymbol{\chi}$ is replaced by $\boldsymbol{\chi} + \boldsymbol{\Lambda}^*_s$, $\boldsymbol{\gamma}$ is replaced by $\boldsymbol{\gamma} + \boldsymbol{D}_3[\boldsymbol{q}]\,\boldsymbol{\Lambda}^*$, and $\boldsymbol{\Lambda}$ is replaced by $\boldsymbol{\Lambda} + \boldsymbol{\Lambda}^*$, where $\boldsymbol{\Lambda}^*$ is an arbitrary function of the arc-length s.

4. Hamiltonian formulations of statics

4.1. Introduction. In this chapter we present various Hamiltonian formulations describing rod statics, for both unconstrained and constrained theories. Each of the Hamiltonian systems in this chapter can be expressed as

$$z_s = J\,\nabla H(z)\,, \tag{4.1.1}$$

where the matrix $\boldsymbol{J}$ is again given by (3.1.2). The structure of equation (4.1.1) is very similar to that of equation (3.1.1) for time dynamics. However in equation (4.1.1), $\boldsymbol{z}$ is a function of s only, H is the energy density at a point s (and not an integral), and the variational derivative δ appearing in equation (3.1.1) is replaced by the gradient ∇ with respect to $\boldsymbol{z}$. Thus the Hamiltonian system given by (4.1.1) is a system of ordinary differential equations which evolve in s. We remark that in statics, in contrast with the situation in dynamics, the analysis of constrained rods is easier than the analysis of unconstrained rods.

This chapter parallels the development of chapter 3. Once again, for simplicity we assume there are no external loads. In sections 4.2 and 4.3 we present Hamiltonian formulations describing the statics of an extensible, shearable rod and an elastica, respectively. In each case we begin with the formulation for a general strain energy density, then specialize to the case of linear elasticity. The two Hamiltonian formulations have very similar structures, and have the same integrals, which are presented in section 4.4.

4.2. Hamiltonian formulation of the statics of an extensible, shearable rod. In the general case of an extensible, shearable rod, where the strain energy is given by $W(\mathsf{u} - \hat{\mathsf{u}}, \mathsf{v} - \hat{\mathsf{v}})$, the Hamiltonian is

$$H(\boldsymbol{r}, \boldsymbol{q}, \boldsymbol{n}, \boldsymbol{\mu}) = W^*(\mathsf{m}, \mathsf{n}) + \hat{u}_j m_j + \boldsymbol{n} \cdot \boldsymbol{d}_3\,. \tag{4.2.1}$$

where W^* denotes the Legendre transform of the strain energy density W, $\mathsf{m} = (m_1, m_2, m_3)$ and $\mathsf{n} = (n_1, n_2, n_3)$. (See [1], for example, for a definition of the Legendre transform.) Notice that now the Hamiltonian (4.2.1) is a scalar-valued function, while in contrast the Hamiltonians (3.2.1) and (3.3.1) are integrals over the rod, yielding the partial differential equations describing the dynamics of rods. In this Hamiltonian formulation, the force $\boldsymbol{n}$ is the variable conjugate to $\boldsymbol{r}$, while the variable $\boldsymbol{\mu}$ conjugate to the Euler parameters $\boldsymbol{q}$ is defined by

$$\boldsymbol{\mu} \equiv W_{u_j}(\mathsf{u} - \hat{\mathsf{u}})\,\frac{2}{|\mathsf{q}|^2}\boldsymbol{B}_j\boldsymbol{q} + \nu\boldsymbol{q}\,, \tag{4.2.2}$$

where u is given in terms of $\boldsymbol{q}$ by equation (2.6.22) and ν has the same meaning as in section 3.2. Accordingly, each component of moment $m_j = W_{u_j}$ is given in terms of the Hamiltonian variables by

$$m_j = \tfrac{1}{2}\boldsymbol{B}_j\boldsymbol{q}\cdot\boldsymbol{\mu}\,. \tag{4.2.3}$$

The associated Hamiltonian system is

$$\boldsymbol{r}_s = \frac{\partial H}{\partial \boldsymbol{n}} = W^*_{n_j}\boldsymbol{d}_j + \boldsymbol{d}_3\,, \tag{4.2.4}$$

$$\boldsymbol{q}_s = \frac{\partial H}{\partial \boldsymbol{\mu}} = \left(W^*_{m_j} + \hat{u}_j\right)\tfrac{1}{2}\boldsymbol{B}_j\boldsymbol{q}\,, \tag{4.2.5}$$

$$\boldsymbol{n}_s = -\frac{\partial H}{\partial \boldsymbol{r}} = \boldsymbol{0}\,, \tag{4.2.6}$$

$$\boldsymbol{\mu}_s = -\frac{\partial H}{\partial \boldsymbol{q}} = \left(W^*_{m_j} + \hat{u}_j\right)\tfrac{1}{2}\boldsymbol{B}_j\boldsymbol{\mu} - W^*_{n_j}\boldsymbol{D}_j[\boldsymbol{q}]\,\boldsymbol{n} - \boldsymbol{D}_3[\boldsymbol{q}]\,\boldsymbol{n}\,, \tag{4.2.7}$$

where W^* is a function of m and n, the matrices $\boldsymbol{B}_j$ and $\boldsymbol{D}_j$ are defined by equations (2.6.9) – (2.6.11) and (2.6.16), and there is an implied summation in the j index.

In the special case of linear elasticity where the strain energy density W is given by (2.4.4), the Legendre transform is

$$W^*(\mathsf{m},\mathsf{n}) = \tfrac{1}{2}K_j^{-1}m_j^2 + \tfrac{1}{2}A_j^{-1}n_j^2\,, \tag{4.2.8}$$

so the Hamiltonian system is

$$\boldsymbol{r}_s = \frac{\partial H}{\partial \boldsymbol{n}} = A_j^{-1}\left(\boldsymbol{n}\cdot\boldsymbol{d}_j\right)\boldsymbol{d}_j + \boldsymbol{d}_3\,, \tag{4.2.9}$$

$$\boldsymbol{q}_s = \frac{\partial H}{\partial \boldsymbol{\mu}} = \left[K_j^{-1}\left(\tfrac{1}{2}\boldsymbol{B}_j\boldsymbol{q}\cdot\boldsymbol{\mu}\right) + \hat{u}_j\right]\tfrac{1}{2}\boldsymbol{B}_j\boldsymbol{q}\,, \tag{4.2.10}$$

$$\boldsymbol{n}_s = -\frac{\partial H}{\partial \boldsymbol{r}} = \boldsymbol{0}\,, \tag{4.2.11}$$

$$\begin{aligned}\boldsymbol{\mu}_s = -\frac{\partial H}{\partial \boldsymbol{q}} &= \left[K_j^{-1}\left(\tfrac{1}{2}\boldsymbol{B}_j\boldsymbol{q}\cdot\boldsymbol{\mu}\right) + \hat{u}_j\right]\tfrac{1}{2}\boldsymbol{B}_j\boldsymbol{\mu} \\ &\quad - A_j^{-1}\left(\boldsymbol{n}\cdot\boldsymbol{d}_j\right)\boldsymbol{D}_j[\boldsymbol{q}]\,\boldsymbol{n} - \boldsymbol{D}_3[\boldsymbol{q}]\boldsymbol{n}\,.\end{aligned} \tag{4.2.12}$$

4.3. Hamiltonian formulation of the statics of an elastica. For an inextensible, unshearable rod, where the strain energy is given by $W(\mathsf{u} - \hat{\mathsf{u}})$, the Hamiltonian is

$$H(\boldsymbol{r},\boldsymbol{q},\boldsymbol{n},\boldsymbol{\mu}) = W^*(\mathsf{m}) + \hat{u}_j m_j + \boldsymbol{n}\cdot\boldsymbol{d}_3\,. \tag{4.3.1}$$

The associated Hamiltonian system is

$$\boldsymbol{r}_s = \frac{\partial H}{\partial \boldsymbol{n}} = \boldsymbol{d}_3\,, \tag{4.3.2}$$

$$q_s = \frac{\partial H}{\partial \mu} = \left(W^*_{m_j} + \hat{u}_j\right) \tfrac{1}{2} B_j q \,, \tag{4.3.3}$$

$$n_s = -\frac{\partial H}{\partial r} = 0 \,, \tag{4.3.4}$$

$$\mu_s = -\frac{\partial H}{\partial q} = \left(W^*_{m_j} + \hat{u}_j\right) \tfrac{1}{2} B_j \mu - D_3[q] n \,, \tag{4.3.5}$$

where the matrices B_j and D_j are defined by equations (2.6.9) – (2.6.11) and (2.6.16), and there is an implied summation in the j index. In the special case where the strain energy density W is given by (2.4.9), the Legendre transform is

$$W^* (\mathrm{m}) = \tfrac{1}{2} K_j^{-1} m_j^2 \,, \tag{4.3.6}$$

so the Hamiltonian system is

$$r_s = \frac{\partial H}{\partial n} = d_3 \,, \tag{4.3.7}$$

$$q_s = \frac{\partial H}{\partial \mu} = \left(K_j^{-1} m_j + \hat{u}_j\right) \tfrac{1}{2} B_j q \,, \tag{4.3.8}$$

$$n_s = -\frac{\partial H}{\partial r} = 0 \,, \tag{4.3.9}$$

$$\mu_s = -\frac{\partial H}{\partial q} = \left(K_j^{-1} m_j + \hat{u}_j\right) \tfrac{1}{2} B_j \mu - D_3[q] n \,. \tag{4.3.10}$$

Notice that the strain energy term (4.2.8) for the extensible, shearable rod differs from the corresponding term (4.3.6) for the elastica only in its second term, which measures the energy due to stretching and shearing. In particular, in the limit as the coefficients A_j approach infinity (i.e., as the rod becomes inextensible and unshearable), the unconstrained Hamiltonian approaches the constrained Hamiltonian.

As in the case of the dynamics discussed in section 3.4, the two Hamiltonian formulations of statics are simplified by the fact that r is an ignorable coordinate. As a consequence, instead of analyzing fourteen equations, one can fix the value of the force n, and then consider the reduced Hamiltonian system consisting of only eight equations given by (4.2.5) and (4.2.7) for an extensible, shearable rod or the analogous system (4.3.3) and (4.3.5) for an elastica.

4.4. Hamiltonian form of integrals for statics. We now describe the integrals arising in the Hamiltonian system of ordinary differential equations describing the statics of rods, together with their associated symmetries. The discussion parallels that given in section 3.4 for the Hamiltonian system of partial differential equations modelling dynamics.

It turns out that the integrals arising for the extensible, shearable rod and for the elastica are identical, so we will only explicitly refer to the extensible, shearable case. Moreover, as was discussed in section 2.5, four of the integrals of the statics of rod are related to the invariants of motion through the conservation laws (2.5.5), (2.5.7), (2.5.11), and (2.5.17). (As was pointed out in section 2.5, the integral arising from the remaining conservation law (2.5.9) is trivial. The reason for this is that the symmetry associated with (2.5.9) is translation in time, which implies the conservation of total energy in the case of dynamics, but has no significance in the case of statics.)

The three components of force $\boldsymbol{n}$ are also constant, due to the translational symmetry of the Hamiltonian system in space. This fact is directly related to the invariance of the three components of total linear momentum (3.4.1), through the conservation law (2.5.5). Similarly the three components of $\boldsymbol{m} + \boldsymbol{r} \times \boldsymbol{n}$ are integrals as a consequence of rotational symmetry about the three inertial axes, which is also associated with the invariance of total angular momentum via equation (2.3.6).

In addition to the integrals listed above which have physical interpretations, there are two integrals which arise from the representation of $SO(3)$ by Euler parameters. The quantity $\boldsymbol{q} \cdot \boldsymbol{q}$ is an integral by the construction of the Hamiltonian formulation, just as in the case of dynamics, and is associated with the invariant of the Hamiltonian under to the transformation in which $\boldsymbol{\mu}$ is replaced by $\boldsymbol{\mu} + \epsilon \boldsymbol{q}$. The quantity $\boldsymbol{\mu} \cdot \boldsymbol{q} + 2\boldsymbol{r} \cdot \boldsymbol{n}$ is also an integral. By contrast, the analogous quantity $\boldsymbol{\zeta} \cdot \boldsymbol{q} + 2\boldsymbol{r} \cdot \boldsymbol{n}$ is *not* an invariant in the case of dynamics; rather we have shown that $\boldsymbol{\zeta} \cdot \boldsymbol{q}$ is invariant. The difference arises from the scaling factor $1/|\boldsymbol{q}|^2$ applied to the Euler parameter representation of the directors. In dynamics, the additional factor was included in the model in order to obtain an additional invariant. But in statics it is not needed, so to conform with common practice in the kinematics literature the factor is omitted. The constancy of $\boldsymbol{\mu} \cdot \boldsymbol{q} + 2\boldsymbol{r} \cdot \boldsymbol{n}$ is associated with the invariance of the Hamiltonian (4.2.1) under the transformation in which $\boldsymbol{q}$ is replaced by $(1+\epsilon)\boldsymbol{q}$, $\boldsymbol{\mu}$ is replaced by $(1+\epsilon)^{-1}\boldsymbol{\mu}$, $\boldsymbol{r}$ is replaced by $(1+\epsilon)^2\boldsymbol{r}$ and $\boldsymbol{n}$ is replaced by $(1+\epsilon)^{-2}\boldsymbol{n}$.

The above integrals arise for all constitutive laws. As before there are two additional, constitutive dependent integrals that arise in each of the cases of a uniform rod, and an isotropic rod. If we assume the rod is translationally symmetric in the arc-length s, then the Hamiltonian H (4.2.1) is an integral. The constancy of H is associated with the invariance of the linear impulse I (2.5.12) through the conservation law (2.5.11). Finally, for isotropic rods $\boldsymbol{m} \cdot \boldsymbol{d}_3$ is an integral in the case of statics, which is linked to the invariance of the isotropic angular momentum (2.5.18) in the case of dynamics through the conservation law (2.5.17).

There is a well developed theory of integrability in Hamiltonian systems of ordinary differential equations. The classic result is that a Hamiltonian system with N degrees of freedom (i.e. $2N$ equations) is completely integrable if there are N linearly independent integrals in involution. Two integrals I_1 and I_2 are said to commute, or be in involution, if the Poisson bracket $\{I_1, I_2\} \equiv \nabla I_1 \cdot \boldsymbol{J} \nabla I_2$ vanishes identically. The N integrals are in involution if all their Poisson brackets vanish identically. We have already shown that there are always eight linearly independent (i.e. different) integrals for our seven degree of freedom Hamiltonian system describing rod statics, and that the symmetries of a uniform, and of an isotropic rod add a ninth and tenth linearly independent integral. A lengthy, but straightforward, calculation (which, because our integrals are all simple polynomials, can be efficiently carried out by a symbolic manipulation package such as Maple or Mathematica) reveals that of the eight integrals only five commute, but that of the ten integrals, seven are in involution. Thus the uniform, isotropic rod is the only (known) case of complete integrability. One of the main consequences of complete integrability is the conclusion that there is a choice of coordinates (unfortunately not defined constructively) in which the initial value problem can in principle be completely solved by quadrature. In the case of a twist–free, unshearable, inextensible rod (and using Euler angles rather than Euler parameters to parameterize $SO(3)$) the appropriate calculations are carried out in [22] (§19). The twisted integrable case and many more calculations are considered in [17] (see also [39] and [38]). The key step is to use cylindrical coordinates to describe the axes of the rod, with the centerline of the cylinder aligned with the direction of the shear force.

The explicit coordinates for integration of the rod equilibrium equations were independently rediscovered by Langer and Singer [23] from a very different, rather geometrical point of view, and subsequently detailed calculations were carried out and applied within the context of DNA modelling by Shi and Hearst [32]. All of these cited works considered only the unshearable, inextensible elastica model. However the Hamiltonian integrability analysis presented here is identical for the elastica and the shearable, extensible rod, so solution by quadrature should be possible in both cases. And indeed in a contemporaneous preprint [31] Shi, Borovik and Hearst have carried out the quadrature for shearable, extensible rods.

It should be understood that a solution by quadrature involves a number of free constants. In an initial value problem, where all the variables are assigned values at $s = 0$, the values of all these constants are typically determined, so that the quadrature is indeed a closed form solution. Initial value problems naturally arise in classic applications of finite dimensional Hamiltonian systems where the independent variable is time. However in the DNA context the independent variable is arc-length, and a two-point boundary value problem arises naturally. In other words, some variables are assigned values at $s = 0$, and some are assigned values at $s = 1$. For such problems a solution by quadrature is not truly explicit, for the values of at least some of the integration constants cannot typically be determined analytically. Accordingly, we are of the opinion that a quadrature solution is of rather limited utility in the solution of the boundary value problems arising in the modelling of DNA. This viewpoint is further expounded in Chapter 5.

In fact the accurate numerical solution of an initial value problem is so straightforward with contemporary machines and software packages, that the use of the solution by quadrature is typically not even an efficient way to numerically generate solutions. In particular, details of various packages developed by K.A. Rogers and R.C. Paffenroth with numerics linked to interactive graphics for examining solutions of the initial value problem for rod equilibria can be found at the World Wide Web URL http://www.glue.umd.edu/~karogers/slinky.html .

5. Equilibria of a closed, twisted elastic loop

5.1. Introduction. In this section we present a brief description of some recent analytical and computational results regarding the analysis of the set of equilibria of a closed, twisted loop. This work is fully described in the sequence of articles [25], [26] and [27] (see also the WWW URL http://jacobi.umd.edu/). The example of a closed elastic rod has now received considerable attention as a very simple, but prototypical model for supercoiling of DNA (see, for example [2], [4], [19], [24], [34], [31], [32], [38], [40], [41], references therein, other articles in this volume, and the extensive survey by Schlick [33]).

The novelty in our treatment is that we exploit the Hamiltonian formulation of the two-point boundary value problem for ordinary differential equations governing equilibria to construct an efficient numerical approach that explicitly takes account of all the symmetries of the problem. In particular, we do not use the Lk = Tw + Wr formula, which would restrict attention to isotropic rods, nor do we use the semi-explicit solution to the initial value problem which, in addition to being computationally inefficient, would restrict attention to the integrable case of a uniform, isotropic rod. As a consequence we can compute equilibria for rather general classes of rods.

Indeed for our approach, which uses parameter continuation, the integrable case is in a sense the hardest to compute. The symmetries of the integrable problem give rise, in both the unshearable, inextensible and shearable, extensible cases, to a closed form 'trivial' solution in which the centerline of the rod is a circle and the cross-sections are uniformly twisted. This closed form solution serves as a convenient starting point for our numerical continuation techniques. The potential difficulty is that the same symmetries which give rise to this special solution also imply that all solutions are nonisolated for a given load parameter (i.e. for prescribed twist, shear force or link). This non-isolation necessitates some care in the numerical continuation approach.

In fact the same symmetries that give rise to the difficulties associated with non-isolation also provide the means to resolve the problem. It was noted in section 3.4 that in a Hamiltonian formulation, or any variational formulation, symmetries generate integrals of the ordinary differential equations. These integrals act to convect information from $s = 0$ to

$s = 1$ with the consequence that some boundary conditions at $s = 1$ are actually implied by boundary conditions at $s = 0$. The upshot is that some boundary conditions are actually implied and can be replaced by additional boundary conditions that factor out the non-isolation in such a way that a numerically well-posed boundary value problem remains.

The benefits of treating the integrable case in this careful fashion are two-fold. First, the complete solution set is obtained for both the shearable, extensible and unshearable, inextensible cases, for arbitrary ratios K_3/K_1 of twisting stiffness to bending stiffness, and for arbitrary values of the unstressed twist $\hat{u}_3$. Second, any solution for the integrable case can be used as a starting point for parameter continuation in the coefficients, with the consequence that all the non-integrable cases can also be computed. In particular, when the model is modified to include a pairwise potential of the form (2.3.7) that prevents self-intersection of the rod, equilibria of a prescribed torus knot type and prescribed linking number can be computed for non-uniform rods with arbitrary unstressed configuration and bend, twist, shear and extension coefficients. We defer to the sequence of papers [25], [26], and [27] and to [13] for a full account of these results, and will here restrict attention to a description of the solution set for the integrable case.

Both the analytic form of the circular solution and the computation of nonplanar equilibria for a shearable, extensible rod we believe to be original. The analytic form of the circular solution in the unshearable, inextensible case is by now well-known, but we believe our computation of the higher modes of the solution set, and of the general pattern of connectivity to be original. The higher modes are of some interest because experiments with highly flexible wires indicate that when effects of self-contact are included, some of the $(M + N)$-foil solutions are stable at least for $(M + N) \leq 14$. We also remark that the rod example described here formed the primary motivation for the development of an interactive graphics package that allows the display of bifurcation sets and associated solutions for two-point boundary value problems [14] (see also the WWW URL http://jacobi.umd.edu/pcr.html).

5.2. The boundary value problem. Consider a uniform, isotropic, linearly elastic rod, so that $\hat{u}_1 = \hat{u}_2 = 0$, $K_2 = K_1$ and for a shearable rod $A_2 = A_1$. In the statics of rods the equilibrium equations can be nondimensionalized with respect to both a length scale and a force scale. Thus we may assume with no loss of generality that the undeformed arc-length is one, $0 \leq s \leq 1$, and that $K_1 = 1$. Note that we can allow an initial natural twist $\hat{u}_3 \neq 0$, but that with our nondimensionalization $\hat{u}_3$ may assume an effective continuum of values that is scaled by the number of base pairs in the segment of DNA. In the case of an unshearable, inextensible rod (i.e. an elastica) there is only one free additional parameter, namely the nondimensionalized value of K_3 (or equivalently the ratio K_3/K_1). In point of fact using the symmetry of isotropy it is straightforward to verify that as $\hat{u}_3$ and K_3 are varied the axis (or centerline) of any equilibrium is unchanged, and the directors merely rotate about the tangent vector at a different (and explicitly determined) constant rate. In this sense a computation of the equilibria for any values of $\hat{u}_3$ and K_3, contains practically all the information for a solution with general values.

The equilibrium conditions are a set of fourteen nonlinear ordinary differential equations, and to be able to straightforwardly apply numerical continuation we must find a set of fourteen boundary conditions and one free parameter, such that at (most) fixed parameter values the solutions of the boundary value problem are locally isolated. The boundary conditions should imply that the centerline closes on itself smoothly, and, in order to obtain local isolation, that the symmetries of translation in space, translation in arc-length, rotations in space, and rotations of the rod about its centerline are all eliminated. As detailed in [25] and [27] for the unshearable and shearable cases respectively, the Hamiltonian form of the equilibrium conditions and the associated integrals can be used to demonstrate that an appropriate set of boundary conditions is:

$$\boldsymbol{r}(0) = (0,0,0)^T , \tag{5.2.1}$$

$$\boldsymbol{r}_s(0)\cdot\boldsymbol{e}_1 = 0 , \tag{5.2.2}$$

$$|\boldsymbol{q}(0)|^2 = 1 , \tag{5.2.3}$$

$$q_1(0) = 0 , \tag{5.2.4}$$

$$q_3(0) = 0 , \tag{5.2.5}$$

$$\boldsymbol{\mu}(0)\cdot\boldsymbol{q}(0) = 0 , \tag{5.2.6}$$

$$\boldsymbol{r}(1) = (0,0,0)^T , \tag{5.2.7}$$

$$\boldsymbol{n}(1) = (\lambda,0,0)^T . \tag{5.2.8}$$

Here

$$\boldsymbol{r}_s(0) = \boldsymbol{d}_3(0) + \frac{\boldsymbol{n}(0)\cdot\boldsymbol{d}_j(0)}{A_j}\boldsymbol{d}_j(0) , \tag{5.2.9}$$

the free parameter is the magnitude of the shear force λ and summation in j is implied. At first sight it might seem more natural to have the free parameter be ΔLk (or equivalently the angle between $\boldsymbol{d}_1(0)$ and $\boldsymbol{d}_1(1)$) or the magnitude of the twisting component of the moment, but it turns out that they are all essentially equivalent for the purposes of constructing one parameter families of solutions. As written (5.2.1) – (5.2.8) are for the shearable case, but to obtain the appropriate unshearable boundary conditions one merely sends the coefficients A_j appearing in (5.2.9) to infinity.

One can show that the boundary conditions (5.2.1) – (5.2.8) are satisfied by the following 'trivial' solutions of the equilibrium equations. The centerline in each solution is a multiply covered circle (covered an integer

N times) lying in the (y, z) plane with $y \leq 0$ and the tangent at $s = 0$ lying along the z-axis. The $\boldsymbol{d}_3$ director makes a constant angle α with the (y, z)-plane, and the other directors uniformly rotate about it with a constant spin rate. The shear angle α must satisfy the relation

$$\sin\alpha = \frac{-2\,(\lambda/A_1)}{1+\sqrt{1+4\,(1-A_1/A_3)\,(\lambda/A_1)^2}}\,. \tag{5.2.10}$$

Then the centerline is given by

$$\boldsymbol{r}(s) = R\,(0,\ \cos 2\pi Ns - 1,\ \sin 2\pi Ns)^T\,, \tag{5.2.11}$$

where

$$R = \frac{1}{2\pi N\cos\alpha}\left(1+\frac{\lambda\sin\alpha}{A_3}\right)\,. \tag{5.2.12}$$

The directors are

$$\boldsymbol{d}_1(s) = \begin{pmatrix} \cos\alpha\,\cos\omega s \\ \sin\omega s\,\cos 2\pi Ns + \sin\alpha\,\cos\omega s\,\sin 2\pi Ns \\ \sin\omega s\,\sin 2\pi Ns - \sin\alpha\,\cos\omega s\,\cos 2\pi Ns \end{pmatrix}, \tag{5.2.13}$$

$$\boldsymbol{d}_2(s) = \begin{pmatrix} -\cos\alpha\,\cos\omega s \\ \cos\omega s\,\cos 2\pi Ns - \sin\alpha\,\sin\omega s\,\sin 2\pi Ns \\ \cos\omega s\,\sin 2\pi Ns + \sin\alpha\,\sin\omega s\,\cos 2\pi Ns \end{pmatrix}, \tag{5.2.14}$$

$$\boldsymbol{d}_3(s) = \begin{pmatrix} \sin\alpha \\ -\cos\alpha\,\sin 2\pi Ns \\ \cos\alpha\,\cos 2\pi Ns \end{pmatrix}, \tag{5.2.15}$$

in which the twist rate ω is related to the force λ through

$$\omega = \frac{\lambda R}{NK_3\cos\alpha} + \left(\frac{K_1}{K_3}-1\right)2\pi N\sin\alpha\,, \tag{5.2.16}$$

and the associated Euler parameters are

$$q_1(s) = \cos\frac{\alpha}{2}\,\sin\pi Ns\,\cos\frac{\omega s}{2} + \sin\frac{\alpha}{2}\,\cos\pi Ns\,\sin\frac{\omega s}{2}\,, \tag{5.2.17}$$

$$q_2(s) = -\cos\frac{\alpha}{2}\,\sin\pi Ns\,\sin\frac{\omega s}{2} + \sin\frac{\alpha}{2}\,\cos\pi Ns\,\cos\frac{\omega s}{2}\,, \tag{5.2.18}$$

$$q_3(s) = \cos\frac{\alpha}{2}\,\cos\pi Ns\,\sin\frac{\omega s}{2} + \sin\frac{\alpha}{2}\,\sin\pi Ns\,\cos\frac{\omega s}{2}\,, \tag{5.2.19}$$

$$q_4(s) = \cos\frac{\alpha}{2}\,\cos\pi Ns\,\cos\frac{\omega s}{2} - \sin\frac{\alpha}{2}\,\sin\pi Ns\,\sin\frac{\omega s}{2}\,. \tag{5.2.20}$$

Finally, the impetuses μ_j associated with the Euler parameters are given by

$$\mu_1(s) = \left(4\pi NK_1\cos\frac{\alpha}{2} + \frac{2\lambda R}{N\cos\alpha}\sin\frac{\alpha}{2}\right)\cos\pi Ns\,\cos\frac{\omega s}{2}$$

$$-\left(4\pi NK_1 \sin\frac{\alpha}{2} + \frac{2\lambda R}{N\cos\alpha}\cos\frac{\alpha}{2}\right)\sin\pi Ns\sin\frac{\omega s}{2}, \tag{5.2.21}$$

$$\mu_2(s) = -\left(4\pi NK_1 \cos\frac{\alpha}{2} + \frac{2\lambda R}{N\cos\alpha}\sin\frac{\alpha}{2}\right)\cos\pi Ns\sin\frac{\omega s}{2}$$
$$-\left(4\pi NK_1 \sin\frac{\alpha}{2} + \frac{2\lambda R}{N\cos\alpha}\cos\frac{\alpha}{2}\right)\sin\pi Ns\cos\frac{\omega s}{2}, \tag{5.2.22}$$

$$\mu_3(s) = -\left(4\pi NK_1 \cos\frac{\alpha}{2} + \frac{2\lambda R}{N\cos\alpha}\sin\frac{\alpha}{2}\right)\sin\pi Ns\sin\frac{\omega s}{2}$$
$$+\left(4\pi NK_1 \sin\frac{\alpha}{2} + \frac{2\lambda R}{N\cos\alpha}\cos\frac{\alpha}{2}\right)\cos\pi Ns\cos\frac{\omega s}{2}, \tag{5.2.23}$$

$$\mu_4(s) = -\left(4\pi NK_1 \cos\frac{\alpha}{2} + \frac{2\lambda R}{N\cos\alpha}\sin\frac{\alpha}{2}\right)\sin\pi Ns\cos\frac{\omega s}{2}$$
$$-\left(4\pi NK_1 \sin\frac{\alpha}{2} + \frac{2\lambda R}{N\cos\alpha}\cos\frac{\alpha}{2}\right)\cos\pi Ns\sin\frac{\omega s}{2}. \tag{5.2.24}$$

In the inextensible, unshearable limit, where the coefficients $\{A_j\}$ go to infinity, the shear angle α given by (5.2.10) goes to zero. In this limit, the expressions (5.2.11) – (5.2.24) above yield the corresponding values for the classic solution of the inextensible, unshearable rod.

The only other planar equilibria of the boundary value problem (5.2.1) – (5.2.8) are found by reflecting the above solution in the (x, z)-plane. The magnitude of the force is unaltered in the reflection, but being a mirror image the sign of the twist is reversed. Accordingly we refer to the original family as + solutions, and to the reflected family as – solutions. (Cf. Figure 5.2.)

5.3. Nonplanar equilibria. Both analytical and computational methods can be applied to show that a number of non-planar equilibria bifurcate (or branch) from the above described solutions at certain critical load parameters. In fact on each trivial branch there is a countable infinity of simple bifurcation points which may therefore be indexed as $(M, N, \pm)$ where M indicates the first, second, third, etc. bifurcation point on the N covered solution with either positive or negative twist. The main point of interest is to follow the branches of non-planar equilibria emanating from the known circular solutions. That procedure must be done numerically, even given the semi-explicit solution of the initial value problem that was discussed at the end of the previous chapter. Indeed Starotsin [38] combined the quadrature solution with a root solving algorithm for the undetermined constants of integration to trace out some (apparently partial) branches in the case of an unshearable rod. Shi & Hearst [32] also use the quadrature solution for an unshearable rod to calculate some isolated solutions, and in [31] an analogous solution by quadrature was found for the shearable extensible rod. However, in this last work no solutions to the two-point boundary value problem are computed due to the complexity of the quadrature formulæ. As remarked previously we regard the *ab initio* numerical generation of solutions to the initial value problem as straightforward computationally, and given the particular solution described in the previous section we also regard the numerical solution of the two-point boundary value problem as routine once the appropriate boundary conditions have been identified. Morever although we will not describe the detailed results here, a direct numerical approach extends to treat non-integrable cases (such as arise for example in the modelling of naturally curved DNA) in both the shearable and unshearable contexts.

We chose to use the package AUTO [12], which uses parameter continuation combined with a discretization based on the collocation method. The known trivial circular solution for the isotropic, uniform rod is used as as a starting point, then a continuation method is applied to numerically trace the bifurcating branches of non-planar equilibria. The result that was rather surprising to us was that the branch emanating from the Mth bifurcation on the N-covered ring with positive twist links to the Nth bifurcation point on the M-covered ring with negative twist, written symbolically as $(M, N, +) \leftrightarrow (N, M, -)$. Moreover, the only instance where the rod has any self-intersection is at a single point on each branch, where an $(M+N)$-foil arises, i.e. a symmetrical configuration with $M+N$ leaves. In fact the centerline of the nonplanar solutions is wound on a torus (or doughnut) in three-space with the cross-section of the torus deforming as the parameter varies. The integers (M, N) determine the torus knot type of the centerline, which is also inherited from the eigenfunction of the eigensolution to the linearized problem at the bifurcation point. The $(M+N)$-foil arises when the hole in the doughnut pinches off at one critical parameter value. The knot type switches at this point, and it is also close to this point that the numerics can be switched to a code that includes a model to preclude self-intersection. For the unshearable rod, this behavior was known in the case $(M + N) = 2$ where the 2-foil is the figure eight, but for $(M + N) > 3$ we believe the conclusion to be original, with the case $(M + N) = 3$ being partially understood previously [34]. The fact that the same behavior also arises for shearable rods was apparently unrealised, but is perhaps unsurprising.

In understanding the results of our bifurcation analysis we have found it extremely valuable to display the results of our computations using an interactive graphics package called PCR [14]. The full benefits of this graphics utility can only be appreciated with the interactive capabilities of a graphics workstation, but we have included some illustrations that depict some of the output. Figures 5.1–5.7 all illustrate certain features of the DNA supercoiling problem. These figures were obtained for the inextensible, unshearable case, but the features are unchanged to the eye for a surprisingly large range of extensible, shearable rods. Figures 5.1, 5.3, 5.4, 5.5, 5.6 and 5.7 each comprise four viewpoints of a single equilibrium of a rod. The four views comprise two elevations, a plan view and a skew-view (the axes reveal the relative orientations of each viewpoint). Each Figure illustrates an equilibrium for a different amount of imposed twist. Figure 5.2 shows a sequence of different configurations on the $(1, 1)$ branch from a single skew perspective. Equilibria of the twisted ring problem arise in one-dimensional families as a load parameter, such as ΔLk, is varied. Figures 5.8 and 5.9 further illustrate the output of the package PCR with displays of certain associated bifurcation diagrams.

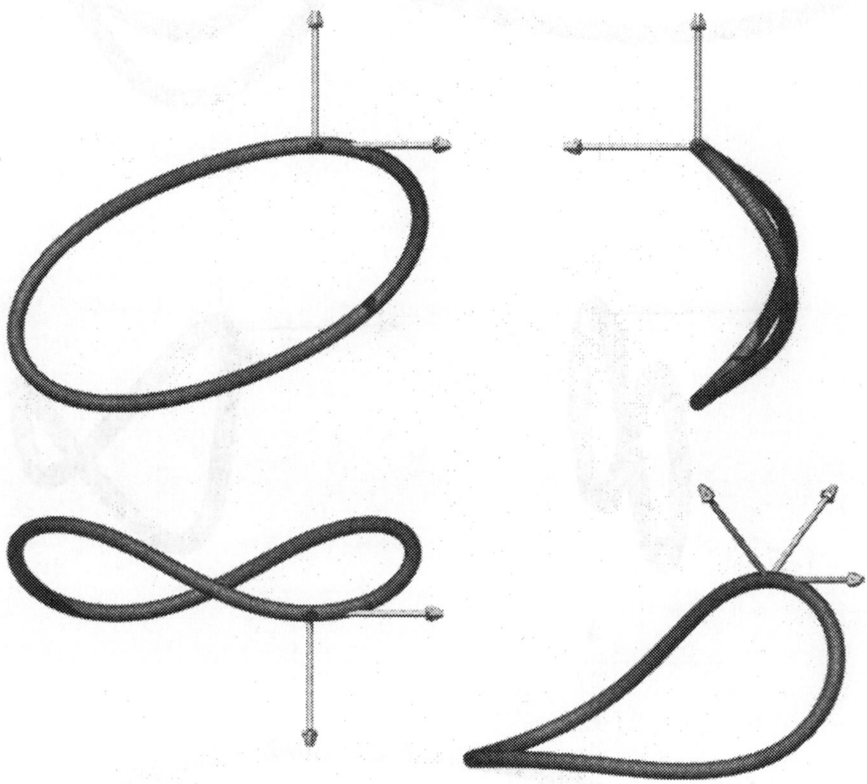

FIG. 5.1 *Four views of a typical configuration on the solution branch emanating from the* $(1, 1, +)$ *bifurcation point. It has left-right symmetry in both side elevations, and period two symmetry in the plan view. Domokos [13] provides a full account of the symmetries of ring equilibria.*

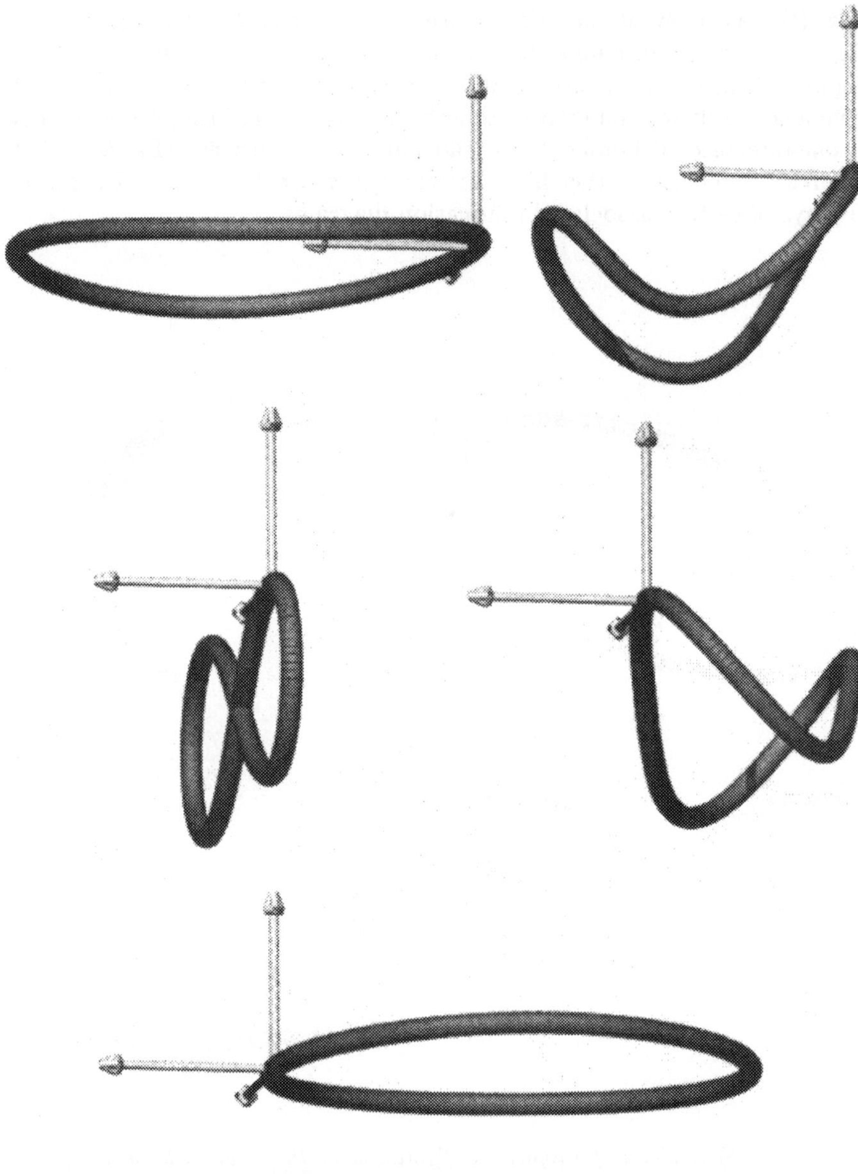

FIG. 5.2 *Different configurations all lying on the branch that links the* $(1,1,-)$ *and* $(1,1,+)$ *bifurcation points. The circle in the upper left is a negative twist solution, and the upper right solution is analogous to Figure 5.1. In the middle left there is a self-intersecting planar figure-eight solution, with another solution analogous to Figure 5.1 in the middle right, and finally the planar positive twist solution. The twist is monotone increasing from a negative to positive value through this sequence. In these calculations effects of self-intersection are ignored.*

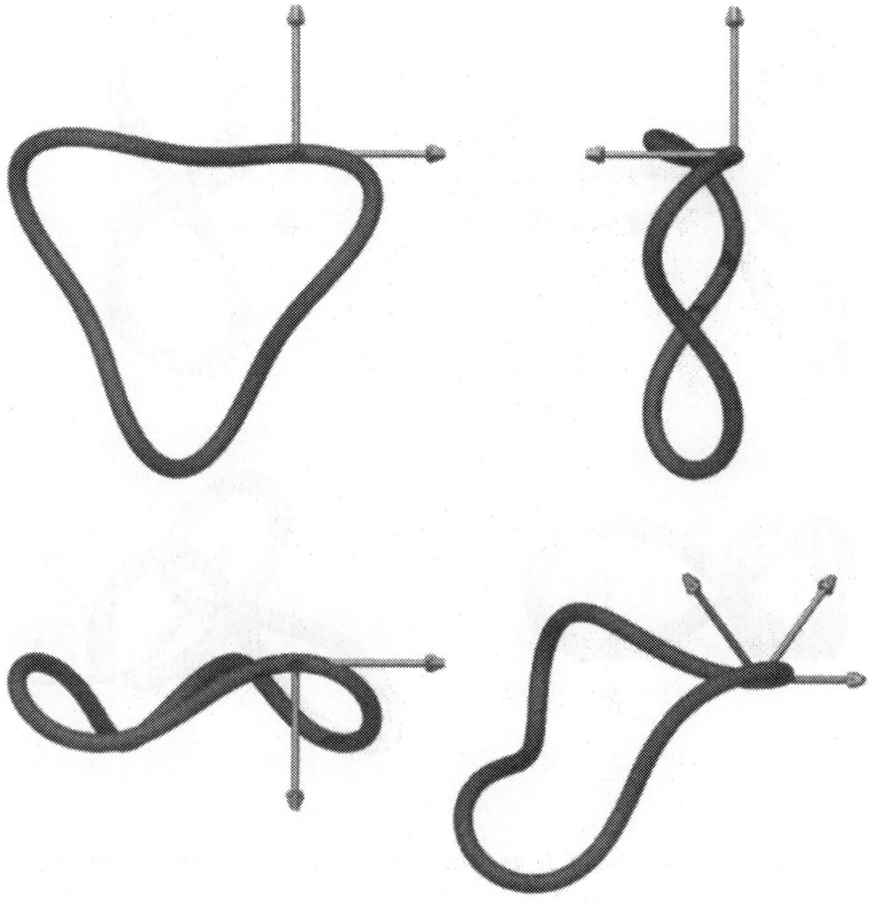

FIG. 5.3 *Figures 5.3, 5.4, and 5.5 show a sequence of three configurations on the branch linking the* $(1,2,-)$ *bifurcation point to the* $(2,1,+)$ *bifurcation point. Each figure depicts four views of the same equilibrium. All three configurations have period three symmetry in the plan view. The analogue of the self-intersecting figure-eight configuration is the self-intersecting trefoil shown in Figure 5.4. For twists slightly greater than the trefoil, the configuration becomes the non-intersecting trefoil knot shown in Figure 5.5, while the configuration shown in Figure 5.3 is unknotted.*

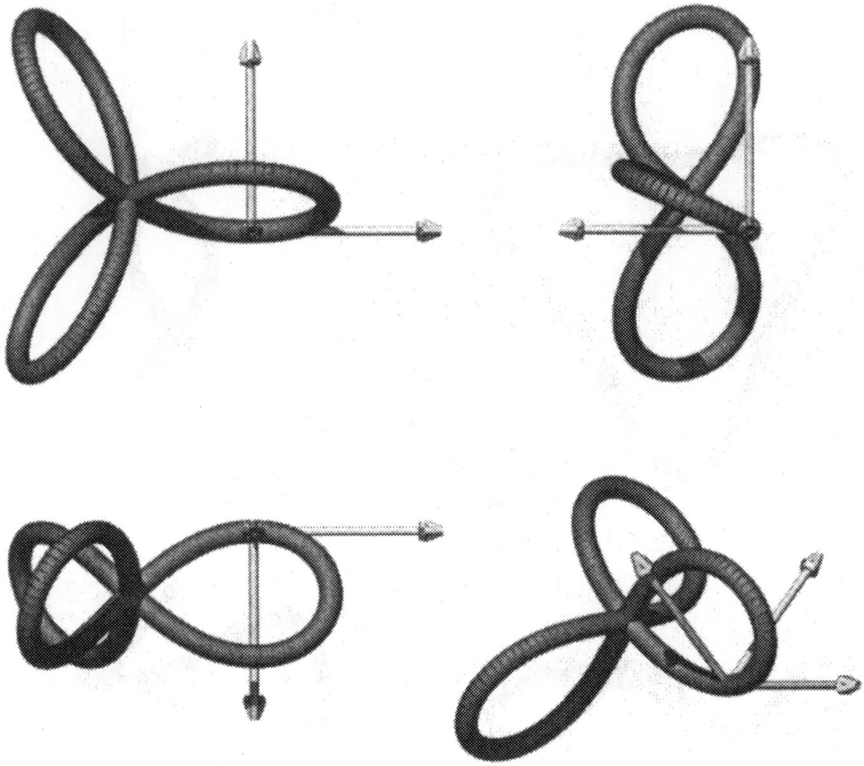

FIG. 5.4 *Figures 5.3, 5.4, and 5.5 show a sequence of three configurations on the branch linking the* $(1, 2, -)$ *bifurcation point to the* $(2, 1, +)$ *bifurcation point. Each figure depicts four views of the same equilibrium. All three configurations have period three symmetry in the plan view. The analogue of the self-intersecting figure-eight configuration is the self-intersecting trefoil shown in Figure 5.4. For twists slightly greater than the trefoil, the configuration becomes the non-intersecting trefoil knot shown in Figure 5.5, while the configuration shown in Figure 5.3 is unknotted.*

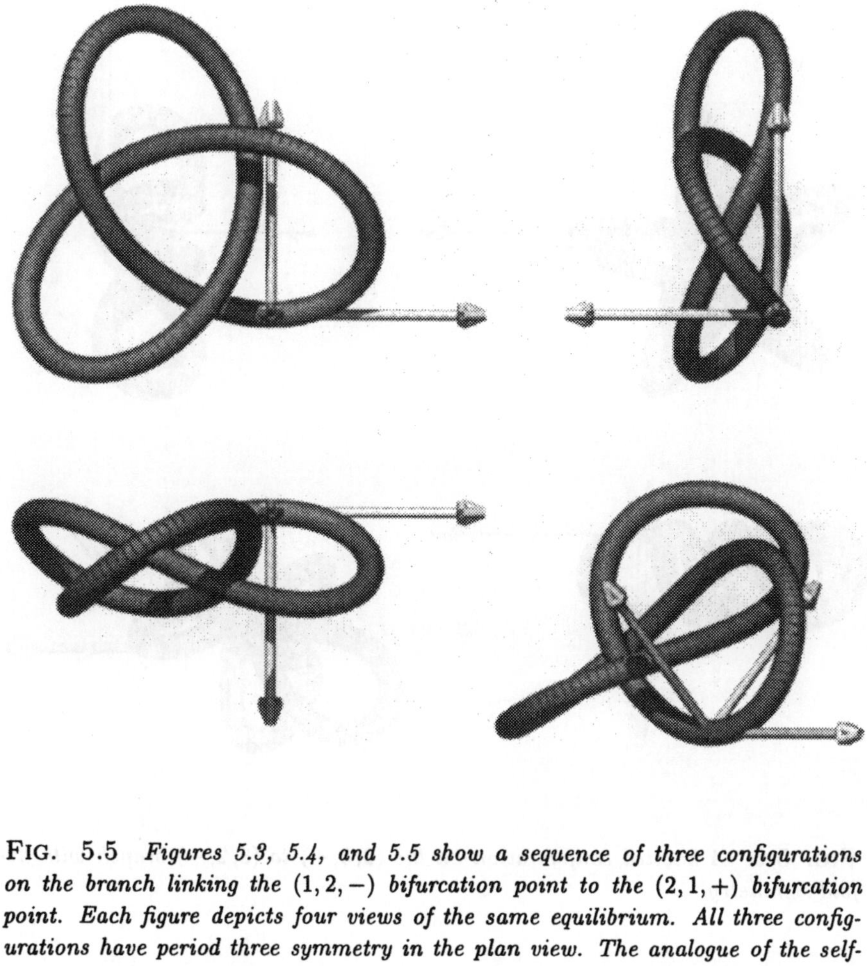

FIG. 5.5 *Figures 5.3, 5.4, and 5.5 show a sequence of three configurations on the branch linking the* $(1, 2, -)$ *bifurcation point to the* $(2, 1, +)$ *bifurcation point. Each figure depicts four views of the same equilibrium. All three configurations have period three symmetry in the plan view. The analogue of the self-intersecting figure-eight configuration is the self-intersecting trefoil shown in Figure 5.4. For twists slightly greater than the trefoil, the configuration becomes the non-intersecting trefoil knot shown in Figure 5.5, while the configuration shown in Figure 5.3 is unknotted.*

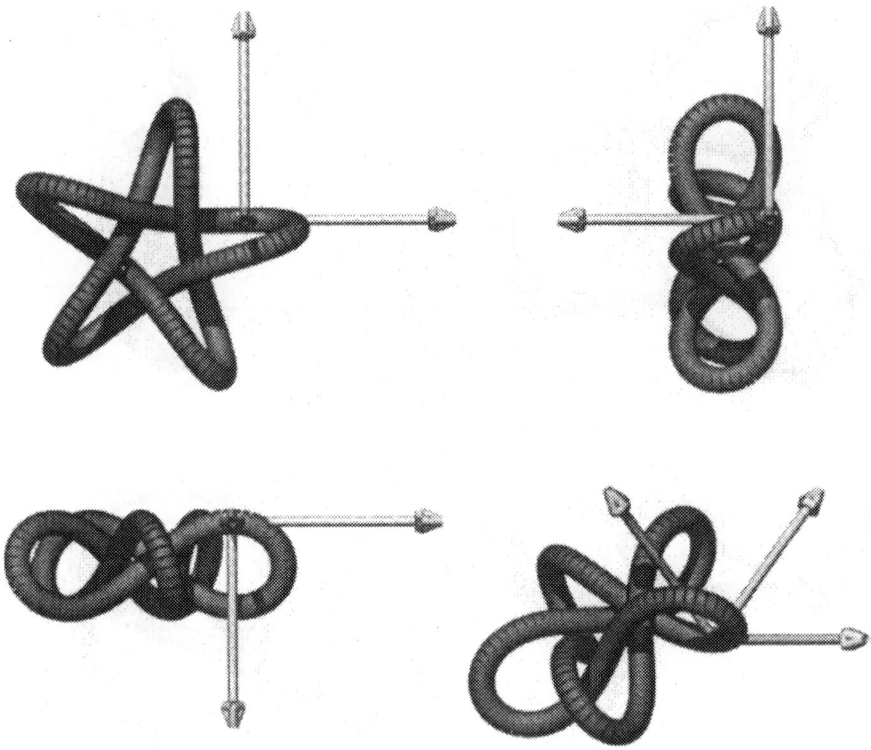

FIG. 5.6 *A knotted configuration on the* $(2,3,+)$ *to* $(3,2,-)$ *branch with five fold symmetry.*

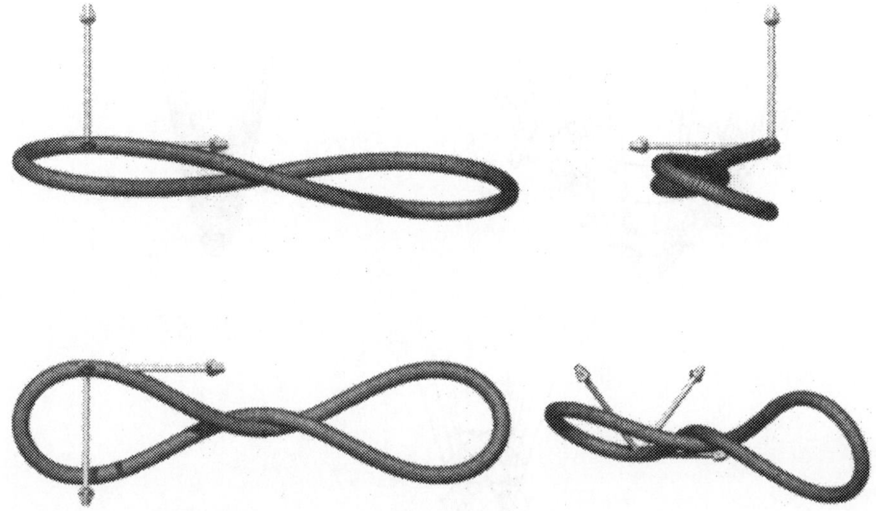

FIG. 5.7 *A configuration calculated with a modified code that models forces of self-contact which forbid self-intersection. Instead of passing through itself after self-contact at the figure eight solution, the rod starts to wrap around itself, forming the start of a plectoneme.*

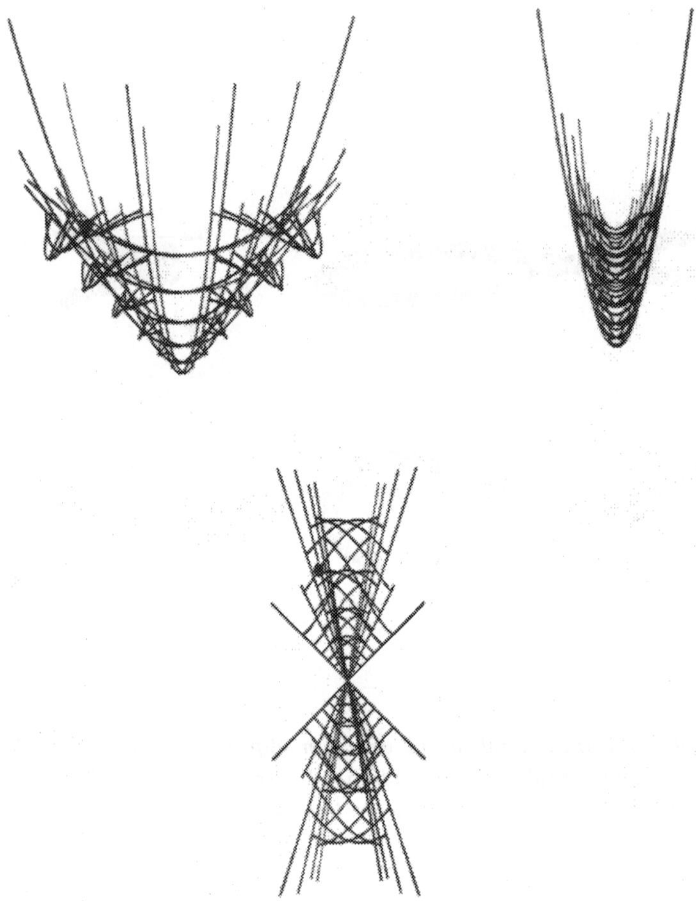

FIG. 5.8 *Three perspective views of the bifurcation diagram that arises when the energy, shear force, twist and writhe are evaluated along families of equilibria. The ordinate and abscissa in the top left, top right and bottom center views of the solution set are respectively energy vs. twist, twist vs. force, and energy vs. force. (The curves can also be colored according to writhe, but this information is lost in the greyscale presented here.) The 'trivial' circular solutions correspond to the straight branches in the bottom projection, and to the parabolic branches in the other two projections. The other solution branches correspond to configurations in which the centerline is a nonplanar curve. The nontrivial branches bifurcate from and link the trivial planar solutions according to the relation* $(M, N, +) \leftrightarrow (N, M, -)$. *All branches for* $(M + N) \leq 7$ *are shown.*

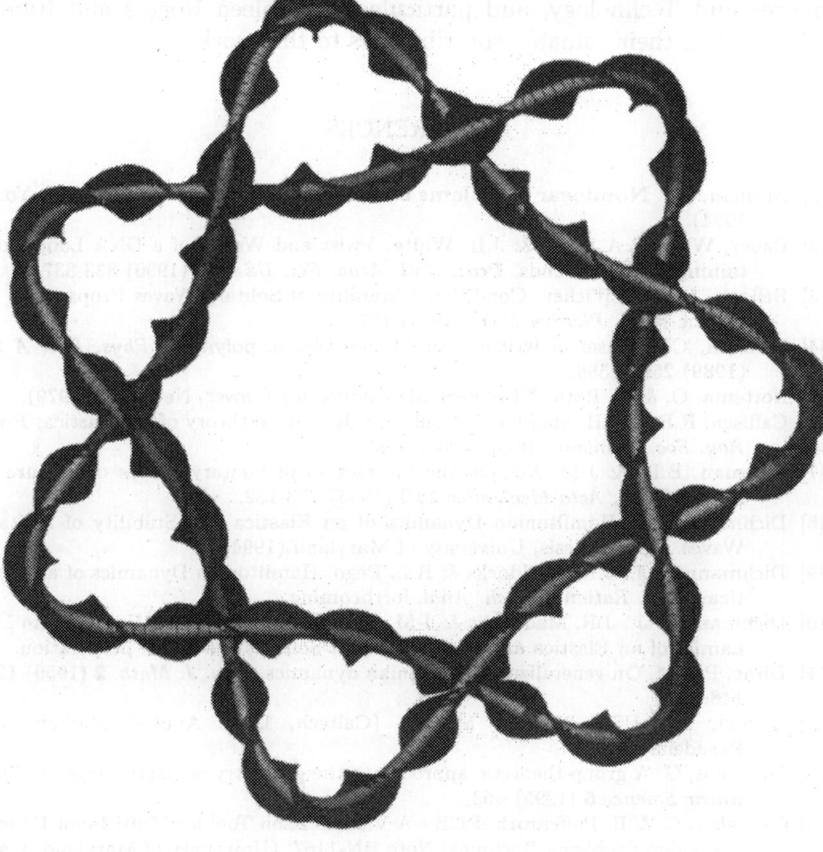

FIG. 5.9 *The reconstructed physical configuration corresponding to the point indicated by the filled circle in the bifurcation diagram of Figure 5.8. The cylinder represents the center line of the configuration, and the ribbon shows how the rod has been twisted. In the actual software package any point on the bifurcation diagram can be selected with a mouse click, and the physical solution is constructed and displayed in a second viewing window.*

Acknowledgements: It is a pleasure for the authors to thank the other members of the Numerical Analysis Laboratory in the Institute for Physical Sciences and Technology, and particularly Kathleen Rogers and Randy Paffenroth for their valuable contributions to this work.

REFERENCES

[1] Antman, S.S. **Nonlinear Problems of Elasticity**. (Springer-Verlag, New York, 1994).

[2] Bauer, W.R., R.A. Lund & J.H. White. Twist and Writhe of a DNA Loop Containing Intrinsic Bends. *Proc. Natl. Acad. Sci. USA* **90** (1990) 833-837.

[3] Beliaev, A. & A. Il'ichev. Conditional Stability of Solitary Waves Propagating in Elastic Rods. *Physica D* **90** (1996) 107.

[4] Benham, C.J. Onset of writhing in circular elastic polymers. *Phys. Rev. A* **39** (1989) 2582-2586.

[5] Bottema, O. & B. Roth. **Theoretical Mechanics** (Dover, New York, 1979).

[6] Caflisch, R.E. & J.H. Maddocks. Nonlinear dynamical theory of the elastica. *Proc. Roy. Soc. Edinburgh* **99A** (1984) 1-23.

[7] Coleman, B.D. & J-M. Xu. On the Interaction of Solitary Waves of Flexure in Elastic Rods. *Acta Mechanica* **110** (1995) 173-182.

[8] Dichmann, D.J. Hamiltonian Dynamics of an Elastica and Stability of Solitary Waves. Ph. D. thesis, University of Maryland (1994).

[9] Dichmann, D.J., J.H. Maddocks & R.L. Pego. Hamiltonian Dynamics of an Elastica. *Arch. Rational Mech. Anal.* forthcoming.

[10] Dichmann, D.J., J.H. Maddocks & J-M. Xu. Three-Dimensional Hamiltonian Dynamics of an Elastica and the Stability of Solitary Waves. In preparation.

[11] Dirac, P.A.M. On generalized Hamiltonian dynamics. *Can. J. Math.* **2** (1950) 129-148.

[12] Doedel, E. AUTO 86 User Manual. (Caltech, Dept. Applied Mathematics, Pasadena)

[13] Domokos, G. A group-theoretic approach to the geometry of elastic rings. *J. Nonlinear Science* **5** (1995) 453.

[14] Domokos, G. & R. Paffenroth. PCR - A Visualization Tool for Multi-Point Boundary Value Problems. Technical Note BN-1167. (University of Maryland, Institute for Physical Science and Technology, College Park, 1994).

[15] Falk, R.S. & J-M. Xu. Convergence of a Second-Order Scheme for the Nonlinear Dynamical Equations of Elastic Rods. *SIAM J. Numerical Analysis* **32** (1995).

[16] Goldstein, H. **Classical Mechanics**, Second Edition (Addison-Wesley, Reading, MA, 1981).

[17] Ilyukhin, A.A. **Spatial problems of non-linear theory of elastic rods**, (Naukova Dumka, Kiev, 1979). [in Russian]

[18] James, R.D. The equilibrium and post-buckling behaviour of an elastic curve governed by a non-convex energy. *J. Elasticity* **11** (1981) 239-269.

[19] Jülicher, F. Supercoiling transitions of closed DNA. *Phys. Rev. E* **49** (1994) 2429-2435.

[20] Kirchhoff, G.R. Ueber das Gleichgewicht und die Bewegung eines unendlich dünnen elastischen Stabes. **Gesammelte Abhandlungen** (Leipzig, 1882).

[21] Klapper, I. & M. Tabor. Dynamics of Twisted Elastic Rods. Preprint.

[22] Landau, L.D. & E.M. Lifshitz. **Theory of Elasticity** (Pergamon Press, New York, 1970).

[23] Langer, J. & D. Singer. Lagrangian Aspects of the Kirchhoff Elastic Rod. *SIAM Review* forthcoming

[24] LeBret, M. Catastrophic Variation of Twist and Writhing of Circular DNAs with Constraint? *Biopolymers* **18** (1979) 1709-1725.

[25] Li, Y. & J.H. Maddocks. On the Computation of Equilibria of Elastic Rods, Part I: Integrals, Symmetry and a Hamiltonian Formulation. Submitted to *J. Comp. Physics*

[26] Li, Y. & J.H. Maddocks. On the Computation of Equilibria of Elastic Rods, part II: Effects of Self-Contact. In preparation.

[27] Li, Y., J.H. Maddocks & D.J. Dichmann. On the Computation of Equilibria of Elastic Rods, part III: Effects of Shear. In preparation.

[28] Love, A.E.H. **A Treatise on the Mathematical Theory of Elasticity**, (Dover, New York, 1944).

[29] Maddocks, J.H. & D.J. Dichmann. Conservation Laws in the Dynamics of Rods. *J. Elasticity* **34** (1994) 83-96.

[30] Olver, P.J. **Applications of Lie Groups to Differential Equations**, (Springer-Verlag, New York, 1986).

[31] Shi, Y., A. Borovik & J.E. Hearst. Elastic Rod Model Incorporating Shear and Extension, Generalized Schrödinger Equations, and Novel Closed-Form Solutions for Supercoiled DNA. *J. Chem. Physics* **103** 3166 (1995)

[32] Shi, Y. & J.E. Hearst. The Kirchhoff elastic rod, the nonlinear Schrödinger equation, and DNA supercoiling. *J. Chem. Phys.* **101** (1994) 5186-5200.

[33] Schlick, T. Modeling Superhelical DNA: Recent Analytical and Dynamical Approaches. **Current Opinions in Structural Biology**, ed. B. Honig. **5** (1995).

[34] Schlick, T., W.K. Olson, T. Westcott & J.P. Greenberg. On Higher Buckling Transitions in Supercoiled DNA. *Biopolymers* **34** (1994) 565-597.

[35] Shuster, M.D. A Survey of Attitude Representations. *J. Astronautical Sciences* **41** (1994) 439-518.

[36] Simo, J.C., J.E. Marsden & P.S. Krishnaprasad. The Hamiltonian Structure of Nonlinear Elasticity: The Material and Convective Representations of Solids, Rods and Plates. *Arch. Rational Mech. Anal.* **104** (1988) 125-183.

[37] Simo, J.C. & L. Vu-Quoc. A Three-Dimensional Finite-Strain Rod Model. Part II: Computational Aspects. *Comput. Meths. Appl. Mech. Engrg.* **58** (1986) 79-116.

[38] Starotsin, E.L. Three-Dimensional Conformations of Looped DNA in an Elastomechanical Approximation. *Proc. 2nd International Conference on Nonlinear Mechanics.* Ed. Chien Wei-zang. (Peking University Press, Peking, 1993).

[39] Tsuru, H. Equilibrium Shapes and Vibrations of Thin Elastic Rods. *J. Phys. Soc. Japan* **56** (1987) 2309-2324.

[40] Wadati, M. & H. Tsuru. Elastic Model of Looped DNA. *Physica D* **21** (1986) 213-226.

[41] Yang, Y., I. Tobias & W.K. Olson. Finite Element Analysis of DNA Supercoiling. *J. Chem. Phys.* **98** (1993) 1673-1686.

STATISTICAL MECHANICS OF SUPERCOILED DNA

JOHN F. MARKO* AND ERIC D. SIGGIA†

The conformation of long supercoiled DNA loops under near physiological conditions has been the subject of a long series of laboratory and Monte Carlo experiments, and impressive agreement between theory and experiment has been achieved [1–2]. A closed unnicked DNA loop has fixed linking number: this constraint leads to a competition between twisting and bending (writhing) elastic energy which drives plectonemic supercoiling. However, entropy, as manifest through a fluctuation induced repulsive force, overwhelms this elastic effect for fractional linking number perturbations $|\sigma| < 0.02$, causing a chiral random-coil conformation. For $|\sigma| > 0.02$, plectonemic supercoils are stable, but thermal fluctuations continue to play an important role, competing with bending and twisting elastic energy to set the superhelical radius. We have constructed an analytic theory for this effect that agrees well with experiment [3], using methods from polymer statistical mechanics.

In addition to describing the basic plectonemic structure, our analysis describes the degree of branching of the plectonemic axis, in good agreement with real and computer experiments. The branches are thermally excited 'defects' in the regular plectonemic structure, and cannot be understood using a purely elastic (zero-temperature) approach. At still larger scales, thermal fluctuations determine the overall size and segment density of a supercoiled loop: we have predicted how these properties depend on the loop length and linking number [3].

The time scale for coil rearrangements or intra-molecular reactions is set by bending modes of the plectoneme as a unit, which is in turn determined by the global supercoil structure. We find that distant-site juxtaposition by random slithering of the DNA through the supercoil is by comparison a very slow process, especially for loops of more than 10 kb [3].

The force vs extension relation for the worm like chain model for a polymer — simply derived from the elastic theory of a uniform rod undergoing thermal fluctuations — can be calculated exactly and agrees well with experiment [4-5]. The effects of variable salt concentration on the entropic elasticity can also be accounted for analytically. When there is an imposed twist on the DNA being extended, the measured force reflects an interesting coexistence of plectonemic and solenoidal states, which we also are able to calculate.

* Center for Theoretical Studies in Physics and Biology, The Rockefeller University, 1230 York Avenue, New York, NY 10021.

† Laboratory of Atomic and Solid State Physics, Cornell University, 538 Clark Hall, Ithaca, NY 14853-2501.

REFERENCES

[1] T.C. Boles, J.H. White, and N.R. Cozzarelli, *Structure of Plectonemically Supercoiled DNA*, J. Mol. Biol. **213** (1990), 931.

[2] A.V. Vologodskii, S.D. Levene, K.V. Klenin, M. Frank-Kamenetskii, and N.R. Cozzarelli, *Conformational and Thermodynamic Properties of Supercoiled DNA*, J. Mol. Biol. **227** (1992), 1224 but a dissenting voice is raised by J. Bednar *et al.*, *ibid.* **235** (1994), 825.

[3] J.F. Marko and E.D. Siggia, *Fluctuations and Supercoiling of DNA*, Science **265** (1994), 506 and Phys. Rev. E, to appear 1995.

[4] S.B. Smith, L. Finzi, and C. Bustamante, *Direct Mechanical Measurements of the Elasticity of Single DNA Molecules by Using Magnetic Beads*, Science **258** (1992), 1122.

[5] C. Bustamante, J.F.Marko, E.D. Siggia, and S. Smith, *Entropic Elasticity of $\lambda-$ phage DNA*, Science **265** (1994), 1555 and J.F. Marko and E.D. Siggia, "Stretching DNA" Macromolecules, submitted.

DETERMINATION OF THE DNA HELICAL REPEAT AND OF THE STRUCTURE OF SUPERCOILED DNA BY CRYO-ELECTRON MICROSCOPY

ANDRZEJ STASIAK*, JAN BEDNAR*, PATRICK FURRER*, VSEVOLOD KATRITCH*†, AND JACQUES DUBOCHET*

Abstract. Cryo-electron microscopy provides an unique possibility to directly observe shape of individual DNA molecules freely suspended in cryo-vitrified liquid media. We used this technique to characterize the superhelical trajectory adopted by linear DNA molecules composed of directly repeated intrinsically bent DNA segments which have 10 or 11 base pair each. Although the DNA helix is not directly discerned by this method, the measured values of diameter, pitch and handedness of the formed superhelices allow to determine the chirality and the number of base-pairs per turn of the constituting DNA. We also used cryo-electron microscopy to study the response of supercoiled DNA molecules to increasing counterions' concentration. We observed that upon substantial neutralization of the negative charge of the DNA, the supercoiled molecules have a tendency to adopt a so called "tight" configuration, whereby the opposing segments of interwound molecules directly approach each other. Metropolis Monte Carlo simulations of shapes of supercoiled DNA molecules revealed that some short range attractive interactions between DNA segments would be needed to compensate for the entropy loss during transition from a loose toward a "tight" configuration of supercoiled DNA molecules. Since earlier studies of DNA helical repeat and DNA supercoiling are not familiar to non-biologists we included in to this chapter a short chronological overview of these studies.

1. Cryo-electron microscopy, its advantages and its limits

Electron microscopy (EM) provided many illuminating insights into structural and functional aspects of DNA. For example, the actual comprehension of DNA supercoiling (see below) required direct EM visualization of supercoiled DNA molecules [1], the sequence dependent DNA bending was widely accepted as a fact only after EM had provided unambiguous micrographs [2], the determination of actual topology of different DNA knots and catenanes formed by DNA topoisomerases or site specific recombination enzymes was only possible thanks to EM [3]. Despite these achievements all classical EM studies suffered from the fact that the investigated biomolecules were not suspended in their liquid media but were adsorbed to supporting films and therefore the intrinsic 3-D structure of observed DNA molecules was lost. In addition interactions with supporting films and subsequent specimen dehydration can change chemical and physical environment of DNA molecules, this can lead to substantial changes of DNA structure [4].

Cryo-electron microscopy overcomes these drawbacks as it allows to observe biomolecules suspended in a thin layer of cryo-vitrified liquid media

* Laboratoire d'Analyse Ultrastructurale, Université de Lausanne, Bâtiment de Biologie, CH-1015 Lausanne, Switzerland.

† Permanent address: Institute of Molecular Genetics, Russian Academy of Science, Moscow, 123182, Russia.

(for a recent review of the method see [5]). Speed of specimen cooling for cryo-vitrification is of the order of 10^6 °K/sec, this eliminates water crystal formation and minimizes possible distortions of biomolecules. Vitrified samples when kept at the temperature below −160°C can be introduced in to a microscope using special cryo-EM specimen holder. At this temperature even at high vacuum of the microscope the vitrified specimens have a negligible rate of sublimation which thus provides enough time (hours) for recording micrographs from different regions of the specimens. However, as the specimens remain very sensitive to electron beam, the photographed objects can not be previewed and assessed before taking pictures since this would damage these previewed regions of the specimen. Minimum dose setting has to be used which ensures that at high magnification the irradiation of the sample is limited only to the dose which is needed to record one or two photographs which are needed for stereo reconstruction[6]. Therefore only after developing of the films one can find out whether molecules of interest are recorded and whether they were properly focused or not. In cryo-EM specimens the DNA is not stained and the contrast is provided only by a different electron density of biomolecules as compared to the surrounding media, therefore optimal use of phase contrast (obtained by strong defocusing) is needed to obtain micrographs on which one can follow the axial trajectory of individual double-stranded DNA molecules. Strong defocusing in turn decreases the resolution of the obtained images.

The micrographs obtained in the EM are strict projection views, however, from two different projections of continuous filamentous objects (like DNA molecules) it is possible to reconstruct the 3-D trajectory of these objects. Two different projections are obtained by tilting the specimen holder between two sequential exposures needed to obtain stero-pair of micrographs. On the basis of such stereo micrographs it is possible to calculate 3-D coordinates of analysed DNA molecules. Computer assisted 3-D reconstruction method was thus developed in order to reconstruct axial trajectories of individual DNA molecules freely suspended in the buffer of interest [6]. We used this methodology to study the shape of linear DNA molecules composed of directly repeated, bent DNA segments which are 10 or 11 base pair long [7] and also the shape of supercoiled DNA molecules suspended in buffers with different concentrations of counter ions [8,9]. The results of these recent studies are presented here in form of overview.

2. DNA helical repeat

In the search for the most direct way to determine the DNA helical repeat. In their famous paper Watson and Crick [10] combined X-ray diffraction patterns of DNA fibers with possible streochemical arrangements of nucleotides (DNA building blocks composed of purine or pyrimidine bases joined via deoxirybose to monophospate groups) to conclude that B-DNA (structure of DNA adopted in aqueous solutions) forms right-handed double helix with about 10 base pairs per turn. However later observation that DNA molecules frequently exist in a circular form (e.g. bacterial chromosomes, DNA of viruses) lead to the realization that chirality of DNA can cause topological difficulties during replication of circular DNA molecules [11]. Therefore alternative model of DNA structure was proposed [12] in which the DNA was composed of 5 bp. long, alternating left- and right-handed segments (this model was consistent with the original low resolution X-ray diffraction patterns and with stereochemistry of nucleotides). This very controversial model in which the overall chirality of DNA was effectively cancelled was called SBS (side by side). Presentation of the SBS model and the fact that DNA can adopt several different configurations [13] including left handed Z-DNA structure [14] prompted researchers to find new ways to analyse helical repeat of DNA by the methods which are more direct and more palpable for nonspecialists than the X-ray diffraction analysis. Of course, nowadays the high resolution X-ray crystallography or NMR data proved beyond any reasonable doubt that most of natural DNA sequences adopt structure of right handed helices with about 10.4 base pairs per turn. However, only narrow group of specialists can comprehend and use X-ray or NMR spectra, therefore new methods were and are sought which can characterize in simple and rather direct way the helicity of DNA. Here some of those methods are reviewed in the chronological order.

Rhodes and Klug [15,16] developed a method where DNA molecules adsorbed to a flat surface were exposed to solutions containing DNase molecules (DNase is an enzyme which cleaves DNA strands). As the most accessible to bulky molecules of DNase are these portions of the DNA helix which are the most distant from the adsorbing surface, it could be expected that within every helical turn of surface adsorbed DNA, one region in each strand will be most likely to be cut and that the distance between such cutting sites on the same strand should thus correspond to the helical repeat of the DNA. Indeed, in an agreement with this expectation it was shown that preferable cutting sites on DNA were spaced by distances ranging between 10.6 and 10.0 bp (depending on the actual nucleotide composition of the analysed DNA). This was consistent with the helical structure of DNA having between 10 and 10.6 bp per turn. However the used method was not supposed to tell if DNA helix is left- or right-handed. In addition the putative SBS (side by side) DNA structure could have produced a very similar DNase digestion pattern and thus SBS model was not refuted by the DNase digestion method. If for example the DNase would have cleaved preferentially within right-handed segment (or left-handed but not in both) then in the surface adsorbed SBS DNA one would still expect to find best exposed regions being spaced also by ca 10 bp i.e. by the complete structural repeat of the putative SBS DNA with ca 5 bp long alternating left and right handed DNA regions.

Wang introduced an ingenious band shift method to assay helical repeat of DNA in solution [17]. This method is based on the earlier observation that when identical linear DNA molecules undergo circularization (covalent closure of both DNA strands by the enzyme ligase) a family of DNA topoisomers is formed [18,19]. Members of this family differ in their linking number (Lk) i.e. in the number of times one DNA strand is topologically linked with the other strand. Since DNA helix has a certain torsional stiffness the most frequent closure of the ends is such which generates the least torsional deformations. However as the available thermal motion randomly undertwists and overtwists constituting segments of DNA molecules, this pushes the DNA molecules out of equilibrium so that at the time of closure they can have some deficit or excess of DNA turns as compared to the lowest energy state. As a result of this one observes a Gaussian distribution of molecules with different linking number (topoisomers). If for example (ideal case) the number of base pairs in circularized DNA molecules can be divided without the rest by the preferable value of DNA helical repeat, then the distribution of the generated topoisomers will be very symmetrical. The most frequent topoisomer will be this corresponding to energetic minimum (topoisomer 0) since this topoisomer arises from the molecules in which the extent of random undertwisting is compensated by the extent of random overtwisting. Less frequent will be molecules in which random thermal motion leads to a slight overall overtwisting or undertwisting (topoisomers +1 and −1 respectively) and then still less frequent will be molecules with increasing deficit or excess of DNA turns (topoisomers +2 and −2 etc.).

Gel electrophoresis of DNA molecules performed in the presence of DNA intercalating drugs like chloroquine or ethidium bromide is a chromatography method which allows separation of individual DNA topoisomers. The discussed above "ideal family" of DNA topoisomers produces upon electrophoresis a characteristic band pattern where the most intense band (topoisomer 0) is flanked on both sides by less intense but equal to each other bands of topoisomers −1 and +1, which in turn are flanked by bands with decreasing intensity (corresponding to topoisomers −2 and +2 and possibly to some further topoisomers). If such "ideal" DNA fragment used for circularization is extended by 5 bp (ca half of a helical repeat), the pattern of topoisomers seen upon gel electrophoresis will change. In the envelope of Gaussian distribution the center will be unoccupied but instead one will see quite symmetrically displaced from the center almost equally intensive bands of topoisomers −0.5 and +0.5 flanked by less intensive bands of topoisomers −1.5 and +1.5 and so on.

Wang used this gel electrophoresis method to find out by how many base pairs the size of circularized DNA would need to be increased (or decreased) to obtain exactly the same distribution of the topoisomers' frequency (within a Gaussian envelope), as the distribution observed for the initial fragment. This method effectively ruled out the SBS model and established that DNA with natural sequence has a helical repeat of about 10.4 bp per turn [17]. However even this method was insensitive to the handedness of DNA and was not indicating if DNA was right- or left-handed.

The more recent phasing methods of determination of helical repeat of nucleic acids are based on the observation that if short linear DNA molecules contain DNA regions with significant intrinsic curvature the electrophoretic migration of such molecules is slowest when the bends are pointing in the same direction that is when they are phased [20]. By varying the size of DNA linkers between the bent regions it was possible to find out what distance phases them best (helical repeat or its multiple). Using this method Drak and Crothers [21] concluded that directly repeated 10 or 11 mers with 5 or 6 bp.long A-tracts have the helical repeat of 10.34 ± 0.04 bp per turn. However also this modern phasing methods could not tell if DNA helix was right- or left-handed.

Scanning tunneling microscopy (STM) seemed to be the method which won the contest for the most direct way to see handedness of DNA helix [22,23]. Atomic resolution of the method (as shown for some inorganic crystals) should have allow without difficulties to see directly the helical structure of DNA. Indeed, in the enthusiastic, initial period of STM usage for studying biological materials several groups worked with DNA and showed images of helical filamentous objects. On some of those objects direction of helical winding and pitch could have been measured [22,23,24]. However these results were met with scepticism since there were serious problems with measured dimensions of DNA helical repeat [25] and it turned out that similar helical objects could have been observed on supporting films without DNA [26]. Therefore it is very likely that such artefacts mislead STM users. Today one knows that with a biological material like DNA the resolution of STM is much poorer than with inorganic crystalline objects [27] and that such phenomena like conductivity problems, weak adsorption and deformability of DNA in contact with a scanning tip may in fact cause that DNA helix will not reveal its structure by the use of this method.

Cryo-electron microscopy approach to determine the sign and the value of DNA helical repeat. To determine helical repeat of DNA we have chosen sequences containing 5 or 6 homopolymeric dA·dT base pairs ("A-tracts") which were shown to deflect significantly the DNA axis [28,29]. When such A-tracts are directly repeated every 10-11 base pairs (bp) the linear DNA fragments adopt a superhelical path with a pitch and a diameter of the order of 10 nanometers [30,31,32]. Although the resolution of electron microscopy does not yet allow to perceive directly the double helical structure of DNA, it is sufficient to measure accurately the pitch and diameter of the formed superhelices. From these two values, the helical repeat of the constituting DNA can be calculated.

Figure 1 explains the principle of this geometric approach to determine the DNA helical repeat. When intrinsically bent right handed helical segments encompassing about one helical turn each are smoothly connected in direct repeat fashion three situations can happen: (a) when each segment makes slightly less than a full rotation, the formed superhelix is left

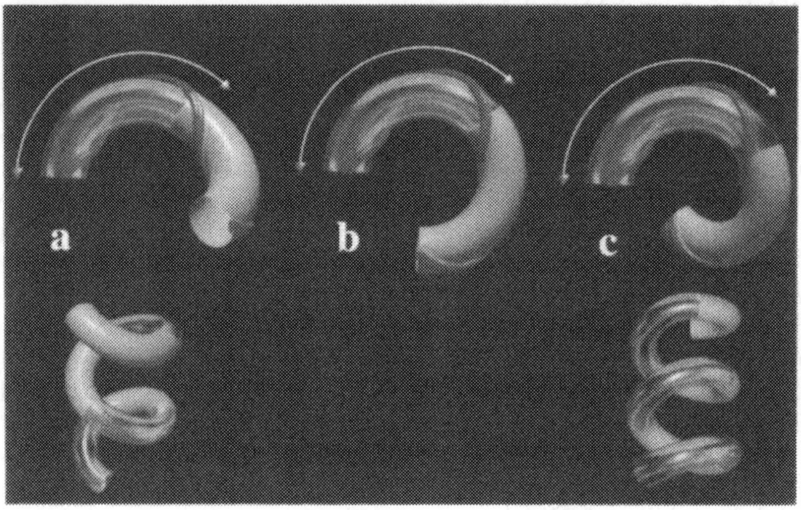

FIG. 1. How torsion contributes to superhelix formation. The tape wrapped around the bent cylinders symbolizes phosphate backbone of one of the complementary strands in a bent DNA fragment. In each case the winding angle of the tape in relation to the axes of bent cylinders is the same, but the composing repeated segments have different length. In (a) each repeated segment is too short to allow for a complete rotation of the tape around the cylinder. Therefore when these cylinders are connected by their "phosphate backbone" the whole construction forms a left-handed superhelix. In (b) the size of repeated bent fragments allows exactly one full rotation of the tape around each repeated segment. Several such identical building blocks with continuous "phosphate backbones" would form a planar arc. In (c) the repeating bent segments are slightly longer than in (b), therefore the total rotation in each segment is bigger than 360° and the formed superhelix is now right-handed. Arcs with constant length are placed above the bent segments in (a), (b), and (c) to accentuate differences in size of the repeating units in each of the cases. This figure is reproduced from [7].

handed, (b) when each segment makes exactly one turn, the resulting construct forms an arc of a circle, and when each segment makes slightly more than one right handed turn (c) the superhelix is right handed. We wanted to mimic situations (a) and (c) with DNA fragments composed of 10 bp (for slightly less than one turn) and 11 bp (for slightly more) and to see if observed superhelices will indeed be left handed for tandem repeats of bent 10 bp segments and right handed for such repeats of 11 bp segments.

Figure 2(a) shows stereomicrographs of 540 bp long DNA molecules composed of tandem repeats of 10 bp segments with the sequence $(A)_5TGCCC$. Stereoscopic viewing of the micrographs allows to see that many molecules form left handed superhelices. The fact that the molecules seen on the micrographs are not very regular is the expected consequence of thermal fluctuations [33]. It has to be reminded here that in cryo-EM preparations the contrast is low since DNA is not stained and therefore in Figure 2(b), selected molecules are traced over for better visualization. Figure 2(c). shows the 3-D path of one $[(A)_5TGCCC]_{54}$ molecule as reconstructed from stereomicrographs [6]. From several micrographs similar to those shown in Figure 2(a) we have selected 41 molecules with well visible superhelical periodicity. Pitch and diameter of those superhelices amounted to 38.4 ± 6.9 nm and 15.5 ± 3.4 nm respectively, as measured on magnified photographic prints.

Figures 3(a) and (b) show a micrograph and the corresponding enhancing tracings of DNA molecules composed of 11 bp segments with the sequence $[(A)_6TGCCC]n$ where n ranges from ca 15 to 30. 3-D reconstruction of the shape of $[(A)_6TGCCC]n$ molecules (Fig. 3(c)), revealed that the formed superhelices were right handed in contrast to the left handed superhelices formed by the sequence $[(A)_5TGCCC]_{54}$. Pitch and diameter of superhelices formed by $[(A)_6TGCCC]_n$ molecules was 30.5 ± 4.3 nm and 7.8 ± 2.0 nm respectively (as measured on 28 molecules with well visible superhelical periodicity).

Thus our direct cryo-EM observations allowed to demonstrate that DNA helix is right-handed. If DNA helix would be left handed then one would expect that a tandem repeat of bent segments which are slightly shorter than one turn would lead to formation of right handed superhelix and conversely a tandem repeat of bent segments which are slightly longer than one turn should result in formation of left handed superhelix. This geometrical method in addition to determining the direction of DNA helix can also provide information about the actual helical repeat value of studied DNA fragments. When helices or superhelices are formed from identical, directly repeated segments, it is possible to derive the torsion contributed by each of those individual segments. In the case of 10 and 11 bp fragments the torsion measures by how much the overall twist angle of DNA in each segment differs from a 360° rotation. The torsion angle (Ψ) per element is expressed by the formula $\Psi = \ell/r \sin\alpha \cos\alpha$ [34] where ℓ is the length of the repeated segment, r is the diameter of the

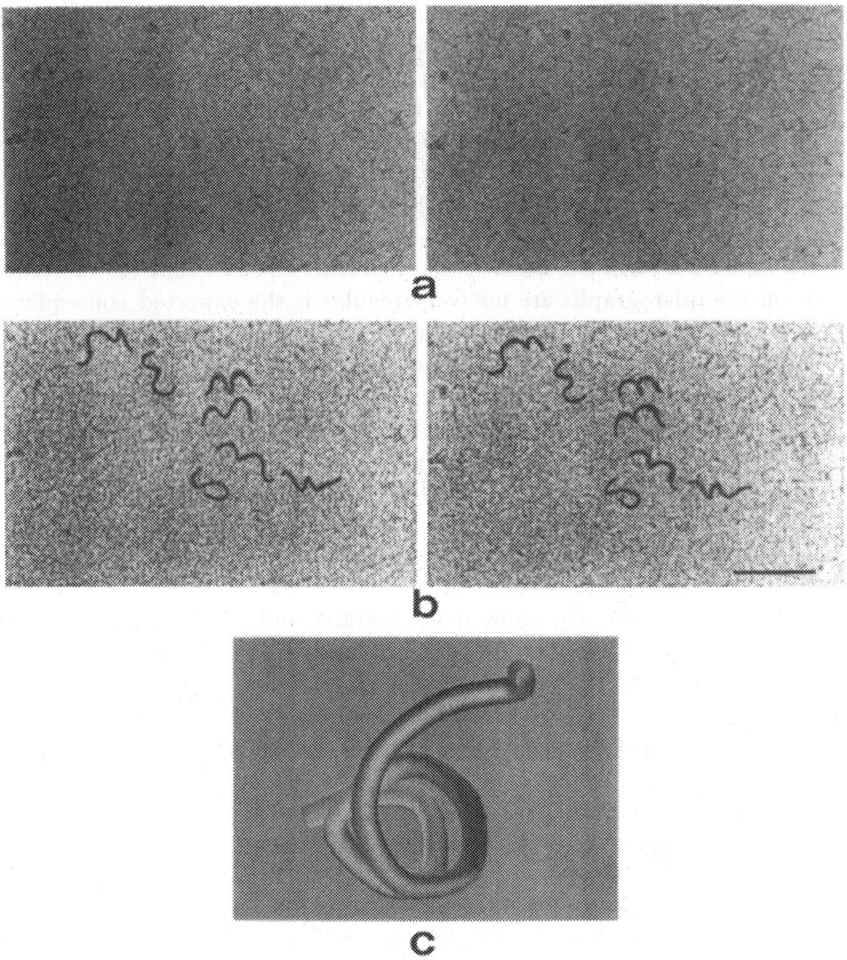

FIG. 2. Left-handed superhelix is formed by direct repeats of 10 bp long bent segments. Stereo micrographs of $[(A)_5TGCCC]_{54}$ DNA molecules and a 3-D reconstruction of one molecule. The DNA molecules are suspended in TE buffer (10 mM Tris-Cl, 1 mM EDTA, pH. 8.0). The molecules, in a thin vitrified layer of buffer are confined to a thickness of about 50 nm [59]. Since axial length of the formed superhelices is longer than 50 nm, they adopt an overall orientation approximate parallel to the plane of the thin layer. They are thus seen in almost lateral projections. The large angular difference between stereo partners (+15° and −15° respectively) improves the precision of 3-D reconstruction by numerical method [6] but makes it difficult to perceive 3-D impression by direct viewing of the stero-pair (a). In (b) some molecules are traced over for a better visualization. The 3-D reconstruction of the superhelical path of one of the photographed $[(A)_5TGCCC]_{54}$ DNA molecules is presented in (c). Magnification bar corresponds to 100 nm. This figure is partially reproduced from [7].

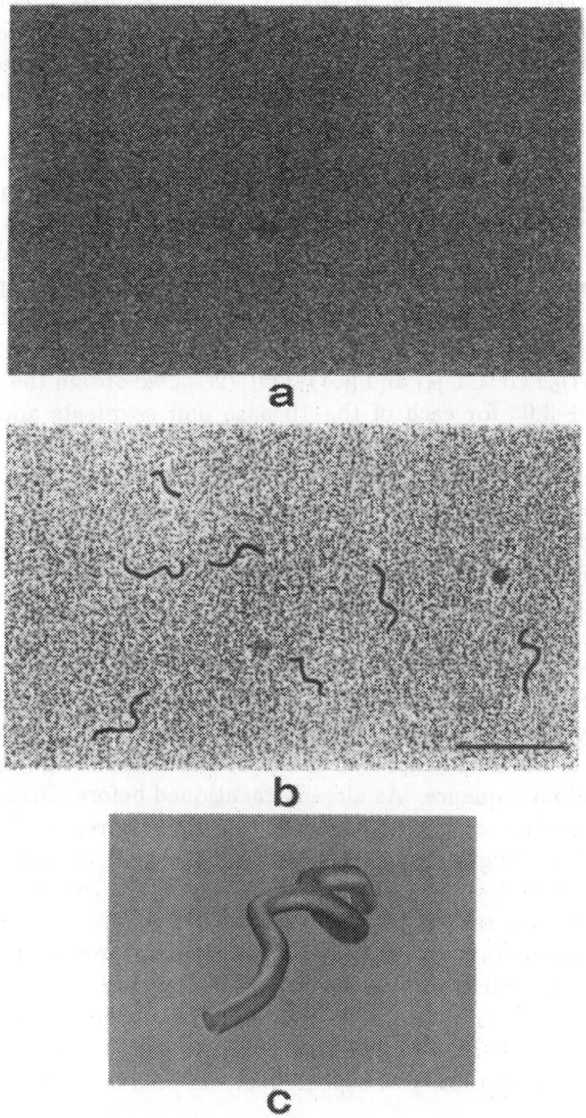

FIG. 3. Right-handed superhelix is formed by direct repeats of 11 bp long bent segments. Micrograph with several $[(A)_6TGCCC]_n$ DNA molecules (a), tracing of the molecules for a better visualization (b) and 3-D reconstruction of one of such molecules (c). Magnification bar corresponds to 100 nm.

superhelix and α is the pitch angle of the superhelix. The diameter of the superhelix (r) and the pitch angle (α) were directly measured values, however the lengths (ℓ) of the repeated elements were not concluded from our micrographs since with weakly contrasted linear molecules it is almost impossible to precisely localise ends of the molecules and thus to reconstruct and measure the entire molecule. Therefore to calculate the length of repeated segments we took new NMR data characterizing the geometry of individual dinucleotide stack [35]. Taking these values and applying the Lavery's Curves program to calculate the path length along the bent axis one obtains the lengths of $30.6\mathring{A}$ and of $33.6\mathring{A}$ for repeated blocks $(A)_5TGCCC$ and $(A)_6TGCCC$ respectively. Applying the formula for the torsion angle $(\Psi)\Psi = \ell/r \sin\alpha\cos\alpha$ to the superhelices formed by the sequences $[(A)_5TGCCC]_{54}$ and $[(A)_6TGCCC]_n$ we obtain the torsion angles of $-11.0° \pm 3.0°$ for each of the 10 base pair segments and $24.0° \pm 5.5°$ for the 11 base pair segments. This means that the total twist of 10 bp fragments is $360° - 11°$ and the twist of 11 bp fragments is $360° + 24°$. From these values it follows that the helical repeat of DNA is 10.32 ± 0.09 bp per turn for the first sequence while it is 10.31 ± 0.15 bp per turn for the second.

Helical repeat of DNA as determined by cryo-electron microscopy agrees well with earlier determinations by other methods. Although the helical repeat of DNA molecules with random sequences was found to be about 10.4 bp per turn [14], the sequences $(dA)_n(dT)_n$ were shown to have a helical repeat very close to 10 bp per turn [36]. Therefore sequences with high proportion of A-tracts are expected to have helical repeat significantly smaller than a random sequence. As already mentioned before, Drak and Crothers [21] used phasing method to conclude that directly repeated 10 or 11 mers with 5 or 6 bp.long A-tracts have the helical repeat of 10.34 ± 0.04 bp per turn, which is indeed very close to the helical repeat values concluded from the direct observation of DNA molecules.

It should be indicated here that the formula we used to calculate the torsion of repeated blocks applies to helices built of planar arcs like those shown in Fig 1. However as the nonplanarity of short 10 and 11 bp long sequences is small, the estimated by us error in the DNA helical repeat due to observed nonplanarity of repeated blocks was not bigger than 0.05 bp per turn.

The presented here Cryo-EM geometric method is probably the most direct and the least distorting way of determining the helicity of DNA molecules. However it should be mentioned that other EM methods also provided information about helical repeat value and about the direction of DNA helix.

Calladine and coworkers [31] used somewhat different geometric method to calculate the DNA helical repeat, though they were unable to determine the direction of DNA helix. Those authors also observed DNA fragments composed of directly repeated bent segments, however as they used classical EM approach the formed superhelix was flattened during adsorption which resulted in the formation of flat sinusoidal trajectory. Observing such flattened DNA one can not say if before the adsorption the original superhelix was left- or right-handed and what was its actual pitch and diameter in solution. In addition upon adsorption (i.e. interaction with a strongly charged surface) DNA can change its properties including the helical repeat value.

Cryo-EM provides 3-D information and this is crucial for solving such questions like direction of DNA helix. However some classical (surface adsorbed) DNA preparation can also provide information about DNA helicity. An illustrious example is the RecA-DNA complex which upon metal coating forms easily discernible in EM right handed helical filaments [37]. Electron microscopy experiments where DNA molecules with different linking number (term explained in the next chapter) were partially covered with RecA gave results which were only consistent with right handed winding of DNA helix [38].

3. DNA supercoiling

Discovery of DNA supercoiling. Studying circular double-stranded DNA isolated from polyoma virus particles Crawford noticed that a sample of these DNA molecules having the same size and base composition can be separated during centrifugation into two different classes with very different sedimentation coefficient [39]. Testing various prediction to explain these differences between seemingly the same DNA molecules Vinograd and coworkers used among other methods also electron microscopy [1]. It turned out that the slowly sedimenting molecules adopt the shape of relaxed, non-contorted circular molecules. In contrast the quickly sedimenting molecules which had the shape of highly contorted, supercoiled circular DNA molecules. To complete their studies Vinograd and coworkers demonstrated that addition of DNase (enzyme which can incise DNA by cleaving one of its strands) converts quickly sedimenting molecules into slowly sedimenting ones. These observations constituted the discovery of DNA supercoiling and the beginning of DNA topology (for reflective overview of this discovery please see [40]). It become clear that DNA behaves as an elastic band with intrinsic twist. When one of the strands in circular DNA molecule is cleaved it provides a freedom of rotation at this point and one can by the use of external force undertwist or overtwist such a band and then close the cleavage site using the enzyme ligase. However, if both DNA strands are uninterrupted the rotation is blocked thus fixing a given linking number (number of times one strand is topologically linked with the other strand). Later studies showed that there are several enzymes (called topoisomerases) which are able to transiently disconnect one or both strands of DNA and thus allow to change the linking number (for a recent short review on this subject please see [41]). Supercoiled DNA isolated from polyoma viruses had a lower linking number than if the molecules would be closed into circles with smallest torsional deformations. Such supercoiling is called negative. Later studies provided growing evidence that the negative DNA supercoiling plays essential functions in living cells. This stirred great interest in DNA supercoiling and many laboratories started studies of supercoiled DNA(for a recent review on DNA supercoiling please see [42]).

Interwound versus solenoidal structure. Theoretical consideration put forward two possibilities of how an elastic polymer like DNA can react to torsional stress. Torsionally stressed thin elastic rod can adopt solenoidal conformation or alternatively can buckle and then adopt a so called interwound structure. Although already the first EM pictures and many later EM studies showed that supercoiled DNA molecules had structure of interwound circles, it was formally possible that in classical EM preparations the interaction of DNA with a supporting surface and the subsequent drying of DNA could have lead to conversion from solenoidal form into interwound supercoiling. Early sedimentation studies of samples containing pure supercoiled DNA showed that depending on the conditions supercoiled molecules showed very different sedimentation properties [43]. This was interpreted as signs of transitions between solenoidal and interwound form of supercoiling [44]. Later studies of the salt effect on supercoiled DNA also indicated some form of transition in supercoiled DNA which was proposed to be a conversion between solenoidal and interwound forms of supercoiling [45,46]. However the indirectness of the methods (X-ray scattering or sedimentation measurements) were leaving open if it was really the transition between solenoidal and interwound form which lead to change of measured experimental value. Thus there was a controversy about the structure of supercoiled DNA in solution. What was needed to resolve this controversy was to directly see supercoiled DNA molecules by the method which could preserve the shape which DNA molecules have in solution. Cryo-EM provided such possibility and it clearly demonstrated that supercoiled DNA in low and high salt buffers adopts interwound configuration, however there were two different forms of interwound supercoiling [8,47].

Two forms of interwound supercoiling; a low salt loose configuration and a high salt tight form. DNA is a charged polymer in aqueous solutions, therefore independent DNA segments repulse each other unless some trivalent or polyvalent counterions are added to screen more than 90% of negative charges of DNA chains [48,49]. Since (in aqueous solutions) divalent or monovalent counterions are unable to achieve this level of DNA charge neutralization [48,49] it was believed for a long time that intersegmental DNA-DNA adhesion should not be observed in physiological range of buffers containing monovalent and divalent counterions. However in a recent paper Shaw and Wang [50] demonstrated that in the buffers containing more than 40 mM $MgCl_2$ some attractive interactions between DNA segments can be detected and that it manifests itself in increased proportion of DNA knots which are formed during circularization of long DNA molecules in such buffers. Using cryo-electron microscopy to study the shape of DNA molecules freely suspended in a buffer of interest we demonstrated recently that even in the presence of monovalent counterions the adhesion between DNA segments can be observed [8]. Importantly, under these ionic conditions this lateral adhesion of DNA segments are observed only in supercoiled DNA molecules and not in the non-constrained DNA molecules. We try to explain below the reasons and implications of such DNA behaviour.

The problem of DNA-DNA repulsion (or attraction) in solution was addressed by many authors. In most of the cases, experimental data were obtained from studies of linear DNA fragments. Frequently studied parameter is the effective diameter of DNA chains, value of this parameter roughly corresponds to the minimal distance at which two independent DNA segments moving with an average thermal motion can approach each other [8,51]. Value of the effective diameter depends very much on the type and concentration of counterions which can screen the negative charge of DNA chains and lower the effective diameter value [50,52]. In the case of linear DNA molecules freely suspended in solution and undergoing a random thermal motion even a weak electrostatic repulsion can impose effective diameter bigger than the actual geometric diameter of double helix. This can effectively prevent charged linear DNA molecules from direct contacts with each other. However, in the case of supercoiled DNA molecules the situation can be different, there the opposing segments of interwound superhelices are pushed toward each other by elastic constraints within the molecules. Thus the residual electrostatic repulsion could be overcome and the segments could directly contact each other.

Using Cryo-EM preparations we can directly observe how concentration and type of counterion affect the shape of supercoiled DNA molecules. We showed earlier that naturally supercoiled DNA molecules change their shape from a loosely to a tightly interwound molecules upon addition of $MgCl_2$ to the solution [47]. In our recent studies we wanted to see if monovalent counterions which are less effective in screening electrostatic repulsion [48,49,50] can also allow formation of the tight association of opposing segments in supercoiled DNA molecules.

Figure 4 shows naturally supercoiled DNA molecules cryo-vitrified in the buffers which had the initial compositions 10 mM Tris-HCl (pH 7.5), 1 mM EDTA (a) or 100 mM NaCl, 10 mM Tris-HCl (pH 7.5), 1 mM EDTA (b). To avoid evaporation of water from thin liquid layers used for Cryo-EM specimen preparation we introduced special humidity controlled chambers for specimen preparation therefore we can expect that final content of water in vitrified specimen is not much lower than in the initial preparation buffer. Despite this uncertainty about the final concentration of salt in used buffers it is visible that molecules observed in low salt buffer have a structure of loosely supercoiled molecules where the distance between

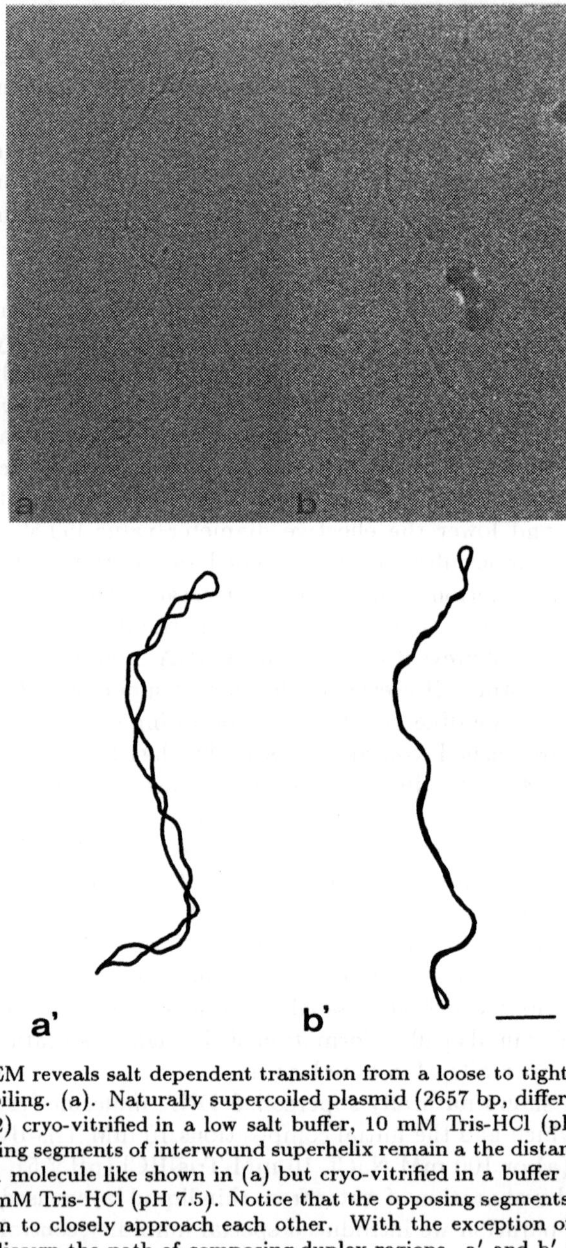

FIG. 4. Cryo-EM reveals salt dependent transition from a loose to tight form of interwound supercoiling. (a). Naturally supercoiled plasmid (2657 bp, difference of linking number ≈ -12) cryo-vitrified in a low salt buffer, 10 mM Tris-HCl (pH 7.5). Notice that the opposing segments of interwound superhelix remain a the distance of about 10 nm. (b). DNA molecule like shown in (a) but cryo-vitrified in a buffer containing 100 mM NaCl, 10 mM Tris-HCl (pH 7.5). Notice that the opposing segments of interwound superhelix seem to closely approach each other. With the exception of ends loops it is difficult to discern the path of composing duplex regions. a′ and b′ tracings of the corresponding molecule s to facilitate their perception. Scale bar = 50 nm.

opposing segments of interwound superhelices amounts to ca 10 nm. It is also visible that addition of 100 mM NaCl to the buffer induced an internal lateral collapse of opposing segments in interwound supercoiled DNA molecules. This collapse looks essentially the same as the one induced by the addition of 10 times lower amounts of $MgCl_2$ to the buffer [47].

Fact that addition of salt can induce transition from loose to tight form of interwound supercoiling can explain the salt induced transition observed in X-ray scattering pattern or sedimentation behaviour of supercoiled DNA without the need to invoke a change from interwound to solenoidal structure. In fact salt induced collapse would be expected to produce a change of X-ray scattering pattern which is consistent with the change which was experimentally observed [46]. Also our observation that collapsed forms have much less branched structure than loosely supercoiled DNA agrees well with the observed changes in radius of gyration of supercoiled DNA upon increase of salt concentration [45].

Comparison of experimental data with some theoretical predictions. Studying the behavior of linear DNA fragments in solution it was concluded that at 100 mM NaCl the effective diameter of DNA is of about 6 nm [51]. It means that two independent DNA segments in solution rarely approach closer than 6 nm. Computations using Metropolis Monte Carlo method allow to simulate the shape of DNA molecules with strict imposition of a given effective diameter [52]. Thus only these conformations where the distance between segments are bigger than the entered effective diameter value are accepted in the simulation procedure (for introduction and methodic details of Metropolis Monte Carlo simulation of DNA conformations see [52]). Figure 5(a) shows a simulated shape of naturally supercoiled DNA molecule with such strictly imposed effective diameter of 6 nm. It is obvious that in such simulation procedure the opposing segments of supercoiled DNA can not come closer than 6 nm. However due to entropic reasons the opposing segments stay separated by a distance which is bigger than the imposed effective diameter of 6 nm. It is clearly visible that such simulated molecule does not resemble the supercoiled DNA molecules directly observed in 100 mM NaCl.

Effective diameter value (or the distance of closest allowable approach) determined for linear DNA fragments should not apply strictly to supercoiled DNA molecules. Approaches of linear DNA molecules in solution are driven solely by a thermal motion. In the case of supercoiled DNA molecules the approaches between opposing segments of interwound superhelices are driven in addition to thermal motion also by the elastic tension within the molecules. Therefore it is to be expected that in supercoiled DNA molecules DNA segments can approach each other closer than the effective diameter value determined for linear DNA fragments. It is difficult to estimate how much closer should be the allowable distance in supercoiled DNA. Therefore we decided to simulate the extreme case where the DNA segments do not repulse each other i.e. where the effective diameter

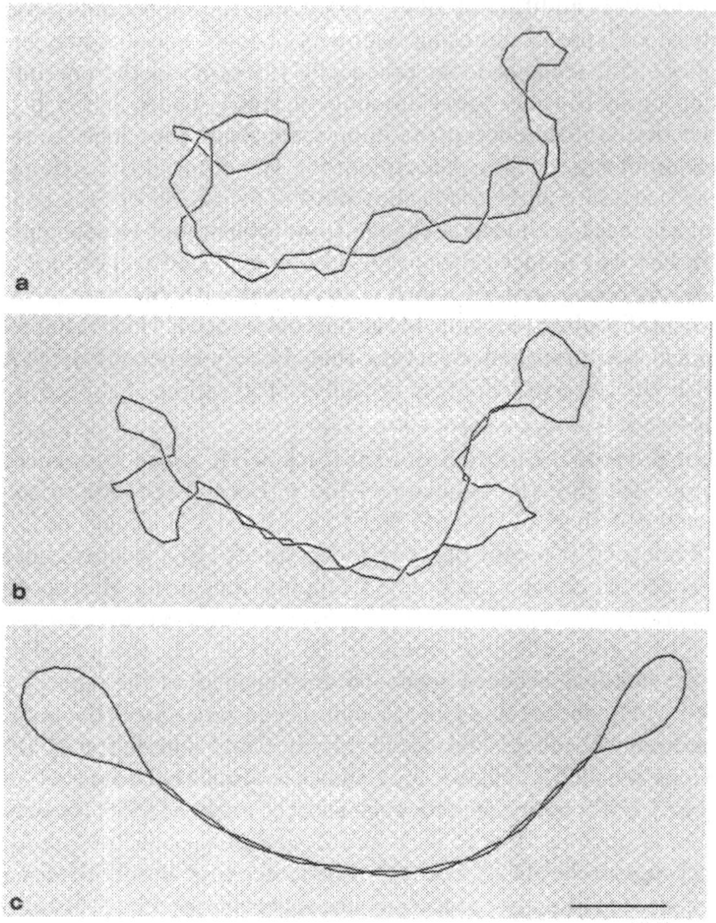

FIG. 5. Monte Carlo simulations of the shape of supercoiled DNA molecules. (a). One of simulated shapes of naturally supercoiled DNA molecules (2750 bp, ΔLk=-12) with an inputted minimal allowable intersegmental distance equal to 6 nm (effective diameter value measured for linear DNA molecules in solutions containing 100 mM NaCl). Notice that this molecule does not resemble the molecule shown in Fig. 4(b). (b). In an attempt to approach conditions leading to formation of tight supercoiling we simulated the case where naturally supercoiled DNA molecules has its electrostatic charge completely neutralized and there is no repulsion nor attraction between the segments (effective diameter was artificially put to zero). Notice that despite the fact that molecules can minimize their bending and twisting energy by tight supercoiling, they do not adopt this state since in the absence of DNA-DNA attractive interactions the entropic cost of collapse is bigger than the enthalpy gain due to the tight supercoiling. (c). Example of the simulated shape of supercoiled DNA molecules under conditions where entropic contribution is brought to zero ($0^\circ K$) and where effective diameter of DNA is equal to geometric diameter of DNA. Under these hypothetical conditions simulated molecules of course adopt tight configuration as now collapse does not lower the entropic contribution to the system. Notice that the drawings only present the axes of DNA molecule, if 2 nm thickness (DNA diameter) would be attributed to drawn lines then in Fig. 5(c) there would be no space left between the opposing segments of this interwound molecule. Scale bar = 50 nm.

is equal to zero. Under such "virtual conditions" DNA chains can not pass through each other but their axes can come in to direct contact without meeting any resistance. One of such simulated supercoiled DNA molecule with an effective diameter value equal to zero is shown in Figure 5(b). It is visible that also in these simulated DNA molecules tight supercoiling is not observed.

Metropolis Monte Carlo simulations account for entropic contribution to the shape of DNA molecules [52]. When the entropic contribution is neglected in simulations then the superhelices adopt tight configuration Fig. 5(c), but see also [53]. This demonstrates that it is the entropic effect which keeps the axes of opposing segments of interwound superhelices separated by a distance bigger than the effective diameter value entered for the simulation.

If the entropic contribution in simulations is not neglected, then to obtain tightly supercoiled molecules one would need to introduce some attractive interaction between laterally associated DNA segments. Indeed recently it was reported that attractive DNA-DNA potential can lead to tight configuration of simulated DNA molecules [54]. Thus we believe that some forms of attractive intersegmental interactions are involved in stabilization of tightly supercoiled DNA molecules.

Why the observed attraction within supercoiled DNA molecules does not manifest itself in linear fragments suspended in 100 mM NaCl? In answering the above mentioned question lets consider the following quite realistic situation

1. At this level of charge neutralization (ca 100 mM NaCl) there is a rather weak long distance Debye-Hückel repulsive interactions between DNA fragments and there is also short range attractive interaction (eg Van der Waals interactions). Combination of the two interactions creates a local minimum in the overall repulsive potential between the interacting DNA segments.

2. Random thermal motion is affected by the residual electrostatic repulsion between linear DNA fragments in 100 mM NaCl. This keeps independent DNA segments spaced by at least 6 nm for most of the time (effective diameter value determined for linear DNA fragments at these conditions). This would normally bar independent DNA segments from approaching each other close enough to sense the attractive interactions.

3. Even when two linear DNA segments are brought into direct longitudinal contact by a thermal motion which had the energy sufficient to reach momentarily the local minimum region of the repulsive potential, this situation will not be stable. The attractive interactions between laterally associated DNA fragments in 100 mM NaCl are most likely weaker than the energy of thermal motions. Thus due to entropic reasons this minimum in the repulsive potential would be to shallow to stabilize doublets of linear DNA fragments.

4. Elastic tension in supercoiled DNA tends to bring two opposing segments of interwound superhelix into a direct contact. The closer two segments are, the higher is the writhe of a molecule and thus the torsional stress gets minimized [8,53]. Lateral adhesion of opposing segments decreases also the bending energy in supercoiled DNA as overall curvature of molecules decreases [8,53]. Therefore in supercoiled molecules elastic energy causes that the opposing segments are pushed toward each other which in energetic terms means that these DNA segments are pushed upwards along the ascending shape of Debye-Hückel repulsive potential. When the supercoiling is strong enough the DNA segments can reach a point in the repulsive potential which is above the local energy minima caused by a short range attractive interactions between DNA segments. If this point is attained the DNA segments can further approach each other by falling into local minima of the repulsive potential.

When the listed above conditions are met then linear fragments will not associate laterally but will remain most of the time separated by at least 6 nm. Therefore someone studying linear fragments by such methods like knotting assay [51] would conclude that at 100 mM NaCl the effective diameter of DNA is 6 nm. However under the same conditions the supercoiled DNA molecules would be expected to show lateral adhesion of the opposing segments of interwound superhelices.

A word of caution may be well placed here. We believe that molecules observed by cryo-electron microscopy remain in the configuration typical for the equilibrium state at room temperature. However it may be that DNA molecules have time to react to changing temperature before their complete immobilization due to growing viscosity of supercooled media. Cooling increases torsional and bending stiffness of DNA and decreases the number of base pair per turn. These changes would promote tightening of supercoiled DNA. On the other hand viscosity of water rapidly grows with cooling and this would effectively prevent DNA from reacting to temperature changes below ca −20°C.

Presented here cryo-electron microscopy determination of the DNA helical repeat suggest that the helical repeat is essentially unchanged as compared to recent measurements of helical repeat of similar DNA sequences by the phasing method [21]. According to literature data on the temperature effect on DNA helical repeat [18], the change from 10.34 bp/turn [21] to 10.31-10.32 bp/turn (our measurements) would correspond to lowering the temperature by 7 to 10°C. If DNA molecules only react to lowering the temperature by the first 10 or so degrees, then the observed molecules reflect quite well the shape of DNA molecules in ambient temperature.

Possible biological implications of lateral adhesion of opposing segments of supercoiled DNA molecules. Supercoiled DNA shows many interesting properties which are absent in relaxed DNA. Well known are such as facilitated opening of promoter regions [55], easiness of association with proteins which promote left handed wrapping or coiling of DNA (e.g. histon octamers), promotion of formation of special DNA structures like cruciforms, Z-DNA or H-DNA (for review see [56,57]). We show here a qualitatively different property namely the selective ability of direct contacts of two segments from the same supercoiled molecule or the same topological domain. We propose that usually the energetic difference between different phasing of the adhering segments is very small. Therefore slithering movements [58] in tightly supercoiled molecules should be still possible as the available thermal energy could cause the adhering segments to move laterally in respect to each other. However the direct contact of opposing segments of supercoiled DNA can in principle lead to sequence specific interactions e.g. formation of additional hydrogen bonds between adhering duplexes. This type of interactions could specifically organize (or phase) supercoiled molecules and be important for the regulation of biological activity of DNA (gene expression).

Concluding remarks. Cryo-electron microscopy provides a method of direct observation of DNA molecules suspended in vitrified buffer. With this method one can observe such higher order arrangements like the shape of supercoiled molecules or a trajectory of unstressed DNA. X-ray crystallographic analysis or NMR do not provide information about interactions at this scale. Classical biophysical methods like sedimentation, X-ray scattering, gel migration, etc. provide a valuable data about behaviour of big DNA molecules in solution however indirectness of those methods require a verification of models and assumptions needed to interpret obtained results. Cryo-EM can very well fulfill this function.

Acknowledgements. We thank Mrs. A. Stasiak for DNA preparations, Prof. A. Vologodskii for discussions and for acquainting us with Metropolis Monte Carlo simulation procedure of DNA chains. We thank also Prof. M. Frank-Kamenetskii for stimulating discussions. This work was supported by Swiss National Science Foundation grant 31-32313 (to J. Dubochet and A. Stasiak).

REFERENCES

[1] Vinograd, J., Lebowitz, J., Radloff, R., Watson, R. & Laipis, P. The twisted circular form of polyoma viral DNA. *Proc. Natl. Acad. Sci. USA* **53**, 1104-1111 (1965).

[2] Griffith, J., Bleyman, M., Rauch, C.A., Kitchin, P.A. & Englund, P.T. Visualisation of the bent helix in kinetoplast DNA by electron microscopy. *Cell* **46**, 717-724 (1986).

[3] Krasnow, M.A., Stasiak, A., Spengler, S.J., Dean, F., Koller, T. & Cozzarelli, N.R Determination of the absolute handedness of knots and catenanes of DNA. *Nature* **304**, 559-560 (1983).

[4] Vollenweider, H.J., James, A. & Szybalski, W. Discrete length classes of DNA depend on mode of dehydration. *Proc. Natl. Acad. Sci. USA* **75**, 710-714 (1978).

[5] Dubochet, J., Adrian, M., Dustin, I., Furrer, P. & Stasiak, A. in *Methods in Enzymology* (eds. Lilley, D.M.J. & Dahlberg, J.E.) 507-518 (Academic Press, Inc., New York, 1992).

[6] Dustin, I., Furrer, P., Stasiak, A., Dubochet, J., Langowski, J. & Egelman, E. Spatial visualisation of DNA in solution. *J. Struct. Biol.* **107**, 15-21 (1991).

[7] Dubochet, J., Bednar, J., Furrer, P., Stasiak, A.Z., Stasiak, A. & Bolshoy, A.A. Determination of the DNA helical repeat by cryo-electron microscopy. *Nature Structural Biology* **1**, 361-363 (1994).

[8] Bednar, J., Furrer, P., Stasiak, A., Dubochet, J., Egelman, E.H. & Bates, A. D. The twist, writhe and overall shape of supercoiled DNA change during counterion-induced transition from a loosely to a tightly interwound superhelix. Possible implications for DNA structure in vivo. *J. Mol. Biol.* **235**, 825-847 (1994).

[9] Dubochet, J., Bednar, J., Furrer, P. & Stasiak, A. in *Structural Biology: The State of the Art* (eds. Sarma, R.H. & Sarma, M.H.) 149-156 (Adenine Press, Albany, NY, 1994).

[10] Watson, J.D. & Crick, F.H.C. Molecular structure of nucleic acid. A structure for deoxyribonucleic acid. *Nature* **171**, 373-738 (1953).

[11] Cairns, J. The bacterial chromosome and its manner of replication as seen by autoradiography. *J. Mol. Biol.* **6**, 208-213 (1963).

[12] Rodley, G.A., Scobie, R.S., Bates, R.H.T. & Levitt, R.M. A possible conformation for double-stranded polynucleotides. *Proc. Natl. Acad. Sci. USA* **73**, 2959-2963 (1976).

[13] Arnott, S., Wilkins, M.H.F., Hamilton, L.D. & Langridge, R. Fourier synthesis studies of lithium DNA, Part III: Hoogsteen models. *J. Mol. Biol.* **11**, 391-402 (1965).

[14] Wang, A.H.-J., Quigley, G.J., Kolpak, F.J., Crawford, J.L., van Boom, J.H., van der Marel, G. & Rich, A. Molecular structure of a left-handed double helical DNA fragment at atomic resolution. *Nature* **282**, 680-686 (1979).

[15] Rhodes, D. & Klug, A. Helical periodicity of DNA determined by enzyme digestion. *Nature* **286**, 573-578 (1980).

[16] Rhodes, D. & Klug, A. Sequence-dependent helical periodicity of DNA. *Nature* **292**, 378-380 (1981).

[17] Wang, J.C. Helical repeat of DNA in solution. *Proc. Natl. Acad. Sci. USA* **76**, 200-203 (1979).

[18] Depew, R.E. & Wang, J.C. Conformational fluctuations of DNA helix. *Proc. Natl. Acad. Sci. USA* **72**, 4275-4279 (1975).

[19] Pulleyblank, D.E., Shure, M., Tang, D., Vinograd, J. & Vosberg, H.- P. Action of nicking-closing enzyme on supercoiled and nonsupercoiled closed circular DNA: formation of a Boltzmann distribution of topological isomers. *Proc. Natl. Acad. Sci. USA* **72**, 4280-4284 (1975).

[20] Bhattacharyya, A., Murchie, A.I.H. & Lilley, D.M.J. RNA bulges and the helical periodicity of double-stranded RNA. *Nature* **343**, 484-487 (1989).

[21] Drak, J. & Crothers, D.M. Helical repeat and chirality effects on DNA gel electrophoretic mobility. *Proc. Natl. Acad. Sci. USA* **88**, 3074-3078 (1991).

[22] Beebe, T.P., Wilson, T. E., Ogletree, D. F., Katz, J. E., Baldhorn, R., Salmeron, M. B. & Siekhaus, W. J. Direct observation of native DNA structures with scanning tunneling microscope. *Science* **243**, 370-372 (1989).

[23] Lee, G., Arscott, P.G., Bloomfield, V.A. & Evans, D.F. Scanning tunneling microscopy of nucleic acids. *Science* **244**, 475-477 (1989).

[24] Arscott, P.G., Lee, G., Bloomfield, V.A. & Evans, D.F. Scanning tunneling microscopy of Z-DNA. *Nature* **339**, 484-486 (1989).

[25] Arscott, P.G., Lee, G., Bloomfield, V.A. & Evans, D.F. Helical period of Z-DNA. *Nature* **346**, 706 (1990).

[26] Clemmer, C.R. & Beebe, T.P. Graphite: a mimic for DNA and other biomolecules in scanning tunneling microscope studies. *Science* **251**, 640-642 (1991).

[27] Arscott, P.G. & Bloomfield, V.A. Scanning tunneling microscopy of nucleic acids. *Methods Enzymol.* **211**, 490-506 (1992).

[28] Wu, H.-M. & Crothers, D.M. The locus of sequence directed and protein-induced DNA bending. *Nature* **308**, 509-513 (1984).

[29] Ulanovsky, L., Bodner, M., Trifonov, E.N. & Choder, M. Curved DNA: design,

synthesis and circularisation. *Proc. Natl. Acad. Sci. USA* **83**, 862-866 (1986).

[30] Ulanovsky, L.E. & Trifonov, E.N. Estimation of wedge components in curved DNA. *Nature* **326**, 720-722 (1987).

[31] Calladine, C.R., Collis, C.M., Drew, H.R. & Mott, M.R. A study of electrophoretic mobility of DNA in agarose and polyacrylamide gels. *J. Mol. Biol.* **221**, 981-1005 (1991).

[32] Furrer, P., Bednar, J., Adrian, M., Stasiak, A., Dubochet, J., Rvet, B., Brahms, S. & Brahms, G.J. in *Structure and Function: Proceedings of Seventh Conversation in Biomolecular Stereodynamics* (eds. Sarma, R.H. & Sarma, M.H.) 229-236 (Adenine Press, Inc., Albany, NY, 1992).

[33] Olson, W.K., Marky, N.L., Jernigan, R.L. & Zhurkin, V.B. Influence of fluctuations on DNA curvature; a comparison of flexible and static wedge models of intrinsically bent DNA. *J. Mol. Biol.* **232**, 530-554 (1993).

[34] Dubochet, J. & Noll, M. Nucleosome arcs and helices. *Science* **202**, 280-286 (1978).

[35] Ulyanov, N.B. & James, T.L. Statistical Analysis of DNA duplex structures in solution derived by high resolution NMR. *Applied Magnetic Resonance* **7**, 21-42 (1994).

[36] Peck, L.J. & Wang, J.C. Sequence dependence of the helical repeat of DNA in solution. *Nature* **292**, 375-378 (1981).

[37] DiCapua, E., Engel, A., Stasiak, A. & Koller, T. Characterization of complexes between RecA protein and duplex DNA by electron microscopy. *J. Mol. Biol.* **157**, 87-103 (1982).

[38] Stasiak, A. & DiCapua, E. The helicity of DNA in complexes with RecA protein. *Nature* **299**, 185-186 (1982).

[39] Crawford, L.V. The physical characteristics of Polyoma virus. *Virology* **19**, 279-282 (1963).

[40] Lebowitz, J. Through the looking glass: the discovery of supercoiled DNA. *Trends in Biochemical Sciences* **15**, 202-207 (1990).

[41] Wang, J.C. DNA topoisomerases: why so many? *J. Biol. Chem.* **266**, 6659-6662 (1991).

[42] Vologodskii, A.V. & Cozzarelli, N.R. Supercoiling, knotting, looping and other large-scale conformational properties of DNA. *Current Opinion in Structural Biology* **4**, 372-375 (1994).

[43] Bauer, W. & Vinograd, J. The interaction of closed-circular DNA with intercalative dyes I. The superhelix density of SV40 DNA in the presence and absence of dye. *J. Mol. Biol.* **33**, 141-171 (1968).

[44] Bauer, W.R., Crick, F.H.C. & White, J.H. Supercoiled DNA. *Sci. Am.* **243** (July), 100-113 (1980).

[45] Campbell, A.M. Conformational variation in superhelical deoxyribonucleic acid. *Biochem. J.* **171**, 281-283 (1978).

[46] Brady, G.W., Satkowski, M., Foos, D. & Benham, C.J. Environmental influences on DNA superhelicity. The effect of ionic strength on superhelix conformation in solution. *J. Mol. Biol.* **195**, 185-191 (1987).

[47] Adrian, M., ten Heggeler-Bordier, B., Wahli, W., Stasiak, A. Z., Stasiak, A. & Dubochet, J. Direct visualisation of supercoiled DNA molecules in solution. *EMBO J.* **9**, 4551-4554 (1990).

[48] Wilson, R.W. & Bloomfield, V.A. Counterion-induced condensation of deoxyribonucleic acid. A light scattering study. *Biochemistry* **18**, 2192-2196 (1979).

[49] Marquet, R. & Houssier, C. Thermodynamics of cation-induced DNA condensation. *J. Biomol. Struct. Dyn.* **9**, 159-167 (1991).

[50] Shaw, S.Y. & Wang, J. Knotted DNA rings: probability of formation and resolution of the two chiral trefoils. *Science* **260**, 533-536 (1993).

[51] Rybenkov, V.V., Cozzarelli, N.R. & Vologodskii, A.V. The probability of DNA knotting and the effective diameter of the DNA double helix. *Proc. Natl. Acad. Sci. USA* **90**, 5307-5311 (1993).

[52] Vologodskii, A.V., Levene, S.D., Klenin, K.V., Frank-Kamenetskii, M. & Cozzarelli, N.R. Conformational and thermodynamic properties of supercoiled DNA. *J. Mol. Biol.* **227**, 1224-1243 (1992).

[53] Schlick, T. & Olson, W.K. Supercoiled DNA energetics and dynamics by computer simulation. *J. Mol. Biol.* **223**, 1089-1119 (1992).

[54] Marko, J.F. & Siggia, E.D. Fluctuations and Supercoiling of DNA. *Science* **265**, 506-508 (1994).

[55] Drew, H.R., Weeks, J.R. & Travers, A.A. Negative supercoiling induces spontaneous unwinding of a bacterial promoter. *EMBO J.* **4**, 1025-1032 (1985).

[56] Frank-Kamenetskii, M.D. in *DNA topology and its biological effects.* (eds. Cozzarelli, N.R. & Wang, J.C.) 185-215 (Cold Spring Harbor Laboratory Press, Cold Spring Harbor, NY, 1990).

[57] Bates, A.D. & Maxwell, A. *DNA Topology*, 1-120 (IRL Press, Oxford, 1993).

[58] Stark, W.M., Boocock, M.R. & Sherratt, D.J. Site-specific recombination by Tn3 resolvase. *Trends Genet.* **5**, 304-309 (1989).

[59] Bednar, J., Furrer, P., Stasiak, A. & Dubochet, J. in *Electron microscopy 92* (eds. Megias-Megias, L., Rodrigez-Garcia, M.I., Rios, A. & Arias, J.M.) 11-12 (University of Granada, Granada, 1992).

DYNAMICS OF TWIST AND WRITHE AND THE MODELING OF BACTERIAL FIBERS

MICHAEL TABOR* AND ISAAC KLAPPER†

Abstract. We discuss a range of issues associated with the dynamics of twist and writhe including some new theoretical and numerical results and techniques. A precise understanding of twist and writhe is important in a variety of physical and biological processes and, in particular, we describe how these ideas can be used to model the dynamics of the self-assembling bacterial fiber, *bacilus subtilis*.

1. Curves and ribbons. The words "twist" and "writhe", in addition to terms like "braiding" and "linking", are rather vivid terms in colloquial usage. They also have a precise physical and mathematical meaning and their correct formulation and accurate computation are needed to explain a range of physical and biological processes that can occur on many different scales. Thus on the microscopic scale, the interplay between twist and writhe is the key to understanding DNA supercoiling and, as we will discuss later, bacterial self-assembly; on a somewhat larger scale, linkage is the key topological idea behind the fluid dynamical notion of helicity; at the macroscopic scale the kinking and twisting of yarn and cables require an understanding of "writhing instabilities"; and on a celestial scale (of the order of 10^5km), solar flaring mechanisms can be related to braided magnetic flux tubes.

We begin by recalling the basic description of a space curve and the notion of an associated ribbon. This will lead to a discussion of the kinematics of curves and a formulation for the kinematics of ribbons. In the usual way we consider a space curve $X(s,t) : [0, L] \to \mathbb{R}^3$ where s is arc-length, which is assumed to be conserved in time, and L is the total length of the curve. In the case of a non arc-length preserving process the curve can be parameterized in terms of a material coordinate and the formalism that follows is easily modified to cover this case. The unit tangent vector along the curve X is given by $T(s,t) = (d/ds)X(s,t)$. A local description of the curve is provided by an appropriate set of coordinate axes. One such set is the Frenet triad consisting of the unit vectors T, N, and B where N is the normal vector and $B = N \times T$ is the binormal vector. This orthonormal triad satisfies the well known Frenet-Serret equations [1]

$$\begin{aligned}\frac{d}{ds}T &= \kappa N\\ \frac{d}{ds}N &= -\kappa T + \tau B\\ \frac{d}{ds}B &= -\tau N\end{aligned} \tag{1.1}$$

* Program in Applied Mathematics, University of Arizona, Tucson, AZ 85721.
† Department of Mathematics, University of California, Los Angeles, CA 90024.

where $\kappa = |dT/ds|$ is the curvature and $\tau = |dB/ds|$ the torsion of X. Note, of course, that N and B are only defined when $\kappa \neq 0$ and in general vary discontinuously through points with $\kappa = 0$. Thus, for the Frenet equations to be well defined κ must be non-vanishing and X must be of class C^3. A convenient way of writing the Frenet equations is in the form

$$\frac{d}{ds}T = \mathbf{\Lambda}_F \times T$$
$$\frac{d}{ds}N = \mathbf{\Lambda}_F \times N$$
$$\frac{d}{ds}B = \mathbf{\Lambda}_F \times B$$

where

$$\mathbf{\Lambda}_F(s,t) = \tau T + \kappa B$$

is called the *Darboux vector*. From this formulation it is seen that τ measures the rotation rate of the normal-binormal plane about T from the point of view of an observer moving along X at unit speed. The total torsional (or Frenet) twist along the curve is then evaluated from the integral

$$Tw = \frac{1}{2\pi}\int_{[0,L]} \tau(s,t)ds \tag{1.2}$$

which is obviously a local additive property of the curve.

Apart from difficulties with points of vanishing κ the Frenet equations give a complete description of X at any given instant in time. The actual temporal motion of X requires a specification of its velocities dX/dt. At this point it is useful to recall the all important distinction between "kinematics" and "dynamics". The former corresponds to the motion of a system with *prescribed velocities* whereas the latter corresponds to the relationship between the velocities of a system and the *prescribed forces* acting on it. For now we will consider the (far simpler) case of kinematics and assume for convenience that the velocity is given in the form

$$\frac{d}{dt}X = UN + VB + WT \tag{1.3}$$

where U, V, W are the normal, binormal and tangential velocity components respectively - each of which can depend on both s and t. Recalling that

$$\frac{d}{ds}X = T$$

self consistency requires equality of the s and t cross derivatives, namely

$$\frac{d^2}{dsdt}X = \frac{d^2}{dtds}X \tag{1.4}$$

In carrying out the computations associated with this relationship it is convenient to write the time dependence of the Frenet triad as

$$\frac{d}{dt}T = \mathbf{\Omega}_F \times T$$
$$\frac{d}{dt}N = \mathbf{\Omega}_F \times N$$
$$\frac{d}{dt}B = \mathbf{\Omega}_F \times B$$

where

$$\mathbf{\Omega}_F(s,t) = \alpha N + \beta B + \gamma T$$

is the rotation vector whose components α, β, γ are related to U, W, W - and hence κ and τ - through equation 1.4. Straightforward computations show these relations to be

$$\alpha_s = -\kappa\gamma + \tau\beta$$
$$\beta_s = \kappa_t - \tau\alpha$$
$$\gamma_s = \tau_t + \kappa\alpha$$

Further calculation leads to the equations of motion for κ and τ in terms of U, V, and W. These take the form

$$(1.5) \quad \frac{d}{dt}\kappa = (U_s + \kappa W - \tau V)_s - \tau(V_s + \tau U)$$

$$(1.6) \quad \frac{d}{dt}\tau = (\frac{1}{\kappa}(V_s + \tau U)_s + \frac{\tau}{\kappa}(U_s - \tau V) + \tau W)_s + \kappa(V_s + \tau U)$$

Various choices of U, V, W give different evolution equations and, of particular interest in recent years [2,3,4,6,5], has been those choices that lead to integrable hierarchies. A rather important example is the local induction approximation, namely

$$\frac{d}{dt}X = \kappa B$$

which describes the motion of a vortex filament in its self-induced velocity field. (Thus, in this approximation one can see that, since the curvature of a ring is inversely proportional to its radius, smaller smoke rings travel faster.) The resulting equations of motion for κ and τ can be transformed via the Hasimoto map

$$\psi \equiv \kappa e^{(i\int^{\sigma}\tau\, d\sigma)}$$

to the nonlinear Schrödinger equation

$$\psi_t + \psi_{ss} + \frac{1}{2}\psi|\psi|^2 = 0.$$

Another example is provided by the choice

$$\frac{d}{dt}X = -\frac{1}{2}\kappa^2 T - \kappa_s N - \kappa\tau B$$

which leads to the (same) Hasimoto function ψ satisfying the complex modified KdV equation

$$\psi_t + \psi_{sss} + \frac{3}{2}\psi_s|\psi|^2 = 0.$$

In fact, Langer and Perline [4] have shown how to recursively generate a whole hierarchy of arc-length preserving velocities for which ψ satisfies the nonlinear Schrödinger equation hierarchy. We shall return to this result later and describe an interesting geometric property of this family.

A particularly useful construction on X is that of a *ribbon* which is defined in terms of a pair of curves $(X, X+\epsilon V)$ where $V(s,t) : [0,L] \to \mathbb{R}^3$ is a unit vector field such that $V \cdot T = 0$, and ϵ is the width of the ribbon. The ribbon can have a real physical significance such as representing a sequence of molecular groups distributed around a molecular "backbone" (represented by X) or the diameter of an elastic cylinder with center-line X. One non-physical, but mathematically useful, choice of ribbon is provided by the *Frenet ribbon* which corresponds to choosing $V = N$ (or B). The orthonormal triad of vectors T, V, W, where $W = T \times V$, provide a convenient coordinate basis for the ribbon. In general the evolution of this triad can be described by

$$\begin{aligned} \frac{d}{ds}T &= \mathbf{\Lambda}_R \times T \\ \frac{d}{ds}V &= \mathbf{\Lambda}_R \times V \\ \frac{d}{ds}W &= \mathbf{\Lambda}_R \times W \end{aligned} \tag{1.7}$$

and

$$\begin{aligned} \frac{d}{dt}T &= \mathbf{\Omega}_R \times T \\ \frac{d}{dt}V &= \mathbf{\Omega}_R \times V \\ \frac{d}{dt}W &= \mathbf{\Omega}_R \times W \end{aligned} \tag{1.8}$$

where $\mathbf{\Lambda}_R(s,t) = \kappa_1 V + \kappa_2 W + \kappa_3 T$ and $\mathbf{\Omega}_R(s,t) = \omega_1 V + \omega_2 W + \omega_3 T$ describe the rotations of the ribbon triad with respect to s and t respectively.

Of particular importance is the twist rate κ_3 which is defined to be the rotation rate of V around T moving along X, i.e.,

$$\kappa_3(s,t) = \mathbf{\Lambda}_R(s,t) \cdot T(s,t) = (V(s,t) \times (d/ds)V(s,t)) \cdot T$$

In the case of the Frenet ribbon this reduces to the torsion τ. For a physical ribbon one can write the twist rate as $\kappa_3(s,t) = \tau + d\theta/ds$ where θ is the angle between V and the Frenet ribbon. The total twist along the ribbon is then given by (cf. (1.2))

$$Tw = \frac{1}{2\pi}\int_{[0,L]} \tau(s,t)ds + \frac{1}{2\pi}\int_{[0,L]} d\theta$$

In the case of a closed ribbon the second contribution to Tw must be integer valued.

In the same way that the cross differentiation relations (1.4) lead to "intrinsic" equations for the motion of a space curve, it is also possible to develop in an analogous manner intrinsic equations of motion for a ribbon. From the relations

$$\frac{d^2}{dsdt}V = \frac{d^2}{dtds}V$$

one finds, with the help of equations (1.7) and (1.8), that

(1.9) $$\theta_t = -\gamma + \omega_3$$

and

(1.10) $$\kappa_{3,t} = -\alpha\kappa + \omega_{3,s}$$

where α and γ are, respectively, the normal and tangential components of $\mathbf{\Omega}_F$ and κ is the curvature of X. It is important to emphasize that the first terms in both of these equations, i.e. $\theta_t = -\gamma$ and $\kappa_{3,t} = -\alpha\kappa$ represent a "parallel transport" effect; namely they represent the time variation of θ and ω as a result of the deformation of the underlying space curve X which can rigidly move the ribbon. This effect can be illustrated by the following simple experiment. Extend your arm horizontally in front of you with wrist bent so that your hand points straight up. Your hand and arm can be thought of as vectors of an untwisted ribbon (with your arm as base curve). Now rotate your arm 90° down to your waist, then 90° out to the side, and then 90° back to its original position, all without any twisting of the wrist. Note that the head-hand "ribbon" has gained a 90° rotation even though no intrinsic twisting has occurred. This experiment also illustrates the idea of a "geometric phase" that occurs upon the circuit of certain classical or quantum systems around a closed path [7]. As will be described later, the twist equation (1.10) plays a very significant role in describing the dynamics of space curves. Noting that $\alpha = -B \cdot T_t$, i.e. the binormal component of the tangential velocity, (1.10) can be written as

$$\kappa_{3,t} = \kappa B \cdot T_t + \omega_{3,s}. \tag{1.11}$$

Since one may easily show that $\alpha = -(V_s + \tau U)$, it follows that for prescribed velocity components U, V, W, equations (1.9), (1.10) and (1.11) define the intrinsic kinematics of a ribbon.

2. Twist and writhe. When a ribbon is closed by gluing its two ends together then its two edges, which may be considered to be two separate space curves, say X and Y, will be found to be linked in various ways. Their linkage Lk is defined to be half the sum of the signed crossing numbers [8]. There are various other ways of computing Lk, the most famous being the formula introduced by Gauss [9] which takes the form

$$Lk(X,Y) = \frac{1}{4\pi} \oint_{S^1} \oint_{S^1} \frac{(T(t) \times S(s)) \cdot (X(t) - Y(s))}{|X(t) - Y(s)|^3} dtds \tag{2.1}$$

where T and S are the tangent vectors associated with X and Y respectively.

More formally the linking number $Lk(X,Y)$ for two disjoint closed space curves is defined as

$$Lk(X,Y) = \frac{1}{4\pi} \int_{S^1 \times S^1} e^* d\Sigma$$

where $e^* d\Sigma$ denotes the pullback of the volume element of S^2 under a map $e : S^1 \times S^1 \to S^2$ defined by

$$(t, s) \mapsto \frac{X(t) - Y(s)}{|X(t) - Y(s)|}. \tag{2.2}$$

Since $S^1 \times S^1$ and S^2 are compact, boundaryless connected spaces and e is smooth [10], it then follows that

$$Lk(X,Y) = \deg(\mathrm{e})$$

where $\deg(e)$ is the degree of the map e. Thus Lk is always an integer and, most significantly homotopy invariant, i.e., it is invariant under smooth deformations of X and Y which do not involve any curve intersections. Expressed explicitly in coordinates Lk takes the form of equation (2.1). From the definition (2.2) of e we see that $Lk = deg(e)$ is the signed crossing number of a generic projection of the curves X and Y.

For the same space curve X, depending on the way the attached ribbon is wound around it, there can be many different values of Tw and Lk. Although this might suggest some lack of invariance associated with X, a global geometric invariant associated with X does in fact exist. This is the *writhe* of the curve and is related to Tw and Lk through the relation

$$Wr(X) = Lk(X, X + \epsilon Y) - Tw(X, Y) \tag{2.3}$$

A history of this important formula can be found in an article by Pohl [11]. (For more history and a connection between (2.3) and conservation of helicity see [12,13].) There are various different, but equivalent, ways of interpreting writhe. For example, one can think of Wr as a measure of the out-of-plane "deformation" of the curve. Alternative, one can think of writhe as a measure of the "helicity" of the curve or, in the biochemical context, the degree of "supercoiling". Writhing has very tangible - and energetically significant - manifestations which can be seen from the following simple experiment. Consider an untwisted rubber tube lying flat on the plane which is then closed by attaching its two ends together by means of a short peg. There is clearly no linkage or twist associated with this configuration. Now "inject" some twist into the tube by twisting the ends around the connecting peg. Depending on the elastic properties of the tube, after a certain amount of twist has been added the tube will in a very real sense "writhe" - deforming into a nonplanar configuration. This process can be quantified through a stability analysis [14,15] which shows that there is a critical twist that triggers this instability. This idea will be discussed further in section 5. The conversion of twist to writhe is constrained by equation (2.3). It plays a crucial role in the knotting and unknotting of circular DNA and, as will be described later, also plays an important part in determining the dynamics of self assembling bacterial fibers.

The actual value of writhe for a curve can be explicitly computed using an integral analogous to that used to compute Lk (Wr can be thought of the "self-linking" of the space curve with itself), namely

$$Wr(X) = \frac{1}{4\pi} \oint_{S^1} \oint_{S^1} \frac{(T(t) \times T(s)) \cdot (X(t) - X(s))}{|X(t) - X(s)|^3} dt ds \tag{2.4}$$

If one makes the identification $Y = X$ the map e (equation (2.2)) is still well-defined on the set $S^1 \times S^1 \backslash \Delta$ where $\Delta = \{(t,t) : t \in S^1\}$. One can extend e to the boundary Δ by

$$e(t,s) = \begin{cases} \frac{X(t)-Y(s)}{|X(t)-Y(s)|} & s \neq t \\ T(t) & s = t \end{cases}$$

and the writhe can then be defined as

$$Wr(X) = \frac{1}{4\pi} \int_{S^1 \times S^1 \backslash \Delta} e^* d\Sigma.$$

However, unlike Lk, the integral is no longer taken over a boundaryless space and the degree is no longer well-defined. Thus writhe is, in general, not an integer and it is not homotopy invariant. In coordinate form, Wr takes the form given in equation (2.4). As with Lk, Wr can also be interpreted as a signed crossing number but now it is necessary to average over all projections.

It is useful to compare and contrast some of the of the basic properties of Tw, Lk and Wr. These follow from their definitions and can be summarized as follows:

Tw: (i) is a conformal invariant; (ii) is additive, i.e., given by a single integral over X; (iii) changes continuously under deformations of the ribbon even if the ribbon is passed through itself.

Lk: (i) is always an integer; (ii) is a topological invariant; (iii) if the ribbon is passed through itself Lk shows a jump of ± 2.

Wr: (i) is a conformal invariant; (ii) changes continuously under deformations of the curve but shows a jump of ± 2 as the curve is passed through itself.

The actual calculation of writhe, either exactly or approximately can be achieved through a number of formulae. Many of these were given by Fuller [16,17] without proof, and in recent work [18] we provided a precise derivation of these formulae. In summary we present three practical routes for computing writhe.

The first involves the idea of an averaged self-linking number in which Wr is determined as the average over all possible projections of the curve. More formally the curve X and a fixed unit vector $\boldsymbol{\delta}$ are said to be in general position if $T \nparallel \boldsymbol{\delta}$; then for small $\varepsilon > 0$

$$Wr(X, \boldsymbol{\delta}) := Lk(X, X + \varepsilon\boldsymbol{\delta})$$

is well-defined and called the *directional writhing number* of X with respect to δ. One may then prove that if $X : [0,1] \longrightarrow \mathbb{R}^3$ is of class C^2 the set

$$M := \{\pm T(t); t \in [0,1]\} \subset S^2$$

has measure zero and

$$Wr(X) = \frac{1}{4\pi} \int_{S^2 \setminus M} Wr(X, \boldsymbol{\delta}) d\boldsymbol{\delta}$$

Our particular proof is based on the formula $Lk = Tw + Wr$ and homotopy invariance whereas, for example, Pohl [11] proved this proposition using the axiomatic definition of $deg(\cdot)$ applied to the linking number. This formula has practical applications in that it can be used to estimate Wr for nearly planar curves such as those studied in electron-microscopy.

Another approach to compute the writhing number is given by the elegant geometrical construction

$$1 + Wr(X) = \frac{A}{2\pi} \, mod\, 2. \tag{2.5}$$

where A is the area enclosed by the curve on the unit sphere traced out by the tangents to the given, closed (non-intersecting) space curve X. This proposition can be proved through use of the Gauss-Bonnet theorem.

Of particular relevance to our studies of filament dynamics is the formula that relates the writhe between two closed curves X_0 and X_1 that can be deformed into each other, namely

$$Wr(X_1) - Wr(X_0) = \tag{2.6}$$
$$\frac{1}{2\pi}\int_0^1 \frac{T_0(t)\times T_1(t)}{1+T_0(t)\cdot T_1(t)}\cdot\frac{d}{dt}\Big(T_0(t)+T_1(t)\Big)dt.$$

where T_0 and T_1 are the tangent vectors associated with X_0 and X_1 respectively. The detailed proof of this formula is quite involved but, in its course, reveals a nice formula for the rate of change of Wr of a curve X under deformation, namely

$$\frac{d}{d\lambda}Wr(X_\lambda) = -\frac{1}{2\pi}\int_0^1\left(\frac{\partial}{\partial\lambda}T(t,\lambda)\times T(t,\lambda)\right)\cdot\frac{\partial}{\partial t}T(t,\lambda)dt \tag{2.7}$$

where λ is the deformation parameter. We point out in particular that although Wr is not an additive quantity, the change in Wr under curve deformation is.

In the case of prescribed kinematics of a closed loop we can ask how the writhe of the curve changes with time. Thus treating time as the deformation parameter in equation (2.7) and assuming the velocities of X are given in the form (1.3), we have

$$\frac{d}{dt}Wr = -\frac{1}{2\pi}\int_0^1 \kappa(V_s+\tau U)ds \tag{2.8}$$

(as long as the decomposition (1.3) makes sense) where subscript s denotes differentiation with respect to s. The reader will recognize the integrand in (2.8); it is precisely the time variation of (ribbon) twist rate, κ_3, due to parallel transport! In the case of the nonlinear Schrödinger equation (1), $U = 0$ and $V = \kappa$ in which case

$$\begin{aligned}\frac{d}{dt}Wr &= -\frac{1}{2\pi}\int_0^1 \frac{1}{2}(\kappa^2)_s ds\\ &= 0\end{aligned}$$

In the case of the next member of the hierarchy, $U = -\kappa_s$ and $V = -\kappa\tau$ which gives

$$\frac{d}{dt}Wr = -\frac{1}{2\pi}\int_0^1 \frac{1}{2}(\kappa^2\tau)_s ds = 0$$

Thus we see that the Wr of closed curves evolving under these two integrable evolution equations is conserved (up to curve crossings). Indeed, further analysis of the Langer-Perline hierarchy shows that all its members, if well-defined, are writhe conserving [19]. (We observe in passing that in terms of the Hasimoto function ψ, the two integrands shown above are $\kappa^2 = \psi\psi^*$ and $\kappa^2\tau = \frac{i}{2}(\psi_s\psi^* - \psi_s^*\psi)$ which have the interpretation of particle "density" and "flux" respectively.)

3. Self-assembling bacterial fibers. The original motivation for many of our studies of twist and writhe is the remarkable behavior of filaments of the bacteria *bacillus subtilis* [20]. The individual cells of this bacteria grow into a cylindrical shape approximately 3 to 4 μm long and $0.8\mu m$ in diameter. When the cell growth occurs, the network of polymers that make up the cell wall are believed to induce twist in the cells [21]. As normal cells reproduce they separate and as each generation of daughter cells reproduce a bacterial colony is formed. However, here we study growth of a bacterial strain in which the cells *fail to separate* after reproduction. Because of this the bacteria cells form an exponentially growing filament in which, as a result of the structure of the cell walls, twist is apparently injected at a steady rate as well. The typical length doubling time for a growing filament is of the order of 70 minutes. The growing filaments undergo a striking "self-assembly" process. In the case of low viscosity media (i.e. of the order of a few centipoise) the growing filaments are observed to flail around and at some point to fold over and make end-to-end contact. In some cases the contact "snaps" open again and in other cases, the resulting loop stays closed. In this latter instance one then observes a dramatic twist-to-writhe conversion in which the loop winds itself up into a tight, plectonemic (i.e. double-stranded helical) structure. This sequence is then repeated as the resulting double helical structure continues to grow exponentially, becomes unstable, folds over and winds up into a quadruple-stranded plectonemic structure, and so on. Eventually macroscopic strands (say 1 cm long) of interwound bacterial filament are produced.

For more viscous media (i.e. of the order of 1000 cp.), most of the twisting and bending motions of the growing filament are suppressed and instead the filament is observed to buckle in the middle but then still undergo a twist-to-writhe conversion (clearly indicating a build up of twist in the cell chain) leading to a double stranded helical structure. As before the sequence is repeated many times resulting in a macro-fiber. A schematic of the two scenarios (i.e. low and high viscosity media) are shown in figures 1 and 2, respectively, and a example of an observed macro-structure is shown in figure 3.

4. The dynamics of twisted rods and the modeling of bacterial fibers. A reasonable physical model of the bacterial filaments described above is that of an elastic rod subjected to twisting forces. An "engineering" approach might be to consider the rod as a truly three di-

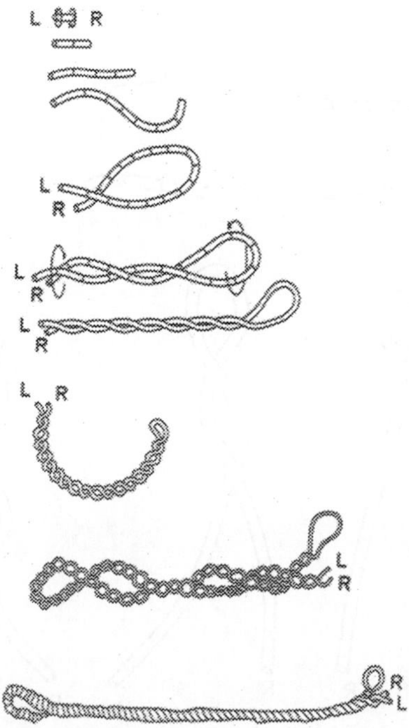

FIG. 1. *Schematic of the writhing instability of bacillus subtilis in a low viscosity medium. This figure is reproduced from [32].*

mensional elastic structure and simulate its evolution as such. However, a simpler approach is possible. As described above the experimental evidence indicates that the twist is generated by the polymer network in the cell walls. This suggests that we can model the filament by a space curve X, representing the center-line of the filament coupled to a ribbon whose twist around X represents the cell wall structure. In this way we can use the standard Kirchhoff equations for a rod (i.e a curve parameterized by a single coordinate, namely the arc-length) provided that the twist evolution of the associated ribbon is correctly incorporated.

The statics of elastic rods has a very long history and is standard fare in many textbooks such as [22,23]. The dynamics of elastic rods is much more complicated and more detailed investigations and formulations are relatively recent (see, for example, [24]). Attempts to develop efficient algorithms for simulation purposes are current research themes [25,26].

Although rod dynamics can be given a rather general formulation in terms of a general set of directors and constitutive relations (the relationship between the local stresses and strains) in what follows we will continue

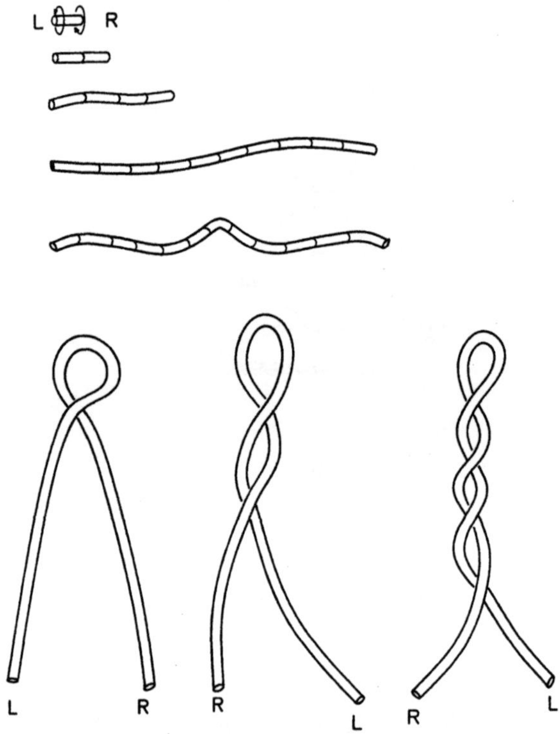

FIG. 2. *Schematic of the writhing instability of bacillus subtilis in a high viscosity medium. This figure is reproduced from [32].*

to work with the ribbon vectors (V, W, T), and assume the approximations of linear elasticity theory. We comment that (V, W) define a plane normal to T, for all time, which is taken to correspond to the cross section of the rod. The assumption that the material cross-section remains normal to T is called unshearability. The basic principle governing the dynamics is the balancing of the local internal stresses, i.e the forces and torques, with the external forces. A complete description of our derivation of the dynamical equations appropriate for modeling the bacterial filaments is given elsewhere [28] so here we give a slightly different, but equivalent, description of the main ideas. If $P = \rho\pi a^2 \dot{X}$ denotes the linear momentum of the filament, where ρ is the filament mass density, a the filament radius and $F(s,t)$ denotes the net internal stress on a perpendicular cross section at $X(s,t)$, we can write

$$\frac{d}{dt}P = \frac{d}{ds}F + g \tag{4.1}$$

The last term $g(s,t)$ represents any external forces acting on the filament

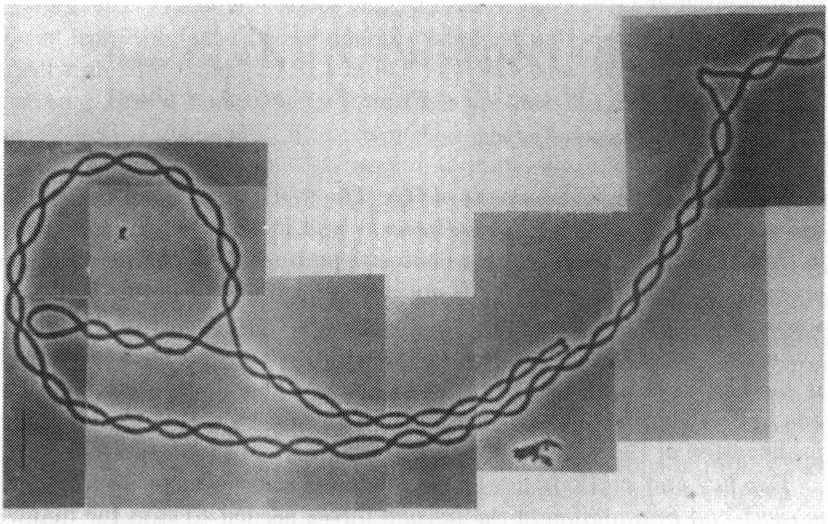

FIG. 3. *Photograph of a filament of bacillus subtilis at an advanced stage of self-assembly. This figure is reproduced from [33].*

such as gravity, viscosity and "contact forces" (to be defined below). Similarly, if Π denotes the angular momentum and $M(s,t)$ the net internal moment on a perpendicular cross section at $X(s,t)$ we have

$$\dot{\Pi} = \frac{d}{ds}M + F \times T + h \tag{4.2}$$

where $h(s,t)$ represents any external moments that might be included such as a torsional damping term. It is fairly straightforward to show (see, e.g [24]) that the angular momentum term is related to the cross-sectional director accelerations, i.e

$$\dot{\Pi} = \rho I(V \times \ddot{V} + W \times \ddot{W}).$$

The pair of momentum balance equations (4.1) and (4.2) are closed by specifying the constituitive relations which in our model are taken to be those of linear elasticity, namely

$$M = EI\kappa_1 V + EI\kappa_2 W + \mu J \kappa_3 T \tag{4.3}$$

where the κ_i are the components of $\mathbf{\Lambda}_R$, E the Young's modulus, μ the torsional rigidity, $I = \pi a^4/4$ the moment of inertia of the circular cross-section about a central axis lying in the plane of the cross section, and $J = \pi a^4/2$ the moment of inertia of the circular cross-section about the tangent T.

If we decompose the force vector into director components, namely $F = F_1V + F_2W + F_3T$, direct computations on equation (4.2) yield (still assuming circular cross-section)

$$\begin{aligned} &EI\kappa_{1,s} - \kappa_2\kappa_3(EI - \mu J) - F_2 = \rho I(\omega_{1,t} + \omega_2\omega_3) \\ &EI\kappa_{2,s} - \kappa_1\kappa_3(EI - \mu J) - F_1 = \rho I(\omega_{2,t} + \omega_1\omega_3) \\ &\mu J\kappa_{3,s} = 2\rho J\omega_{3,t} \end{aligned} \tag{4.4}$$

where the ω_i are the components of $\mathbf{\Omega}_R$. The first two of these three equations give explicit relationships between F_1 and F_2 and the κ_i and ω_i which can then be used in the linear momentum equation (4.1). We note that the tangential force component, F_3, is not determined at this point. Under the approximation of unshearability ω_1 and ω_2 are components of $\dot{T}$. For the simulations presented here $\dot{T}$ is small and thus we neglect ω_1 and ω_2 without further consideration of the consistency of this approximation. This leads to particularly simple expressions for F_1 and F_2 with a corresponding simplification of the steps required in the numerical simulations.

Two key and subtle issues in our modeling are the correct evolution of the twist and formulation of the tension forces needed to stop the filament from flying apart. In the case of the twist evolution we recall from our analysis of ribbon motion in section 1 that $\kappa_{3,t} = -\alpha\kappa + \omega_{3,s}$. Thus, by eliminating ω_3 between this relation and equation (4.4) we obtain the *twist evolution equation*

$$\frac{d^2\kappa_3}{dt^2} - \frac{\mu}{\rho}\frac{d^2\kappa_3}{ds^2} = \frac{d}{dt}\left(\kappa B \cdot \frac{dT}{dt}\right) \tag{4.5}$$

The parallel transport term on the right hand side of this equation provides the essential coupling between the curve motion and the twist. We note that without it the equation reduces to the familiar torsional wave equation for a fixed rod [23].

An essential part of any numerical scheme for filament simulations is *tension*. Its role is to keep the points that represent the discretization of X from flying apart. This can be accomplished by using springs to hold adjacent points together (e.g., [27]). However, for stiff filaments like those considered here, resolving the (stiff) spring time scale may require overly-small time steps and/or implicit integrators. To avoid these difficulties we have derived a method to simulate inextensible filaments by introducing the tension which enforces arc-length preservation to round-off accuracy. In our current formulation this can be identified as the tangential force component

F_3. There are various routes to determining the equation governing the tension but, by direct computation on equation (4.1), one finds that

$$
\begin{aligned}
F_{3,ss} - ({\kappa_1}^2 + {\kappa_2}^2)F_3 &= 2\kappa_2 F_{1_s} - 2\kappa_1 F_{2_s} + (\kappa_{2_s} - \kappa_1\kappa_3)F_1 \\
&\quad -(\kappa_{1_s} + \kappa_2\kappa_3)F_1 - \rho(\omega_1^2 + \omega_2^2).
\end{aligned}
\tag{4.6}
$$

The right-hand side of the equation for F_3 depends only on known quantities and hence the tension is determined in such a way that arc-length is preserved. (That the above equation is a statement of inextensibility follows from the fact that arc-length conservation requires that $T \cdot T_t = 0$ for all time and hence $T \cdot T_{t,t} = -T_t \cdot T_t$ which is precisely our equation governing F_3.) The boundary conditions are periodic for closed filaments and $F_3(0,t) = F_3(L,t) = 0$ for open filaments (unless external loading is present in which case the load may determine the boundary conditions). In practice, however, a naive discretization of this equation fails because errors tend to set off numerical instabilities. This problem was overcome by finding a discretization of (4.6) which works to round-off error [28]

5. Some numerical results for open and closed filaments. In our numerical simulations the curve X and twist κ_3 are discretized into a number of points x_i and the corresponding $\kappa_{3,i}$ which are evolved according to second order centered difference schemes for (4.1), (4.2), (4.3), and (4.5). Time integration is done using leapfrog or 4th order Runge-Kutta methods. In equation (4.1), the term g sometimes includes a viscous damping term $-\gamma_1 \dot{X}$ or contact force Ψ. To prevent the curve passing through itself, contact forces are introduced in the form of a stiff inverse power law ($\Psi \sim r^{-10}$) with a "charge" density 1 concentrated at the discretization points. Ψ is scaled so that the filament radius a is approximately two to three discretization lengths. The requirement $a/L \ll 1$ is a necessary one for numerical stability and convergence. The stiff power law is chosen in order to approximate a sharp boundary contact force. The issue of sensitivity of results to choice of power law has not been investigated. The function h in equation (4.2) sometimes includes a torsional damping term $-\gamma_2\dot{\theta}$. The tension is determined in the manner described in the previous section.

The main method for checking numerical errors was a convergence study using decreasing discretization lengths. Other diagnostics included conservation of the energy for non-dissipative runs and conservation of the linking number (2.3) for closed rods. For sufficient resolution these two quantities are conserved to within one percent. (An additional check, mentioned below, comes from the verification of the Zajac instability criterion for closed filaments and provides credence to both the numerical results and the Zajac and Benham analyses.)

An inextensible rod closed into a circle in which twist is uniformly distributed is easily seen to be in equilibrium. However this equilibrium is known to be unstable for sufficiently large twist. The critical total twist Ω has been calculated [14,15] to be $\Omega = \sqrt{3}(EI/\mu J) = \sqrt{3}\Gamma$ and this result is found to be born out rather well in our numerical studies (not shown here). For the special case $\mu J = 0$ (i.e., no contribution to the twist energy) it is in fact known that, in the absence of contact forces, the circle is the only stable equilibrium [29]. In general however the question of identification of stable equilibria is open, with or without contact forces. Much work has been done on generating equilibrium configurations using the static rod equations, usually without contact forces, but little is known about the stability of these configurations (see e.g. [30]).

In the circular (unstable) equilibrium conformation, the rod has twist and no writhe. As discussed earlier, when instability occurs, twist energy is decreased by the conversion of twist to writhe. We have investigated the dynamics of this instability with and without contact forces. Although the latter, albeit unphysical case, gives some interesting insights into the instability, we report here only an illustrative example of the twist to writhe conversion with contact forces.

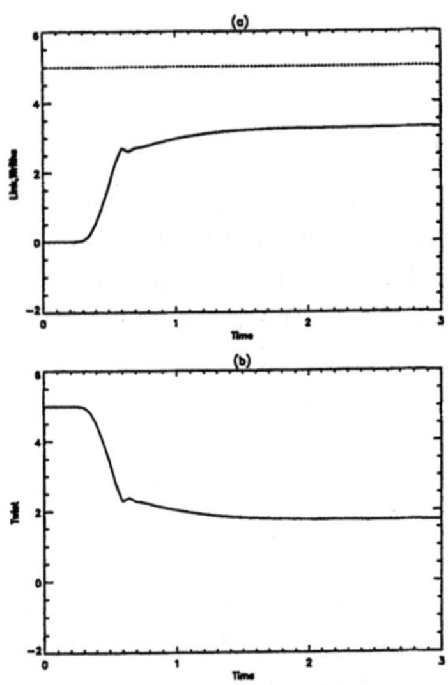

FIG. 4. *Evolution of Tw and Wr for the clover-leaf example. (a) Lk (dashed curve) and Wr (solid curve) versus time. (b) Tw versus time.*

In figure 4 the behaviors of Lk, Wr, and Tw are shown for certain initial conditions (the "clover" conditions, see below). In this case contact forces prevent the filament from crossing through itself so Lk is constant and Tw and Wr evolve continuously. A sequence of snapshots of the dynamics of approach to equilibrium for two examples ($\Gamma = 1$, $r_0 = 1/2$) are shown in figures 5 and 6. In both examples we choose as initial conditions a circular conformation with $Tw = 5$. The four-leaf clover in the final frame of figure 5 is the eventual equilibrium when the initial twist is distributed

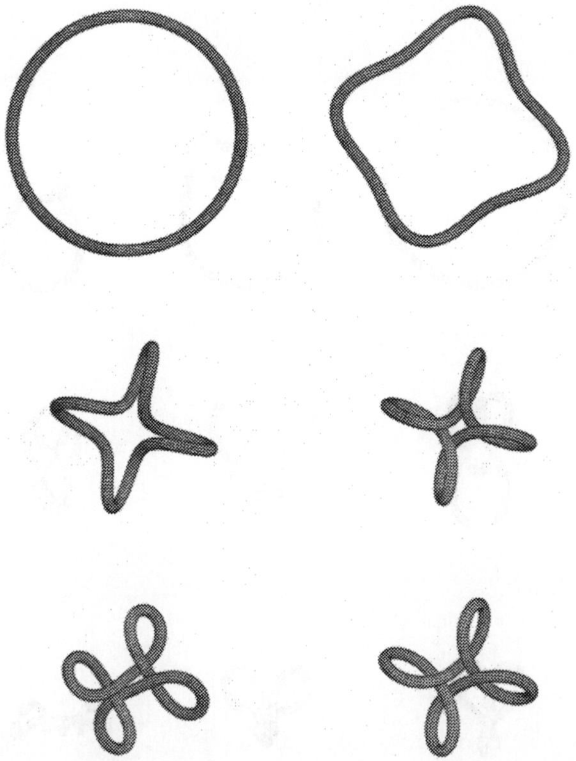

FIG. 5. *Approach to equilibrium "clover" conformation. First row:* $t = 0$, $t = .40$. *Second row:* $t = .55$, $t = .75$. *Third row:* $t = .95$, $t = 3.0$.

uniformly with a small localized perturbation. The final frame of figure 6 illustrates the eventual equilibrium conformation for exactly the same parameters and intial conditions as the first example except that the initial twist includes an order one non-localized perturbation from uniformity, in particular $\kappa_3(s,0) = 2\pi TwL^{-1}(1 + .5\sin(2\pi s/L))$. The equilibrium of figure 6 is sometimes called the plectonemic conformation and is familiar from DNA studies (e.g., [31]). We stress that the final conformations of figures 5 and 6 are distinct, obtainable, local energy minima of unknotted closed rods with identical linking number. For the particular case shown here the plectonemic structure has roughly 2/3 the energy of the clover structure.

Our formalism and algorithm can be used to study the dynamics of open elastic rods subject to twisting forces. To begin with we comment on the behavior of a rod of *fixed length L*. A significant point revealed by our simulations is the role of inertia. In a relatively undamped medium, interference and interaction effects of non-linear twist and curvature waves can produce large amplitude and seemingly random "writhing" motions. This

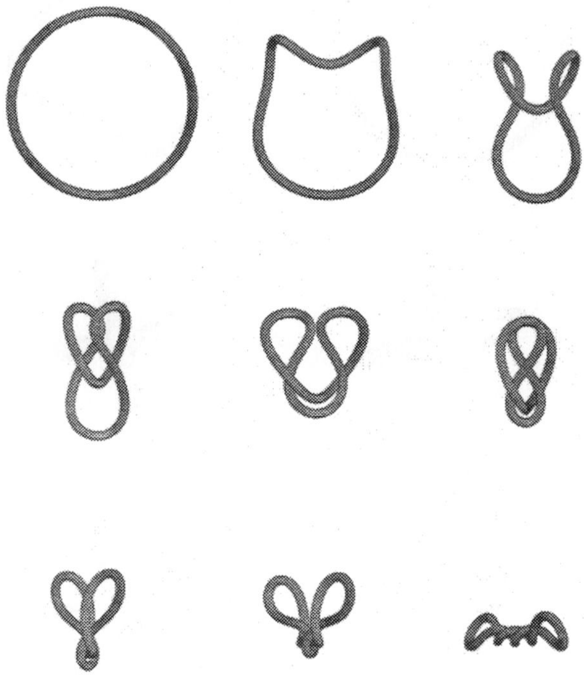

FIG. 6. *Approach to equilibrium "plectonemic" conformation. First row:* $t = 0$, $t = .35$, $t = .45$. *Second row:* $t = .55$, $t = .75$, $t = .95$. *Third row:* $t = 1.25$, $t = 1.55$, $t = 3.50$.

behavior however is dramatically suppressed in a highly viscous medium (even without internal damping). Because of its highly dynamic nature, this effect is best illustrated with a movie but, as described in [28], studying surfaces of $\kappa(s,t)$ (the geometric curvature) and $\kappa_3(s,t)$ (the twist rate) of an evolving twisted filament in low and high viscosity media ("low" being defined to mean that inertial effects are noticeable), reveals clear difference in curvature and twist wave amplitudes. These are consistent the experimentally observed dichotomy of "writhing" versus "buckling" instabilities exhibited by the bacteria in low versus high viscosity regimes respectively.

A complete study of the bacterial filaments requires that the simulation includes exponential growth of the filament. This is clearly a significant computational challenge that is now being addressed. In the mean time, however, we believe that the fundamentals of the bacterial filament evolution are captured by recognizing that it is the growth of the filament that is injecting twist into the structure. We believe that it is this injection that is responsible for stabilizing the plectonemic structures that are formed.

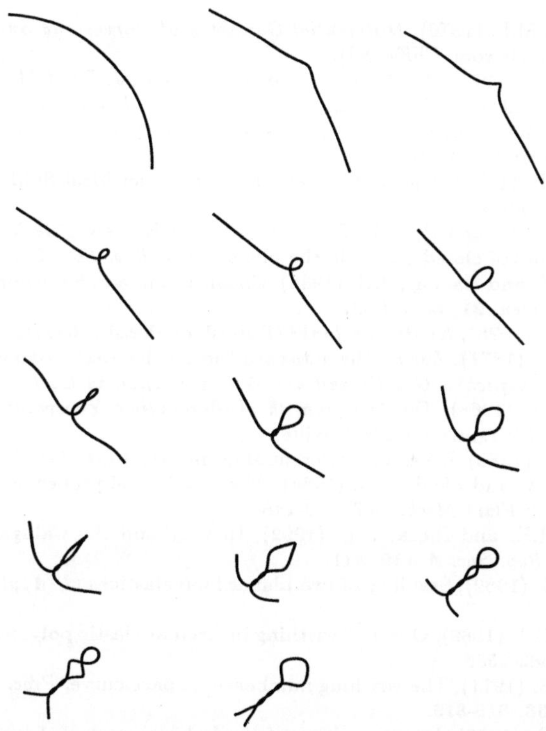

FIG. 7. *Evolution of an open filament in viscous medium with continuously injected twist.*

Thus for a filament of fixed length a rough equivalent to the growth process can be provided by continuously "injecting" twist into the filament. This effect is accomplished in actual fact by modifying the elastic moment constitutive law (4.3) to be

$$M = EI\kappa_1 V + EI\kappa_2 W + \mu J(\kappa_3 - \hat{\kappa})T$$

where $\hat{\kappa}$, the relaxed twist rate, is a given, and in this case time dependent, function of arc-length. In figure 7 we show the results of a simulation using this more general constitutive law using a $\hat{\kappa}$ with Gaussian-like profile (in arc-length) and amplitude increasing linearly in time. The pictures are striking and clearly show the buckling phenomenon followed by plectonemic wind-up.

Acknowledgements. The authors would particularly like to thank A. Goriely and A. Rao for their considerable assistance with the computer graphics. The authors also benefited from conversations with N. Mendelson and J. Maddocks. This work was supported by DOE grant DE-FG03-93-ER25174 and an NSF post-doctoral fellowship.

REFERENCES

[1] Do Carmo, M.P. (1976), *Differential Geometry of Curves and Surfaces* (Prentice-Hall, Englewood Cliffs, NJ).

[2] Hasimoto, H. (1972), A soliton on a vortex filament, *J. Fluid Mech.* **51**, 477-485.

[3] Lamb, G.L. (1977), Solitons on moving space curves, *J. Math. Phys.* **18**, 1654-1661.

[4] Langer, J. and Perline, R. (1991), Poisson geometry of the filament equations, *J. Nonlinear Sci.* **1**, 71-93.

[5] Keener, J.P. (1990), Knotted vortex filaments in an ideal fluid, *J. Fluid Mech.* **211**, 629-651.

[6] Goldstein, R.E. and Petrich, D.M. (1991), The Korteweg de Vries hierarchy as dynamics of closed curves in the plane, *Phys. Rev. Lett.* **67**, 3203-3206.

[7] Berry, M.V. and Hannay, J.H. (1988), Classical non-adiabatic angles, *J. Phys. A: Math. Gen.* **21**, L325-L331.

[8] Rolfsen, D. (1976), *Knots and Links* (Publish or Perish, Berkeley, CA).

[9] Gauss, C.F. (1877), Zur mathematischen theorie der electrodynamischen wirkungen, *Koniglichen Gesellschaft der Wissienshaften zu Gottingen* **5**, 602-629.

[10] Milnor, J.W. (1965), *Topology from the Differentaible Viewpoint* (The University Press of Virginia, Charlottesville).

[11] Pohl, W.F. (1980), DNA and differential geometry, *Math. Intelligencer* **3**, 20-27.

[12] Berger, M.A. and Field, G.B. (1984), The topological properties of magnetic helicity, *J. Fluid Mech.* **147**, 133-148.

[13] Moffatt, H.K. and Ricca, R.L. (1992), Helicity and the Calugareanu invariant, *Proc. Roy. Soc. A* **439**, 411.

[14] Zajac, E.E. (1962), Stability of two planar loop elasticas, *J. Appl. Mech.* **29**, 136-142.

[15] Benham, C.J. (1989), Onset of writhing in circular elastic polymers, *Phys. Rev. A* **39**, 2582-2586.

[16] Fuller, F.B. (1971), The writhing number of a space curve, *Proc. Natl. Acad. Sci. USA* **68**, 815-819.

[17] Fuller, F.B. (1978), Decomposition of the linking number of a closed ribbon: A problem from molecular biology, *Proc. Natl. Acad. Sci. USA* **75**, 3557-3561.

[18] Aldinger, J., Klapper, I., and Tabor, M. (1995), Formulae for the calculation and estimation of writhe, *J. Knot Theory Ramifications* **4**, 343-372.

[19] Klapper, I. and Tabor, M. (1994), A new twist in the kinematics and elastic dynamics of curves and ribbons, *J. Phys. A: Math. Gen.* **27**, 4919-4924.

[20] Mendelson, N.H. (1990), Bacterial macrofibers: the morphogenesis of complex multicellular bacterial forms, *Sci. Progress Oxford* **74**, 425-441.

[21] Thwaites, J.J. and Mendelson, N.H. (1991), Mechanical behavior of bacterial cell walls, *Adv. Microbiol. Physiol.* **32**, 174-222.

[22] Love, A.E.H. (1927), *A Treaty on the Mathematical Theory of Elasticity*, Fourth Edition (Cambridge University Press, Cambridge, reprinted by Dover Publications, New York).

[23] Landau, L.D. and Lifschitz, E.M. (1959), *Theory of Elasticity* (Pergamon Press, Oxford).

[24] Coleman, B.D., Dill, E.H., Lembo, M., Lu, Z., and Tobias, I. (1993), On the dynamics of rods in the theory of Kirchhoff and Clebsch, *Arch. Rational Mech. Anal.* **121**, 339.

[25] Simo, J.D., Marsden, J.E., and Krishnaprasad, P.S. (1988), The Hamiltonian structure of nonlinear elasticity: the material and convective representations of solids, rods and plates, *Arch. Rational Mech. Anal.* **104**, 125-183.

[26] Maddocks, J.H., et al., this volume.

[27] Schlick, T. and Olson, W.K. (1992), Trefoil knotting by molecular dynamics simulations of supercoiled DNA, *Science* **257**, 1110-1115.

[28] Klapper, I. (1994), Biological applications of the dynamics of twisted elastic rods, *J. Comp. Phys.* in press.

[29] Langer, J. and Singer, D.A. (1985), Curve straightening and a minimax argument for closed elastic curves, *Topology* **24**, 75-88.

[30] Shi, Y. and Hearst, J.E. (1994), The Kirchhoff elastic rod, the nonlinear Schrödinger equation, and DNA supercoiling, to appear, *J. Chem. Phys.*.

[31] Tsuru, H. and Wadati, M. (1986), Elastic model of highly supercoiled DNA, *Biopolymers* **25**, 2083-2096.

[32] Mendelson, N.H. and Thwaites, J.J. (1990), Bending, folding and buckling processes during bacterial macrofiber morphogenesis, *Mat. Res. Soc. Symp.* **174**, 171-178.

[33] Mendelson, N.H. (1976), Helical growth of *Bacillus subtilis*: A new model of cell growth, *Proc. Natl. Acad. Sci.* **73**, 1740-1744.

INTEGRATION METHODS FOR MOLECULAR DYNAMICS*

BENEDICT J. LEIMKUHLER†, SEBASTIAN REICH‡, AND ROBERT D. SKEEL§

Abstract. Classical molecular dynamics simulation of a macromolecule requires the use of an efficient time-stepping scheme that can faithfully approximate the dynamics over many thousands of timesteps. Because these problems are highly nonlinear, accurate approximation of a particular solution trajectory on meaningful time intervals is neither obtainable nor desired, but some restrictions, such as *symplecticness*, can be imposed on the discretization which tend to imply good long term behavior. The presence of a variety of types and strengths of interatom potentials in standard molecular models places severe restrictions on the timestep for numerical integration used in explicit integration schemes, so much recent research has concentrated on the search for alternatives that possess (1) proper dynamical properties, and (2) a relative insensitivity to the fastest components of the dynamics. We survey several recent approaches.

Key words. leapfrog method, Verlet method, symplectic method, multiple-timestep methods, symplectic integrator, molecular dynamics simulation, constrained dynamics, SHAKE, multiple time scales, long-time integration.

1. Introduction.

For classical molecular dynamics (MD) of macromolecules such as nucleic acids, proteins and polymers, atomic trajectories obey Newton's second law of motion:

$$M \frac{d^2}{dt^2} q = -\nabla_q V(q), \tag{1.1}$$

where q is vector containing all positions (in cartesian coordinates), M is a diagonal matrix of atomic masses with each mass replicated thrice, and $V(q)$ is the (empirical) potential energy function. Standard numerical schemes for simulating the dynamical behavior of a macromolecule are based on discrete timestepping. When the number of atoms is large, we need a fast method to calculate electrostatic interactions at each step, and we need parallelism, preferably scalable parallelism, but it is not enough to be able to take fast time steps for microsecond to millisecond dynamics: ways must be found to increase the length of the timesteps.

These systems are highly nonlinear and have the hallmarks of chaos. In particular, they exhibit sensitive dependence on perturbations. For example, numerical experiments in [1] show that tiny perturbations introduced

* The work of the first author was performed under grant number NSF Grant DMS-9303223 and relied on resources of the Kansas Institute for Theoretical and Computational Science. The work of the second and third authors was performed at the Beckman Institute of the University of Illinois and supported in part by DOE/NSF grant DE-FG02-91-ER25099/DMS-9304268, by NIH Grant P41RR05969, and by NSF/ARPA Grant ASC-9318159

† Department of Mathematics, University of Kansas, Lawrence, KS 66045, U.S.A.

‡ Konrad-Zuse Zentrum, Heilbronner Str. 10, D–10711 Berlin, Germany.

§ Department of Computer Science, University of Illinois, 1304 West Springfield Avenue, Urbana, IL 61801–2987, U.S.A.

into the trajectories are doubled about every picosecond meaning that even with double precision (about 15 digits of precision) the trajectory would be overwhelmed by the effects of roundoff error after just fifty picoseconds of simulation. Moreover, it is typical to use random values for initial velocities. Thus for long-time integrations, accuracy with respect to a particular trajectory is neither desired nor obtainable. Instead, what is sought is generic or representative behavior of the system, or in some cases good sampling of phase space.

The variety of types and strengths of terms in the potential function for a biomolecule complicates numerical simulation. These problems involve *multiple time scales.* In most cases, the primary restriction on timestep comes from the presence of the strong chemical bonds (which contribute high-frequency components to the motion). Occasionally, the word "stiff" is used to describe the problem with multiple scales, but this usage is nonstandard, and may confuse the issue. Strictly speaking, the term "stiff" implies that we are seeking particular trajectories of the system that are smoother than nearby trajectories. However, due to the ergodic nature of the differential equations for MD, no trajectory can be smoother than its neighbors for more than a short time interval. In the sequel, we will sometimes use the terms "fast modes" or "high frequency components," to refer to components of *linearized* dynamics.

The numerical integration schemes of molecular dynamics roughly divide into four categories according to their handling of these critical "fast" modes:

(i) Methods that accurately resolve the highest frequencies, e.g. explicit methods (with small timesteps), including multiple time stepping and the method LIN [47].
(ii) Methods which inaccurately resolve the highest frequencies: e.g. the implicit midpoint method with large timestep Δt. Although to date apparently not used in MD, this method has recently been employed for other problems in nonlinear dynamics, e.g. for rigid and elastic bodies [38].
(iii) Methods based on removal of the fast components through the introduction of constraints, including SHAKE [34] and RATTLE[2].
(iv) Methods which attempt to suppress the fast components, e.g. the method LI [29], which introduces dissipation via the implicit Euler scheme to enable larger timesteps and balances this with a random noise term to maintain the desired temperature.

Traditionally, numerical integrators have been analyzed in terms of their *stability* and *accuracy.* In standard usage, "stability" means boundedness with respect to perturbations, and "accuracy" means accuracy with respect to a given trajectory; these concepts are of limited practical value for the treatment of nonlinear dynamics on long time intervals, due to the sensitive dependence on initial conditions. Although it is found that classical stability and local accuracy are prerequisite for a given integrator to function well in molecular dynamics, these standard concepts are inadequate for gauging its effectiveness.

An alternative point of view is to attempt to show that the numerically computed trajectory is close to *some* true trajectory of the physical problem (i.e., starting from perturbed initial values), but obtaining such a ***shadowing*** property even for relatively simple systems has so far proven very difficult, despite many attempts. Still another approach is to attempt to show that the numerical trajectory is close to some true trajectory of a ***nearby system of differential equations***, preferably one with dynamical properties similar to those of the original system. This is similar to the viewpoint of ***backward error analysis*** introduced by J.H. Wilkinson to study the algorithms of numerical linear algebra.

An important feature of Newton's equations (1.1) is that they can be viewed as a Hamiltonian system, i.e. they can be put in the form $\dot{q} = \nabla_p H$, $\dot{p} = -\nabla_q H$, for some smooth energy function $H = H(q,p)$. A certain category of differential invariants, the *symplectic* invariants, is associated with a Hamiltonian system. It is sensible to consider a ***symplectic discretization*** which preserves the same class of invariants. Much recent interest in symplectic discretization schemes for Hamiltonian systems emanates from fundamental results establishing the "backward error" point of view (i.e., the numerical flow generated by a symplectic discretization of a Hamiltonian system is the exact flow of a nearby Hamiltonian system). The practical ramifications of symplecticness for general Hamiltonian systems or for molecular dynamics are still not completely clear, but symplectic methods have proven to be important in related problems, such as long term simulation of the solar system [46] and for the design of particle accelerators [13].

The main goals of this article are first to discuss current approaches to time discretization, in particular the popular leapfrog/Verlet integrator, and, second, to sketch some alternatives that may permit the taking of longer timesteps. The first part of the paper is concerned primarily with the issue of numerical stability and correct long time behavior, particular attention being given to symplectic integration:

2. Numerical time-stepping schemes: leapfrog/Verlet
3. Symplectic discretizations
4. Alternatives to leapfrog/Verlet

The second part of the paper addresses the computational costs associated with high frequency components of trajectories. The idea of multiple time steps is to use an overall long time step with shorter time steps only for selected interactions. We can manage with even less computation if we are willing to accept less in terms of accuracy—e.g., some kind of smoothed solution—and this is apparently possible with constrained dynamics. These ideas constitute the remaining sections:

5. Multiple time stepping
6. Constrained dynamics

2. The leapfrog/Verlet scheme. For the molecular dynamics problem, the Hamiltonian takes the form $H(q,p) = T(p) + V(q)$ where the kinetic energy $T(p) = \frac{1}{2}p^T M^{-1}p$, so the corresponding equations of motion are

$$\frac{d}{dt}q = M^{-1}p, \quad \frac{d}{dt}p = -\nabla_q V(q). \tag{2.1}$$

The pair $(q(t), p(t))$ specifies a point in *phase space*, a $6N$ dimensional Euclidean space where N is the number of particles. These values determine the state of the system at a time t and from these one can in principle uniquely determine the values $(q(t+\Delta t), p(t+\Delta t))$ at some later time $t + \Delta t$. This mapping from the earlier to later set of values we call the Δt-flow of the system, and it will be handy to denote it by $\phi_{\Delta t,H}$. Hence,

$$\phi_{\Delta t,H}(q(t), p(t)) = (q(t+\Delta t), p(t+\Delta t)),$$

where (q,p) solves the Hamiltonian system (2.1).

The purpose of an integration method is to approximate the flow $\phi_{\Delta t,T+V}$. Many popular explicit methods are obtained by *splitting*: e.g., the Strang [41] splitting gives

$$\phi_{\Delta t,T+V} = \phi_{\Delta t/2,V} \circ \phi_{\Delta t,T} \circ \phi_{\Delta t/2,V} + O(\Delta t^3). \tag{2.2}$$

where $\circ$ denotes composition of mappings. The idea here is to replace the generally intractable computation of the flow of the original system by a sequence of systems that are analytically integrable. For example the flow $\phi_{\Delta t/2,V}$ is obtained by solving

$$\frac{d}{dt}q = 0, \quad \frac{d}{dt}p = -\nabla_q V(q).$$

If we let

$$q(t_n) = q_n, \quad p(t_n) = p_n,$$

denote the current location in phase space, the solution is trivially

$$q(t) \equiv q_n, \quad p(t) = p_n + (t - t_n)(-\nabla_q V(q_n)),$$

and integration for an interval $\Delta t/2$ yields the new value

$$p_{n+1/2} = p_n + (\Delta t/2)(-\nabla_q V(q_n)).$$

The approximation suggested by eq. (2.2) is merely a formal way of expressing the very popular leapfrog/Verlet method: We begin the step with values q_n, p_n, and $F_n = -\nabla_q V(q_n)$ obtained from the previous step. Then we compute

$\phi_{\Delta t/2,V}$, "half a kick":

$$p_{n+1/2} = p_n + \frac{\Delta t}{2} F_n,$$

$\phi_{\Delta t,T}$, "drift":

$$q_{n+1} = q_n + \Delta t M^{-1} p_{n+1/2},$$

$\phi_{\Delta t/2,V}$, "half a kick":

$$F_{n+1} = -\nabla_q V(q_{n+1})$$
$$p_{n+1} = p_{n+1/2} + \frac{\Delta t}{2} F_{n+1}.$$

The physical interpretation is that two impulses of equal size are imparted to the particles just before and just after each time point t_n and that between these time points the particles just drift with constant velocity.

A contribution to the error of order Δt^3 is introduced at each step of leapfrog integration. A standard stability result shows that the accumulated error over a fixed time interval is then bounded by $K\Delta t^2$ for Δt sufficiently small. The constant K in this formula grows exponentially with the length of the time interval; for molecular dynamics applications it may be very large.

Numerical integration methods can be analyzed to determine the extent to which they possess correct long-time behavior. A detailed analysis is possible for the harmonic oscillator:

$$\frac{dq}{dt} = p, \tag{2.3}$$

$$\frac{dp}{dt} = -\omega^2 q. \tag{2.4}$$

Correct behavior for the harmonic oscillator seems to be a necessary property for success in MD simulations, even if it is not sufficient. The oscillator gives insight into the vibrational behavior of bond lengths and bond angles in MD problems.

Analytically, if we plot $p(t)$ versus $\omega q(t)$ we get a circular orbit in phase in which we have a rotation by $\omega\Delta t$ for every timestep of the solution. One step of Verlet gives instead [39]

$$\begin{bmatrix} \omega q_{n+1} \\ p_{n+1} \end{bmatrix} = D \begin{bmatrix} \cos\theta & \sin\theta \\ -\sin\theta & \cos\theta \end{bmatrix} D^{-1} \begin{bmatrix} \omega q_n \\ p_n \end{bmatrix}$$

where

$$\theta = 2\arcsin\frac{\omega\Delta t}{2}$$

and

$$D = \mathrm{diag}(1, \cos\frac{\theta}{2}),$$

provided that $\omega\Delta t \leq 2$. Thus we have for the Verlet method (i) a slight stretching (ii) followed by a rotation in error by $O((\omega\Delta t)^3)$ (ii) followed by a slight shrinking.

In a sequence of Verlet steps the intermediate stretchings and shrinkings all cancel. This does not happen though if the method or the timestep is varied. The restriction $\omega\Delta t \leq 2$, which is necessary for stability, is the weakest restriction of any conventional explicit method [11].

An important property of the Verlet method is that it is *symplectic*. We will now describe this property.

3. Symplectic discretizations. In two dimensions, a mapping is symplectic if it is *area preserving*: any bounded region is mapped to another region of equal area. In higher dimensions, symplecticness is still a geometrical property, although somewhat more difficult to state. Denote a point in $\mathbf{R}^{2m}$ by $(q^1, q^2, \ldots, q^m, p^1, p^2, \ldots, p^m)$ Consider the m projections of a two-dimensional oriented surface D onto the $q^i p^i$-coordinate planes, and then sum the oriented areas of these projections to get a number $\mu(D)$. A mapping ϕ of $\mathbf{R}^{2m}$ is symplectic if $\mu(\phi(D)) = \mu(D)$ for every 2 dimensional oriented surface D. It can be demonstrated that a consequence of this property is preservation of volume.

The Δt-flow $\phi_{\Delta t,H}$ of a Hamiltonian system is always μ-preserving, hence symplectic. In the context of flows of Hamiltonian systems, the invariant $\mu(D)$ is referred to as the Poincaré integral invariant.

The geometric definition of symplecticness can be replaced by an algebraic one. Let $\Phi = \Phi(x)$ denote the Jacobian matrix of a mapping $\phi : \mathbf{R}^{2n} \to \mathbf{R}^{2n}$. ϕ is symplectic if

$$\Phi(x)^T J \Phi(x) \equiv J,$$

where J is the $2m \times 2m$ skew-symmetric "symplectic structure matrix:"

$$J = \begin{bmatrix} 0 & I \\ -I & 0 \end{bmatrix}.$$

For demonstrating symplecticness of complicated mappings, it is sometimes convenient to use the language of differentials: a mapping is said to be symplectic if it preserves the "2-form" $dp \wedge dq = \sum_{i=1}^{m} dp^i \wedge dq^i$. (The algebra of exterior products of differentials can then be used to simplify the calculations.)

A discretization of (2.1) can be viewed as a mapping $\psi_{\Delta t}$ of phase space that approximates the flow map $\phi_{\Delta t,H}$. Since, as we have seen, the symplecticness of the flow map of an Hamiltonian system is such a strong condition, it is natural to seek discretizations which also possess this property. Recently, a great deal of work on symplectic discretizations has been undertaken (see e.g. [36] for a survey). Much of the motivation for this work comes from the following backward error argument: if a symplectic discretization is applied to approximate the Δt-flow $\phi_{\Delta t,H}$, then the numerical map $\psi_{\Delta t}$ is exactly the Δt-flow of a perturbed Hamiltonian:

$$\psi_{\Delta t} \equiv \phi_{\Delta t,\tilde{H}},$$

where (with r the order of the method)

$$\tilde{H} = H + \Delta t^r H_r + \Delta t^{r+1} H_{r+1} + \cdots$$

for some functions H_i, $i \geq r$.

The sense in which $\tilde{H}$ is "close" to H is complicated. In particular, the formal series expansion for $\tilde{H}$ need not converge.

An interesting theorem due to Ge and Marsden [16] states roughly that for a generic Hamiltonian system, a symplectic method cannot exactly conserve the energy unless it is a reparameterization in time of the true flow. In the Verlet method, the perturbed Hamiltonian ($\tilde{H}$) corresponding to the discretization is affected in both the kinetic energy and potential energy terms, causing the mass matrix M to cease to be diagonal. The magnitude of energy fluctuations in actual simulation is caused by this perturbation and is thus a useful indicator of accuracy.

The Verlet method also conserves the angular momentum [48].

Another strong property possessed by the flow of a Hamiltonian system such as (2.1) is time-reversibility. This means that if we integrate forward τ units in time from point A to point B, then replace t by $-t$ and p by $-p$ in the differential equations and integrate τ units starting from B, we arrive back at point A. A similar property can be defined for a discretization scheme (for example, the leapfrog/Verlet scheme is reversible), and may also be of importance for long term dynamics [9].

From the form of typical methods such as leapfrog, there is no reason why the stepsize Δt may not vary with the timestep number. For example, we could imagine adjusting the Δt from step to step according to some estimate of the local discretization error. However, the results are less than satisfactory in practical implementations [10]. This can be explained in terms of the nearby Hamiltonian: when we vary the stepsize in a symplectic integrator the effect is the same as if we would solve a different modified Hamiltonian system over each timestep. Thus it is no longer possible to say that the numerical solution coincides with the solution of one and the same nearby Hamiltonian.

4. Alternatives to leapfrog/Verlet. Various alternatives to leap frog for numerical integration in molecular dynamics have been proposed and implemented over the years. These include multistep schemes and Runge-Kutta methods.[1] A survey of MD experiments using various methods is found in [5]. One can also look for efficient splitting methods based

[1] The term "multistep method" should not be confused with "multiple time stepping" discussed in the following section.

on alternatives to the Strang splitting. One general approach is based on the idea of multiple time stepping, which we discuss in section 5.

Multistep methods make use of the previously computed timesteps and function values to improve the approximation. The explicit "Gear methods" implemented in many molecular dynamics codes are an example of a type of multistep method. These methods are not symplectic or reversible, and it is not clear that the extra order obtained is relevant, since they exhibit relatively poor long term stability of energy. Also of interest in the family of multistep methods are the higher order symmetric schemes (the cousins of position Verlet/Störmer's rule) [21]. These have recently been applied with great success in long-term simulations of the motion of the solar system [30]. The chief advantage of using these multistep methods is that high accuracy can be obtained with minimal cost measured in terms of force evaluations. It has not been shown that multistep methods can be symplectic in the strict sense of the term, but the symmetric methods are time-reversible.

Higher order multistep schemes will likely offer only moderate improvements in efficiency *for the accuracies needed in MD*, e.g., 40% decrease in computing time [25].

Runge-Kutta methods involve a sequence of intermediate "stage" calculations which are then combined in such a way as to obtain desirable features such as high order of accuracy, stability or symplecticness. The standard explicit fourth order method uses four stages. It is well known that to achieve order $p > 4$ with an explicit RK scheme, more than $p + 1$ stages are necessary [17]. Since each stage involves a force evaluation, it is probably not advantageous to consider very high order RK methods for molecular dynamics calculations, particularly since high-order of accuracy is not necessarily the critical feature.

Implicit methods require the solution of a system of nonlinear equations at each timestep. For example, the implicit midpoint method can be written

$$q_{n+1} = q_n + \Delta t M^{-1} \frac{p_n + p_{n+1}}{2}, \tag{4.1}$$

$$p_{n+1} = p_n - \Delta t \nabla_q V\left(\frac{q_n + q_{n+1}}{2}\right). \tag{4.2}$$

It can be shown that this method is symplectic, time-reversible and second order accurate. Implicit methods are typically employed with the hope of circumventing stability restrictions and enabling much larger timesteps. A family of higher order implicit methods (Gauss-Legendre Runge-Kutta schemes [17]) with similar properties have been used, although for molecular dynamics applications the advantages of these higher order schemes appear to be outweighed by increased implementation costs. Some work on the implementation of these methods for molecular dynamics has been undertaken by [19].

The stability of the implicit midpoint method on the harmonic oscillator (2.3)-(2.4) can be thoroughly analyzed. A short calculation shows that with q_{n+1} and p_{n+1} computed from (4.1)-(4.2), they satisfy

$$\begin{bmatrix} \omega q_{n+1} \\ p_{n+1} \end{bmatrix} = \begin{bmatrix} \cos\theta & \sin\theta \\ -\sin\theta & \cos\theta \end{bmatrix} \begin{bmatrix} \omega q_n \\ p_n \end{bmatrix},$$

with θ defined by

$$\theta = 2\arctan\frac{\omega\Delta t}{2}.$$

As for the leapfrog scheme, we have a rotation through an angle θ which is in error by $O(\Delta t^3)$. In this case, there is no stretching or shrinking (the implicit midpoint method preserves quadratic invariants).

Of interest is the behavior of this method for large $\omega\Delta t$. We can see that as $\Delta t\omega \to \infty$, $\theta \to \pi$. This means that for large $\omega\Delta t$, the frequency gets completely distorted. There is no stability restriction on Δt, but the high frequencies are not accurately represented unless Δt is proportionately small.

In the case of generic linear systems, the foregoing analysis applies independently to each of the fundamental modes. This means that the equations can be decoupled into essentially the equations of a set of independent harmonic oscillators. This means that if we are only interested in the slowest modes, we can resolve these by using a large timestep. The high frequency components will be incorrectly resolved, but this causes no difficulty since they are decoupled from the lower modes.

On the other hand, the situation for nonlinear problems is much more complex. No simple decoupling of the modes of the problem is possible, and the possibility exists that errors induced in the high frequency components of the linearization may contaminate the numerical solution—particularly over longer time intervals.

Actually, difficulties can arise even in relatively simple planar nonlinear problems for which only one linear mode is present. In Figure 4.1 we have indicated the result of an experiment with a simple modification of the harmonic oscillator with potential $V(q) = \omega^2 q^2/2 + \epsilon q^4/4$, where $\omega = 10$ and $\epsilon = 1$. Applying the implicit midpoint method for various values of the stepsize, we compute the energy error at each timestep and then plot the energy error vs. both time and the stepsize. This results in a jagged surface that becomes smoother for smaller timesteps. Note that for values of the stepsize near $\Delta t = .2$, one observes very large variations in the total energy. We refer to this phenomena as "stepsize resonance." Although a detailed analysis of these resonances is beyond the scope of the current article, the existence of the resonances should raise a red flag regarding the use of implicit methods in order to lengthen the stepsize.

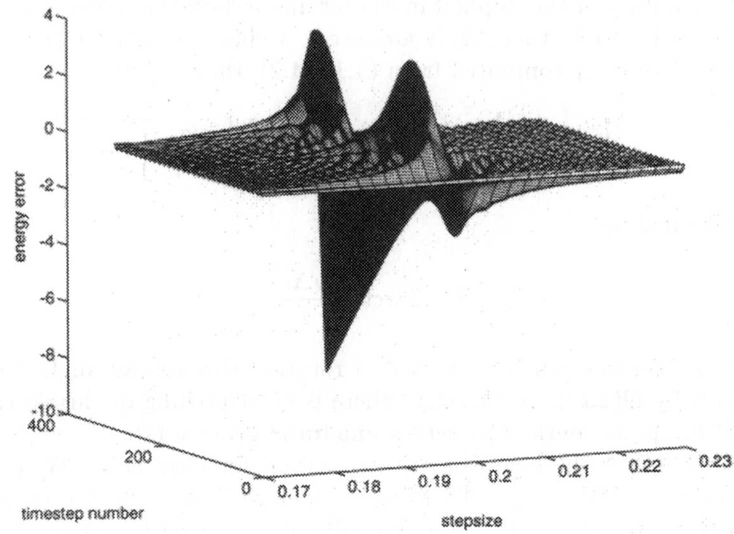

FIG. 4.1. *Stepsize resonance for a simple model problem. The graph indicates the energy error in the computed solution with a given stepsize and at a given timestep number.*

5. Multiple time stepping. In molecular dynamics simulations, the highest frequencies in the system are due to the bond stretching and bond-angle bending potentials, which can be written in the form

$$V_{\text{bond}}(q) = \frac{1}{2}\sum_i k_i g_i(q) g_i(q). \tag{5.1}$$

Here the functions g_i stand for $g_i(q) = r - r_0$ in case of bond stretching and for $g_i(q) = \phi - \phi_0$ in case of bond-angle bending. The constants k_i are the force constants which typically take values of a few hundred kcal mol^{-1} Å^{-2} for the bond-stretching potentials and a few tens up to a hundred kcal mol^{-1} deg^{-2} for the bond-bending potentials.

Denoting the remaining entries in the potential function by $V(q)$ and using matrix notation, the Hamiltonian function of a molecular system can be written as

$$H(q,p) = \frac{p^T M^{-1} p}{2} + V(q) + \frac{g(q)^T K g(q)}{2}, \tag{5.2}$$

where g is the column vector of all the g_i's and K is a diagonal matrix with diagonal entries $K_{ii} = k_i$.

The corresponding equations of motion are then given by

$$\begin{aligned} \frac{d}{dt}q &= M^{-1}p, \\ \frac{d}{dt}p &= -\nabla_q V(q) - g_q(q)^T K g(q). \end{aligned} \tag{5.3}$$

The impediment to taking longer time steps is the high frequency components of the motion due to the potential (5.1).

The idea of multiple timesteps (MTS) is to evaluate different force terms at different interval lengths; positions and velocities are updated in principle every time any force is evaluated. For example, for the interactions represented above by V and for those represented by $V^{\mathrm{hard}} = \frac{1}{2}g(q)^T Kg(q)$ the time scales are considerably different, and it would be natural to use different timesteps for V than for V^{hard}. We might try the following time-splitting integration method:

$\Delta t/2, V$
$\Delta t, T + V^{\mathrm{hard}}$ numerically—N steps of Verlet with stepsize $\Delta t/N$
$\Delta t/2, V$

Such *symplectic* multiple time stepping methods were discovered independently at Illinois [6,15] and Columbia [43]. Nonsymplectic MTS methods go back over 25 years in the astrophysics literature [18] and 16 years in the MD literature [42]. Typical nonsymplectic MTS methods will exhibit poor behavior on long enough time intervals. On the other hand, it seems that there may be serious accuracy and stability problems with symplectic MTS methods. For example, the possibility of resonance is reported in [15,6]. And there are other concerns that do not appear in the literature. Unless these can be cleared up, it seems more prudent to use the better nonsymplectic MTS methods proposed in [15,6,37].

The use of MTS with a timestep fixed for each bonded interaction is appropriate and probably optimal, but it is inadequate for nonbonded interactions because softness or hardness can vary greatly depending on the interparticle distance r. It is highly desirable to vary the stepsize for such an interaction depending on the distance between two atoms.

Variable stepsize is possible in a way that still employs the machinery of MTS: For each nonbonded interaction $U(r)$ we introduce an artificial splitting

$$U(r) = U^{\mathrm{hard}}(r) + U^{\mathrm{soft}}(r)$$

such that $U^{\mathrm{hard}}(r)$ vanishes for $r \geq r_{\mathrm{cut}}$ and $U^{\mathrm{soft}}(r)$ is soft for all r. So $U^{\mathrm{soft}}(r)$ never requires a small timestep. The effect of this is to permit a large timestep whenever r exceeds the cutoff.

For example with $U(r) = 1/r$ we suggest

$$U^{\mathrm{soft}}(r) = \begin{cases} \frac{3}{2}r_{\mathrm{cut}}^{-1} - \frac{1}{2}r_{\mathrm{cut}}^{-3}r^2, & r \leq r_{\mathrm{cut}}, \\ r^{-1}, & r \geq r_{\mathrm{cut}}, \end{cases}$$

and $U^{\mathrm{hard}} = U - U^{\mathrm{soft}}$. Preliminary experiments with this method appear in [40].

6. Constrained dynamics. A traditional approach to taking larger timesteps has been to freeze the bonded motions, thus reducing the stability restriction. This leads to constrained equations of motion which can be discretized, e.g., by the well-known SHAKE/RATTLE method. In this section we wish to discuss various recent results on the numerical and analytical aspects of constrained dynamics. In the following subsection, we introduce the concept of the *smoothed dynamics* [31] of a molecule and show that, for example, constraining the bond-lengths to their equilibrium value leads to a first approximation of the corresponding smoothed dynamics. The second subsection is devoted to numerical aspects of constrained dynamics. In particular, we discuss new efficient techniques for solving the nonlinear equations at each SHAKE/RATTLE step. Finally, in the last subsection, we come back to the concept of the smoothed dynamics of a molecule and derive a better approximation. This will lead us to a new type of constraints which we call *soft* or *flexible constraints* [8],[31]. It seems possible that the introduction of soft constraints will allow one to constrain bond-angles as well as bond-lengths and thus to effectively reduce the motion of a molecule to its torsion dynamics. (It has been shown in [44] that constraining the bond-angles by "hard" constraints to their equilibrium value leads to qualitatively incorrect results.)

6.1. Analytical aspects I. In this section we introduce the notion of the *smoothed dynamics* [31] of a Hamiltonian system (5.3). By this we mean the following:

With a suitable scaled Hamiltonian H, the shortest period in the motion of (5.3) due to the potential $V(q)$ is of order $O(1)$. In contrast to this, the potential $g(q)^T K g(q)/2$ contributes high-frequency terms with period of order $O(1/\sqrt{||K||})$. To separate these high frequency components from the slowly varying parts, we introduce the smoothing operator

$$(6.1) \qquad \langle w\rangle_\alpha(t) \;:=\; \frac{1}{\alpha}\int_{-\infty}^{+\infty} \rho(\frac{t-t'}{\alpha})\, w(t')\, dt',$$

with $0<\alpha<1$ and $w:\mathbf{R}\to\mathbf{R}$. Here $\rho:\mathbf{R}\to\mathbf{R}$ is an appropriate weight function such that for any smooth function w we have

$$(6.2) \qquad \langle w\rangle_\alpha(t) - w(t) \;=\; O(\alpha^s),$$

and for any (bounded) continuous function w we have

$$(6.3) \qquad \langle\langle w\rangle_\alpha\rangle_\alpha(t) - \langle w\rangle_\alpha(t) \;=\; O(\alpha^s),$$

where s is a fixed integer with $s \gg 1$. One could, for example, chose for ρ the Meyer scaling function [12]. From now on we will always assume that the terms of order $O(\alpha^s)$ are sufficiently small and can be neglected.

We call a function $w(t)$ *smooth w.r.t. the time scale* α, if $\langle w\rangle_\alpha(t) \approx w(t)$. The idea is now to replace the rapidly varying solutions $q(t)$ of (5.3) by $\langle q\rangle_\alpha(t)$ with

$$\sqrt{||K^{-1}||} \;<\; \alpha \;<\; 1,$$

and then to seek numerical approximations to the smooth $\langle q \rangle_\alpha$ rather then to the rapidly varying $q(t)$. We call the functions $\langle q \rangle_\alpha(t)$, corresponding to solutions $q(t)$ of (5.3), the *smoothed dynamics* of (5.3) [31].

For a given scaling of the Hamiltonian H, $k_B T$, k_B the Boltzmann constant and T the temperature, becomes a quantity which we call δ and which we assume to satisfy

$$||K^{-1}|| \leq \delta < 1.$$

For example, in a molecular dynamics simulation where the potential V contains the torsion, Lennard-Jones, and electrostatic potentials, the corresponding scaling of the Hamiltonian by $\approx 5\,\text{kcal/mol}$ yields at room-temperature $\delta \approx 0.1$. Let us also introduce the quantity ϵ by

$$\epsilon := ||K^{-1}||.$$

We typically have $\epsilon \approx 0.1$ for the bond stretching and bond-angle bending potentials.

In the sequel, we will show how the application of the smoothing operator (6.1) to (5.3) leads to constrained equations of motion that approximate the smoothed dynamics of (5.3). By such an approximation we mean a differential equation in the variables Q and P such that the corresponding solutions $(Q(t), P(t))$ satisfy

$$\langle q \rangle_\alpha(t) - Q(t) = O(\delta^k \epsilon^l),$$

and

$$\langle p \rangle_\alpha(t) - P(t) = O(\delta^k \epsilon^l),$$

over bounded intervals of time with appropriate integers $k \geq 0$ and $l > 0$. Here $(q(t), p(t))$ is a solution of (5.3) and $(Q(0), P(0))$ is chosen such that $\langle q \rangle_\alpha(0) - Q(0) = O(\delta^k \epsilon^l)$ and $\langle p \rangle_\alpha(0) - P(0) = O(\delta^k \epsilon^l)$.

Let us introduce now the new variable λ by

$$\lambda := Kg(q).$$

Then the equations of motion (5.3) can be rewritten as

$$\begin{aligned} \frac{d}{dt}q &= M^{-1}p, \\ \frac{d}{dt}p &= -\nabla_q V(q) - g_q(q)^T \lambda, \qquad (6.4) \\ K^{-1}\lambda &= g(q). \end{aligned}$$

In a naive approach, one would now look for the smooth solutions of (6.4) hoping that this yields an order $O(\epsilon)$ approximation to the smoothed dynamics of (5.3). Since a smooth $p(t)$ implies that $(dp/dt)(t) = O(1)$, λ has to remain bounded as $\epsilon \to 0$. Thus

$$g(q) = O(\epsilon).$$

In the limit $\epsilon \to 0$, this suggests replacing the last equation in (6.4) by the holonomic constraint $g(q) = 0$ and we obtain the constrained equations of motion

$$\begin{aligned} \frac{d}{dt}Q &= M^{-1}P, \\ \frac{d}{dt}P &= -\nabla_q V(Q) - g_q(Q)^T \Lambda, \\ 0 &= g(Q). \end{aligned} \tag{6.5}$$

Note that the new variable Λ is implicitly defined by twice differentiating the constraint $g(Q) = 0$ with respect to time. As a result we obtain

$$\Lambda = [g_q M^{-1} g_q^T]^{-1} [g_{qq}(M^{-1}P)(M^{-1}P) - g_q M^{-1} \nabla_q V],$$

where, for notational convenience, we suppressed the variable Q in $V(Q)$, $g_q(Q)$, and the second derivative $g_{qq}(Q)$. The equations (6.5) constitute a constrained Hamiltonian system with Hamiltonian

$$H(Q,P) = \frac{P^T M^{-1} P}{2} + V(Q) + g(Q)^T \Lambda,$$

and constraints $g(Q) = 0$. The flow of (6.5) can be shown to be symplectic [23]. The solutions of (6.5) are also time-reversible.

The approximation (6.5) has been used in MD simulations to remove the bond stretching/bond-angle bending modes [34], [44]. Since $g(Q) = 0$ constrains the bond-lengths/bond-angles to their equilibrium values r_0, ϕ_0 respectively, we call $g(Q) = 0$ *hard constraints*.

We will discuss in Subsection 3 how well the constrained equations (6.5) reflect the correct smoothed dynamics of the original system (5.3). In particular, we will see that the equations (6.5) are appropriate for constraining the bond-lengths, but are doomed to fail when applied to the bond-angle bending.

6.2. Numerical aspects. Ryckaert, Ciccotti, Berendsen (1977) proposed an algorithm SHAKE [34] for the constrained equations (6.5), which can be formulated as

$$\begin{aligned} Q_{n+1} &= Q_n + \Delta t\, M^{-1} P_{n+1/2}, \\ P_{n+1/2} &= P_{n-1/2} - \Delta t\, [\nabla_q V(Q_n) + g_q(Q_n)^T \Lambda_n], \\ 0 &= g(Q_{n+1}). \end{aligned} \tag{6.6}$$

Together with the starting procedure

$$P_{1/2} = P_0 - \frac{1}{2}\Delta t\,[\nabla_q V(Q_0) + g_q(Q_0)^T \Lambda_0],$$

this can be viewed as a constrained variant of leap-frog. Strictly speaking, the scheme is implicit since nonlinear equations must be solved at each timestep, but it is partially explicit in the sense that only one $\nabla_q V$-evaluation is needed per time step. The equations can be reduced to nonlinear equations for Λ_n only:

$$0 = F(\Lambda) := g(\bar{Q}_{n+1} - \Delta t^2 M^{-1} g_q(Q_n)^T \Lambda_n), \tag{6.7}$$

where $\bar{Q}_{n+1}$ is the *unconstrained step* based on Verlet approximation. Thus the dimension of the nonlinear system is the number of constraints.

An alternative scheme, RATTLE, was proposed by Anderson in 1983. RATTLE is a modification of SHAKE such that P_{n+1} also satisfies the "hidden constraint" $g_q(Q)M^{-1}P = 0$. However, both methods lead to identical results in terms of the Q variable [22].

It has been shown [22] that (6.6) is a second-order, time-reversible, symplectic discretization of (6.5). In addition, the method preserves angular momentum [48] and can be viewed as the exact solution of a perturbed constrained Hamiltonian system [32].

But SHAKE requires an efficient technique for solving the nonlinear equations at each step. In fact, the original paper [34] describing the SHAKE discretization presented an iterative solver for the nonlinear equations, and the term SHAKE typically is used to refer to the overall procedure consisting of discretization by (6.6) together with the iterative solver used in satisfying the constraints (*coordinate resetting*). For clarity, we will refer to the iterative solver as *SHAKE iteration* and the discretization scheme as *SHAKE discretization*.

In SHAKE iteration, we cycle through the constraints one by one, adjusting one multiplier at each step. Let g_i be the ith component of g, and let G_i represent the ith row of g_q, then the iteration proceeds as follows: First, we initialize $Q = \bar{Q}_{n+1} = Q_n + \Delta t\, M^{-1}[P_{n-1/2} - \Delta t\, \nabla_q V(Q_n)]$ which corresponds to the initial guess $\Lambda_n = 0$. Next, we compute the correction of the ith component $\Delta\Lambda^i$ needed to satisfy the ith linearized constraint equation:

$$\Delta\Lambda^i \leftarrow \frac{g_i(Q)}{G_i(Q)M^{-1}G_i(Q_n)},$$

and update Q by

$$Q \leftarrow Q - \Delta t^2 M^{-1} G_i(Q_n)^T \Delta\Lambda_i.$$

This cycle should be repeated until all constraint residuals $g_i(Q)$ are below a prescribed tolerance.

As shown in [4], SHAKE can be viewed as a 1-step nonlinear Gauss-Seidel-Newton iteration in the framework developed by Ortega and Rheinboldt [26]. Local convergence is guaranteed precisely when standard Gauss-Seidel iteration converges for the linear system of equations

$$A_n x = b,$$

where $A_n = g_q(Q_{n+1})M^{-1}g_q(Q_n)$. For unconstrained approximations $\bar{Q}_{n+1}$ sufficiently close to Q_{n+1} (i.e., for sufficiently small timestep Δt), SHAKE iteration converges with asymptotic rate of convergence approximately

$$\rho((D+L)^{-1}U),$$

where $A_n = L + D + U$ is the splitting of A_n into strictly lower triangular, diagonal, and strictly upper triangular parts.

The matrix $B_n = g_q(Q_n)M^{-1}g_q(Q_n) = A_n + O(h)$ has a very special structure which can be easily analyzed. In the current setting where g consists only of bond length constraints, it can be shown that the matrix is nearly constant along solutions, so also the matrix A_n is nearly constant along solutions. Thus SHAKE iteration will converge with a similar rate at each timestep. It can also be seen that SHAKE iteration will converge very slowly for chainlike structures unless successive bonds are nearly perpendicular, and degraded convergence would be anticipated for molecules with high connectivity.

A natural improvement to Gauss-Seidel iteration is based on the use of overrelaxation. In this method, we update the iterate at each step by forming a weighted combination of the previous iterate and the G-S step: in essence, we exaggerate each G-S correction by a *relaxation parameter* ω, changing the update of Λ from

$$\Lambda^i \leftarrow \Lambda^i + \Delta\Lambda^i$$

to

$$\Lambda^i \leftarrow \Lambda^i + \omega\Delta\Lambda^i.$$

The parameter ω can be a fixed value obtained through some preliminary experiment, or it can be obtained automatically during the integration by a simple adaptive algorithm [4].

This adjustment is essentially free of cost. And it can mean substantial speedups in coordinate resetting. Since the method is just an alternative nonlinear equation solver to SHAKE iteration, the converged numerical solution will be identical (up to rounding error) to that computed by SHAKE iteration.

A second alternative to SHAKE iteration was also considered in [4]: we could use a Newton iteration, or variant thereof, to compute successive updates to the vector Λ_n. In particular, we could apply a true Newton iteration to the original nonlinear equations and use *sparse matrix techniques* to solve the resulting linear equations. This turns out to be somewhat expensive for most problems because the factorization of a matrix—even a sparse matrix—is relatively costly. A better alternative is to use an adaptive Newton iteration based on a factorization of the symmetric matrix B_n which is close to A_n and changes little from step to step.

These methods iteratively improve the whole vector of multipliers:

$$\Delta\Lambda \leftarrow R^{-1}F(\Lambda),$$

where for the true Newton iteration (NIP) we have $R = g_q(Q)M^{-1}g_q(Q_n)^T$ with Q the latest approximation to Q_{n+1}. For the symmetric, adaptive iteration we have $R = GM^{-1}G^T$, with $G = g_q(Q_k)$ for some previous step k (G is updated as needed for convergence). The Newton-based techniques are robust as long as the initial data lie on or very near the constraint manifold; in practice, SHAKE (or SOR) should probably be used at the first timestep to insure that the initial iterate is very close to the constraint manifold.

The techniques have been integrated into the CHARMM software package and successfully applied to several examples: C_{60}, a box of water, Myoglobin, BPTI; detailed results are reported in [4]. A MATLAB "toolbox" has been developed by Barth for studying the algorithms [3].

Roughly speaking, we can summarize the results as follows: (1) the use of an SOR parameter can improve SHAKE iteration convergence by a factor of two to three in protein dynamics simulations for no additional cost or loss of robustness, (2) the adaptive scheme for determining optimal ω in [4], though not robust, does demonstrate the feasibility of computing the SOR parameter, and (3) the symmetric adaptive Newton method may show improvement over SHAKE iteration at large step size Δt for problems with high bond connectivity, e.g. the *buckminsterfullerene* C_{60}.

Some higher-order extensions of SHAKE/RATTLE have been proposed by Reich [32] and independently by Jay [20]. However, competitive practical implementations for molecular dynamics applications have not been demonstrated.

6.3. Analytical aspects II. Let us see now how well the constrained equations (6.5) reflect the correct smoothed dynamics of the original system (5.3). To do so, we introduce (for theoretical purposes) local coordinates (q_1, q_2) by

$$\begin{aligned} q_1 &= g(q), \\ q_2 &= b(q), \end{aligned}$$

where $b(q)$ is a vector valued function such that $g_q(q)M^{-1}b_q(q)^T = 0$ and the composed matrix $[g_q(q)^T\ b_q(q)^T]$ is invertible. The existence of such a coordinate system follows, at least locally, from the Frobenius theorem [7].

To obtain a transformed system that remains Hamiltonian, we define *conjugate momenta* [23] by

$$[g_q(q)^T \, b_q(q)^T] \begin{bmatrix} p_1 \\ p_2 \end{bmatrix} = p,$$

which results in the Hamiltonian

$$H = \frac{p_1^T g_q \, M^{-1} g_q^T p_1}{2} + \frac{p_2^T b_q \, M^{-1} b_q^T p_2}{2} + V + \frac{q_1^T K q_1}{2}. \tag{6.8}$$

The equations of motion are now given by

$$\begin{aligned} \frac{d}{dt} q_1 &= g_q \, M^{-1} g_q^T p_1, \\ \frac{d}{dt} p_1 &= -\nabla_{q_1} V - K q_1 - \nabla_{q_1} \frac{p_1^T g_q \, M^{-1} g_q^T p_1 + p_2^T b_q \, M^{-1} b_q^T p_2}{2} \end{aligned} \tag{6.9}$$

and

$$\begin{aligned} \frac{d}{dt} q_2 &= b_q \, M^{-1} b_q^T p_2, \\ \frac{d}{dt} p_2 &= -\nabla_{q_2} V - \nabla_{q_2} \frac{p_1^T g_q \, M^{-1} g_q^T p_1 + p_2^T b_q \, M^{-1} b_q^T p_2}{2}, \end{aligned}$$

where, for notational convenience, we suppressed the arguments in the mappings $V(q_1, q_2)$, $g_q(q_1, q_2)$, and $b_q(q_1, q_2)$.

The constrained equations (6.5) are obtained by replacing the first set of equations by $Q_1 = P_1 = 0$. In terms of the variables (Q_2, P_2), this results in

$$\begin{aligned} \frac{d}{dt} Q_2 &= b_q(Q_2) \, M^{-1} b_q(Q_2)^T P_2, \\ \frac{d}{dt} P_2 &= -\nabla_{q_2} V(Q_2) - \nabla_{q_2} \frac{P_2^T b_q(Q_2) \, M^{-1} b_q(Q_2)^T P_2}{2}. \end{aligned}$$

By applying the smoothing operator (6.1) to the second equation in (6.9), we obtain

$$\begin{aligned} \langle \frac{d}{dt} p_1 \rangle_\alpha &= -\langle \nabla_{q_1} V \rangle_\alpha - K \langle q_1 \rangle_\alpha - \\ &\quad \langle \nabla_{q_1} \frac{p_1^T g_q M^{-1} g_q^T p_1 + p_2^T b_q M^{-1} b_q^T p_2}{2} \rangle_\alpha, \end{aligned} \tag{6.10}$$

since each term in (6.10) has to remain bounded as $\epsilon \to 0$ [31], it follows that

$$\langle q_1 \rangle_\alpha = O(\epsilon).$$

Furthermore, equipartition of energy [27] implies [31]

$$\langle q_1^T q_1 \rangle_\alpha = O(\delta \, \epsilon).$$

Now, by Taylor expansion of V about $\langle q_1\rangle_\alpha$ and applying the smoothing operator (6.1) to this expansion, one can show [31] that

$$\langle V(q_1, q_2)\rangle_\alpha \;=\; V(\langle q_1\rangle_\alpha, q_2) + O(\langle q_1^T q_1\rangle_\alpha).$$

(Here we used facts like $\langle q_2\rangle_\alpha \approx q_2$, $\langle V(\langle q_1\rangle_\alpha, q_2)\rangle_\alpha \approx V(\langle q_1\rangle_\alpha, q_2)$, $\langle \Delta q_1\rangle_\alpha := \langle q_1 - \langle q_1\rangle_\alpha\rangle_\alpha \approx 0$, and $\langle \Delta q_1^T \Delta q_1\rangle_\alpha = \langle q_1^T q_1\rangle_\alpha + O(\epsilon^2)$.) Now, per degree of freedom, we have $\langle \phi^2\rangle_\alpha \approx 0.01\,\text{deg}^2$ for the bond-angle bending and $\langle r^2\rangle_\alpha \approx 0.001\mathring{A}$ for the bond stretching. Thus

$$\langle V(q_1, q_2)\rangle_\alpha \approx \; V(\langle q_1\rangle_\alpha, q_2),$$

and

$$\langle \nabla_{q_2} V(q_1, q_2)\rangle_\alpha \;\to\; \nabla_{q_2} V(0, q_2),$$

as $\epsilon \to 0$, and a similar result holds for the term

$$\langle \nabla_{q_2} \frac{p_2^T b_q M^{-1} b_q^T p_2}{2}\rangle_\alpha.$$

However, by equipartition of energy, we have [31]

$$\langle \frac{p_1^T g_q M^{-1} g_q^T p_1}{2}\rangle_\alpha \;\approx\; \frac{m}{2}\delta,$$

where m is the number of constraints. Thus the force term

$$\langle \nabla_{q_2} \frac{p_1^T g_q M^{-1} g_q^T p_1}{2}\rangle_\alpha$$

is not necessarily small even in the limit $\epsilon \to 0$. It has been pointed out before [14] that the appropriate correction to the constrained dynamics (6.5) is given by the Fixman potential

$$V_F \;=\; \frac{\delta}{2} \ln \det g_q M^{-1} g_q^T. \tag{6.11}$$

Thus an $O(\epsilon)$ approximation to the correct smoothed dynamics of (5.3) is given by the modified constrained equations

$$\begin{aligned} \frac{d}{dt}Q &= M^{-1}P, \\ \frac{d}{dt}P &= -\nabla_q V(Q) - \nabla_q V_F(Q) - g_q(Q)^T \Lambda, \\ 0 &= g(Q). \end{aligned} \tag{6.12}$$

The variable Λ satisfies now

$$\Lambda = [g_q M^{-1} g_q^T]^{-1}\, [g_{qq}(M^{-1}p)(M^{-1}p) - g_q M^{-1}[\nabla_q V + \nabla_q V_F]]. \tag{6.13}$$

For large systems, the computation of the Fixman potential (6.11) becomes rather time-consuming. However, a few simplifications seem possible: (i) In case that we constrain only the bond-lengths, the matrix $g_q\, M^{-1} g_q^T$ can be shown to be a function of the bond-angles ϕ and the bond-lengths r which we keep constant. Since the corresponding force $\nabla V_F(\phi)$ amounts to a few $k_B T\,\mathrm{deg}^{-1}$ [45] and is thus small compared to the contributions stemming from the bond-angle bending potentials, the Fixman potential can safely be ignored in this situation. (ii) The situation differs if we include bond-angle constraints. In this case the matrix $g_q\, M^{-1} g_q^T$ depends on the torsion angles ψ (and, of course, on the bond-lengths and the bond-angles which, however, we keep constant). The Fixman potential $V_F(\psi)$ can now no longer assumed to be small compared to the torsion potentials [45].

To keep the computational costs low, we suggest [33] to approximate the Fixman potential by

$$V_F(\psi) \;\approx\; \sum_j V_F^j(\psi_j),$$

where $V_F^j(\psi_j)$ stands for the Fixman potential due to the three bond-length and two bond-angle constraints defining the torsion angle ψ_j. (In this case, each matrix $g_q\, M^{-1} g_q^T$ becomes a 5×5 matrix that depends only on ψ_j.). This approximation is justified by the fact that

$$\frac{\partial^2}{\partial \psi_i\, \partial \psi_j} V_F(\psi) \;\approx\; 0,$$

for all $i \neq j$ [33].

Remark. The Fixman potential can easily be derived in the following way [31]: Let $Q(q)$ be an orthogonal matrix such that $Q^T g_q\, M^{-1} g_q^T Q$ is a diagonal matrix $D(q)$ with entries $d_{ii}(q)$. Let π_i denote the ith entry in the m dimensional vector $Q(q)^T p_1$. Then

$$\langle \nabla_{q_2} \frac{p_1^T g_q\, M^{-1} g_q^T p_1}{2} \rangle_\alpha \;\approx\; \sum_i \frac{\langle \pi_i^2 \rangle_\alpha}{2} \nabla_{q_2} \langle d_{ii} \rangle_\alpha,$$

and, by equipartition of energy,

$$\langle \pi_i^2 \rangle_\alpha \;\approx\; \frac{\delta}{\langle d_{ii} \rangle_\alpha}.$$

Thus we obtain

$$\sum_i \frac{\langle \pi_i^2 \rangle_\alpha}{2} \nabla_{q_2} d_{ii}(Q) \;\approx\; \frac{\delta}{2} \nabla_{q_2} \ln \det D(Q),$$

which, in terms of the original matrix $g_q\, M^{-1} g_q^T$, leads to the potential (6.11).

The constrained equations (6.12) yield correct results only in the limit $\epsilon \to 0$. While this condition is approximately satisfied for the force constants corresponding to bond stretching, one has to take finite size effects of ϵ into account when looking at bond-angle bending [44]. In other words, the estimate

$$\langle q_1 \rangle_\alpha \;=\; 0$$

has to be replaced by a more accurate one. As suggested in [8],[31], this can be achieved by using (6.10) together with $\langle dp_1/dt \rangle_\alpha = O(\epsilon)$ (which follows from the fact that $\langle q_1 \rangle_\alpha = O(\epsilon)$ and $\langle p_1 \rangle_\alpha = O(\epsilon)$) to derive

$$\langle q_1 \rangle_\alpha \;=\; -K^{-1} \nabla_{q_1} V(\langle q_1 \rangle_\alpha, q_2) + O(\delta\,\epsilon).$$

Here we also used that

$$\nabla_{q_1} \frac{p_1^T g_q M^{-1} g_q^T p_1 + p_2^T b_q M^{-1} b_q^T p_2}{2} \;=\; O(\delta),$$

by equipartition of energy [31] and that $\epsilon^2 \leq \delta\,\epsilon$.

The corresponding constrained equations of motion are now given by

$$\begin{aligned}
\frac{d}{dt}Q &= M^{-1}P, \\
\text{(6.14)}\quad \tfrac{d}{dt}P &= -\nabla_q V(Q) - \nabla_q V_F(Q) - g_q(Q)^T \tilde{\Lambda}, \\
0 &= g(Q) + K^{-1}\left[g_q(Q) M^{-1} g_q(Q)^T\right]^{-1} g_q(Q) M^{-1} \nabla_q V(Q).
\end{aligned}$$

One can show that this formulation provides an $O(\delta\epsilon)$ approximation [31].

In contrast to the formulation (6.5), the system (6.14) can no longer be derived from a Hamiltonian principle. However, the solutions of (6.14) are still time-reversible. For a Hamiltonian reformulation of (6.14) see [31].

As opposed to the *hard constraint* $g(Q) = 0$, we call the constraint $\tilde{g}(Q) = 0$ with

$$\tilde{g}(Q) \;:=\; g(Q) + K^{-1}\left[g_q(Q) M^{-1} g_q(Q)^T\right]^{-1} g_q(Q) M^{-1} \nabla_q V(Q),$$

a *soft constraint.*

Finally a few words on the discretization of (6.14). Any constrained Hamiltonian system of the form (6.5) can efficiently be discretized by (6.6). The same scheme can also be applied to the constrained system (6.14) with soft constraints. This time we obtain

$$\begin{aligned}
Q_{k+1} &= Q_k + \Delta t\, M^{-1} P_{k+1/2}, \\
\text{(6.15)}\quad P_{k+1/2} &= P_{k-1/2} - \Delta t\left[\nabla_q V(Q_k) + \nabla_q V_F(Q_k) + g_q(Q_k)^T \tilde{\Lambda}_k\right], \\
0 &= \tilde{g}(Q_{k+1}).
\end{aligned}$$

The method is time-reversible but not symplectic.

The scheme (6.15) is computationally expensive. If the nonlinear system of equations in $\tilde{\Lambda}_k$ is solved by Newton's method, it requires the computation of $\nabla_q V(q)$ in each Newton iteration. To avoid this, one can use [8] the following modification

$$
\begin{aligned}
\bar{Q}_{k+1} &= \bar{Q}_k + \Delta t\, M^{-1} P_{k+1/2}, \\
P_{k+1/2} &= P_{k-1/2} - \Delta t\, [\nabla_q V(Q_k) + \nabla_q V_F(Q_k) + g_q(\bar{Q}_k)^T \tilde{\Lambda}_k^1], \qquad (6.16) \\
0 &= g(\bar{Q}_{k+1}),
\end{aligned}
$$

and

$$
\begin{aligned}
Q_k &= \bar{Q}_k + M^{-1} g_q(\bar{Q}_k)^T \tilde{\Lambda}_k^2, \\
0 &= g(Q_k) + K^{-1} [g_q(\bar{Q}_k) M^{-1} g_q(\bar{Q}_k)^T]^{-1} g_q(\bar{Q}_k) M^{-1} \nabla_q V(\bar{Q}_k).
\end{aligned}
\qquad (6.17)
$$

This scheme is time-reversible and differs from the method (6.15) by a term of order $O(\epsilon^2)$ [8]. Now the force field $\nabla_q V(q)$ has to be computed only twice per integration step. Both nonlinear systems in $\tilde{\Lambda}_k^1$ and $\tilde{\Lambda}_k^2$ can be solved by the same methods as proposed for (6.6) in Section 6.2.

A further simplification seems possible [8]: If we split the potential into $V = V^{\text{hard}} + V^{\text{soft}}$, one can replace the constraint function $\tilde{g}$ by

$$
\tilde{g}(Q) \;\approx\; g(Q) + [g_q(Q) M^{-1} g_q(Q)^T]^{-1}\, g_q(Q)\, M^{-1} \nabla_q V^{\text{hard}}(Q),
$$

where V^{hard} would include perhaps only the Lennard-Jones and the torsion potentials.

Example. In this example we consider a four-bead-three-bond structure [28] where the structure is restricted to move in a finite volume by the potential

$$
V_r(q) \;=\; \sum_i K_r \left(\frac{r_i}{\sigma}\right)^6 .
$$

Here r_i denotes the distance of each of the four beads to the origin, $\sigma = 2$, and $K_r = 0.1$. We set the mass of all four beads equal to $m = 1$ and choose $r_0 = 1$ as the equilibrium bond-length and $\phi_0 = 90^o$ as the equilibrium bond-angle. For simplicity, we included no torsion potential. The force constant for the harmonic bond-angle bending potentials was $K_a = 60\,\text{deg}^{-2}$ and $K_b = 600$ for the corresponding bond stretching potentials. Note that these values correspond to force constants typically found in molecular dynamics simulations for a scaling of the Hamiltonian by approximately 1.0kacl/mol [44].

We started the structure from its equilibrium position with the initial velocities in x-direction equal to $p_x = 1.0$. The impact of the structure clashing with the potential-wall V_r can be seen in Figure 6.1. We computed the trajectory of one bond-angle (plotted as $\cos(\phi)$) for the unconstrained formulation and compared this trajectory with the one obtained by constraining the bond-lengths by hard constraints and the bond-angles by flexible constraints.

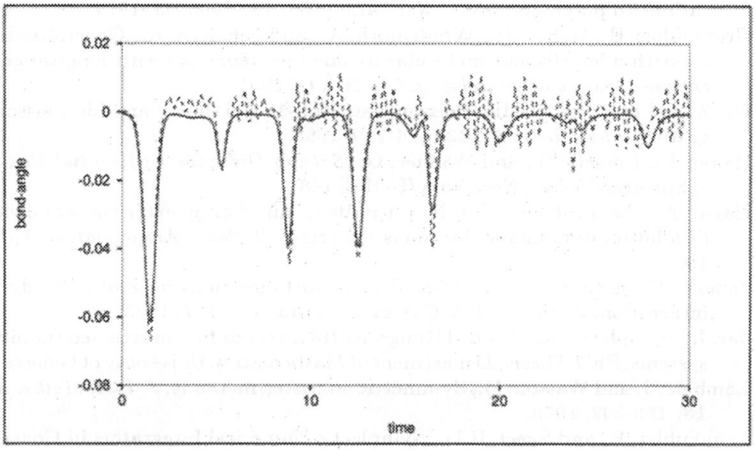

FIG. 6.1. ***Trajectory of one of the bond-angles for the unconstrained formulation (dotted line) compared to the one for the formulation with hard constraints on the bond-lengths and flexible constraints on the bond-angles (solid line).***

REFERENCES

[1] Allen, M.P. and Tildesley, D.J., *Computer Simulation of Liquids*, Oxford Science, New York, 1987.

[2] Andersen, H.C., Rattle: a 'velocity' version of the shake algorithm for molecular dynamics calculations, *J. Comput. Phys.* **52**, 24–34, 1983.

[3] Barth, E., A user's guide to MBLab, *1993 Matlab Conference Proceedings*, electronic publication available via anonymous ftp from ftp.mathworks.com.

[4] Barth, E., Kuczera, K., Leimkuhler, B., and Skeel, R.D., Algorithms for constrained molecular dynamics, *J. Comput. Chem.*, **16**, 1192–1209, 1995.

[5] Beeman, D., Some multistep methods for use in molecular dynamics calculations, *J. Comput. Phys.* **20**, 130–139, 1976.

[6] Biesiadecki, J.J. and Skeel, R.D., Dangers of multiple-time-step methods, *J. Comput. Phys.* **109**, 318–328, 1993.

[7] Boothby, W.M., *An Introduction to Differentiable Manifolds and Riemannian Geometry*, 2nd edition, Academic Press, New York, 1986.

[8] Brooks, B.R., Zhou, J., and Reich, S., Elastic molecular dynamics with flexible constraints, in preparation.

[9] Buneman, O., Time-Reversible Difference Procedures, *J. Comput. Phys.* **1**, 517–535, 1967.

[10] Calvo, M.P. and Sanz-Serna, J.M., The development of variable step symplectic integrators, with application to the two-body problem, *SIAM J. Sci. Comp.* **14**, 936-952, 1993.

[11] Chawla, M.M., On the order and attainable intervals of periodicity of explicit Nystrom methods for $y'' = f(t,y)$, *SIAM J. Numer. Anal.* **22**, 127–131, 1985.

[12] Daubechies, I., *Ten Lectures on Wavelets*, SIAM, Philadelphia, 1992.

[13] Dragt, A., Neri, F., Rangarajan, G., Douglas, D., Healy, L., and Ryne, R., "Lie algebraic treatment of nonlinear beam dynamics", *Ann. Rev. Nucl. Part. Sci.* **38**, 455-496, 1988.

[14] Fixman, M., Classical statistical mechanics and constraints: A theorem and applications to polymers., *Proc. Nat. Acad. Sci.* **71**, 3050–53, 1974.

[15] Grubmüller, H., Heller, H., Windemuth, A., and Schulten, K., Generalized Verlet algorithm for efficient molecular dynamics simulations with long-range interactions, *Molecular Simulation* **6**, 121–142, 1991.

[16] Ge, Z. and Marsden, J., Lie-Poisson Hamilton-Jacobi theory and Lie-Poisson integrators, *Phys. Lett. A* **133**, 134-139, 1988.

[17] Hairer, E., Norsett, S., and Wanner, G., *Solving Ordinary Differential Equations, I*, Springer-Verlag, New York/Berlin, 1987.

[18] Hayli, A., Le problème des N corps dans un champ extérieur application a l'évolution dynamique des amas ouverts-I, *Bulletin Astronomique* **2**, 67–89, 1967.

[19] Janezic, D. and Orel, B., Implicit Runge-Kutta methods for molecular dynamics integration, *J. Chem. Inf. Comput. Sci.* **33**, 252–257, 1993.

[20] Jay, L., Symplectic partitioned Runge-Kutta methods for constrained Hamiltonian systems, PhD Thesis, Department of Mathematics, University of Geneva, 1993.

[21] Lambert, J. and Watson, D., Symmetric multistep methods, *J. Inst. Maths. Applic.* **18**, 189-202, 1976.

[22] Leimkuhler, B. and Skeel, R.D., Symplectic Numerical Integrators in Constrained Hamiltonian Systems, *J. Comput. Phys.* **112**, 117–125, 1994.

[23] Marsden, J.R. and Ratiu, T., *An Introduction to Mechanics and Symmetry*, Springer-Verlag, New York, 1994.

[24] McCammon, J.A. and Harvey, S.C., *Dynamics of Proteins and Nucleic Acids*, Cambridge University Press, Cambridge, 1987.

[25] Okunbor, D. and Skeel, R.D., Canonical numerical methods for molecular dynamics simulations, *J. Comput. Chem.* **15**, 72–79, 1994.

[26] Ortega, J.M. and Rheinboldt, W.C., *Iterative Solution of Nonlinear Equations in Several Variables*, Academic Press, New York, 1970.

[27] Pathria, P.K., *Statistical Mechanics*, Pergamon Press, Oxford, 1972.

[28] Pear, M.R. and Weiner, J.H., Brownian dynamics study of a polymer chain of linked rigid bodies, *J. Chem. Phys.* **71**, 212–224, 1979.

[29] Peskin, C.S. and Schlick, T., Molecular dynamics by the backward Euler method, *Comm. Pure and Appl. Math.* **42**, 1001–1031, 1989.

[30] Quinlan, G.D. and Tremaine, S., Symmetric multistep methods for the numerical integration of planetary orbits, *Astron. J.* **100**, 1694–1700, 1990.

[31] Reich, S., Smoothed dynamics of highly oscillatory Hamiltonian systems, Physica D 89, 28–42, 1995.

[32] Reich, S., Symplectic integration of constrained Hamiltonian systems by composition methods, *SIAM J. Numer. Anal.*, **33**, 475–491, 1996.

[33] Reich, S., Molecular dynamics with soft constraints, Technical Report UIUC-BI-TB-94-12, Beckman Institute, Urbana, 1994.

[34] Ryckaert, J.P., Ciccotti, G., and Berendsen, H.J.C., Numerical integration of Cartesian equations of motion of a system with constraints: Molecular dynamics of n-alkanes, *J. Comput. Phys.* **23**, 327–342, 1977.

[35] Sanders, J.A. and Verhulst, F., *Averaging Methods in Nonlinear Dynamical Systems*, Springer-Verlag, New York, 1985.

[36] Sanz-Serna, J.M. and Calvo, M.P., *Numerical Hamiltonian Problems*, Chapman and Hall, London, 1994.

[37] Scully, J.L. and Hermans, J., Multiple time steps: Limits on the speedup of molecular dynamics simulations in aqueous systems, *Mol. Simulation* **11**, 67–77,

1993.

[38] Simo, J., Rarnow, N., and Wong, K.K., Exact energy-momentum conserving algorithms and symplectic schemes for nonlineaer dynamics, *Comp. Meth. Appl. Mech. Eng.* **1**, 1528–1544, 1992.

[39] Skeel, R.D., Variable step size destabilizes the Stormer/leapfrog/Verlet method, *BIT* **33**, 172–175, 1993.

[40] Skeel, R.D. and Biesiadecki, J.J., Symplectic integration with variable stepsize, *Annals of Numer. Math.* **1**, 191–198, 1994.

[41] Strang, G., On the construction and comparison of difference schemes, *SIAM J. Numer. Anal.* **5**, 506–517, 1968.

[42] Streett, W.B., Tildesley, D.J., and Saville, G., Multiple time step methods in molecular dynamics, *Mol. Phys.* **35**, 639–648, 1978.

[43] Tuckerman, M., Berne, B.J., and Martyna, G.J., Reversible multiple time scale molecular dynamics, *J. Chem. Phys.* **97**, 1990–2001, 1992.

[44] van Gunsteren, W.F. and Karplus, M., Effects of constraints on the dynamics of macromolecules, *Macromolecules* **15**, 1528–1544, 1982.

[45] van Gunsteren, W.F., Constrained dynamics of flexible molecules, *Molecular Physics* **40**, 1015–1019, 1980.

[46] Wisdom, J. and Holman, M., Symplectic Maps for the N-body problem, *Astron. J.* **102**, 1528–1538, 1991.

[47] Zhang, G. and Schlick, T., LIN: A new algorithm to simulate the dynamics of biomolecules by combining implicit-integration and normal mode techniques, *J. Comput. Chem.* **14**, 1212–1233, 1993.

[48] Zhang, M.Q. and Skeel, R.D., Symplectic integrators and the conservation of angular momentum, *J. Comput. Chem.* **16**, 365–369, 1995.

ON THE PARALLELIZATION OF CHARMM ON THE CM-5/5E

JILL P. MESIROV*, PABLO TAMAYO†, AND ROBERT J. NAGLE‡

Abstract. We describe a new port of CHARMM to the CM-5 supercomputer. This port is based on the original B-H parallelization of CHARMM[2] for MIMD machines but it has an improved communications library.

1. Introduction. CHARMM (Chemistry at HARvard Macromolecular Mechanics)[1] is a flexible multi-functional molecular simulation program based on semi-empirical energy functions. The program has the capability to apply different molecular modeling and simulation methods such as energy minimization, normal modes, *ab initio* and molecular dynamics with or without constraints. It allows for the analysis of many equilibrium static and dynamic properties of molecular configurations such as energy comparisons, time series, correlation functions, kinetic and thermodynamic properties. The energy functions or "force-fields" are based on lumping nuclei and electrons into atom-like particles. Bonded interactions and non-bonded interactions are considered. The program can read and store parameters, coordinates and additional molecular information or model-build the structure of interest. Typical applications of CHARMM are simulations of macromolecular systems with large molecules and assemblies such as biopolymers, peptides, polysaccharides, nucleic acids, and lipids in vacuum, implicit, or explicit solvent environments.

Since its introduction more than ten years ago CHARMM has been widely used by the academic community and its development continues with the addition of new capabilities and features.

2. The B-H parallelization of CHARMM. In 1992, B. Brooks and M. Hodošček developed a version of CHARMM for MIMD computers[2]. This version of CHARMM had the following features,

- All features of CHARMM were supported; however, not all of them were parallelized.
- The Molecular Dynamics and Energy Minimization were parallelized by distributing the work load across processors.
- The source code was compatible with the standard version of CHARMM for serial machines and workstations.
- The parallelization scheme was general enough to work on most MPP machines and clusters of workstations.

The parallelization strategy taken was based on the following principles: i) coordinates are replicated in each processing node; ii) the non-

* Computer Science Department, Boston University, Boston, MA 02215.
† Thinking Machines Corp., Cambridge, MA 02142.
‡ Chemistry Department, Harvard University, Cambridge, MA 02138.

bonded list is generated by a hash coding procedure that distributes interacting pairs across processors (the load imbalance is less than 10%); iii) the integration of the equations of motion and other calculations are made by each processor working on a contiguous block of atom coordinates (each block is nearly equal in size but the boundaries are chosen in such way that the atom groups are not divided); and iv) communications between nodes was done by vector distributed broadcast and sum operations provided by a communications library.

The communications library[2] supports the following primitives,

- *Vector Distributed Global Sum.*- This is used to accumulate force components, which are distributed across the nodes, on the corresponding node responsible for the integration of the equations of motion for each particular block of atoms.
- *Distributed Vector Broadcast.*- This is used to broadcast new coordinates, velocities and other parameters across all nodes. It is a type of all-to-all broadcast where each node contributes a different block of data.
- *Vector Global Sum.*- Internally this function is implemented by combining the distributed sum and broadcast operations.
- *Global Input/Output and Timing Functions.*- These are used to read parameters, print results and measure the performance of the program and are not time-critical operations.

The algorithm to implement the distributed sum and broadcast is based on the global combine operations as discussed by Fox et al. [3] and van de Geijn[4]. This approach is based on using minimal spanning trees to exchange power-of-two subsets of the data vector in such a way that the total times are nearly constant for large number of processors and grow linearly with the size of the vector.

The B-H parallelization of CHARMM was initially implemented on the Intel Delta and was subsequently ported to the CM-5 (as part of release C23F3). However, the CM-5 port had a number of limitations and performance problems. The main performance problem was high communications overhead and limited scaling with larger numbers of processors. This motivated us to look for improvements.

3. Improving the communications library. A new implementation of the communications library was done by using *active messages*[5,6]. Active messages allow for the setting up of "communications ports" that perform fast data transfers across processors without the usual overhead of point-to-point message-passing communications. We sought to reduce the initial overhead, increase the bandwidth by overlapping send and receive operations, and ultimately improve the scaling behavior so the communications time would not rise significantly when more processors were used. This goal was achieved and the communications performance improved by about 50%. To test the new communications code and measure its performance outside of CHARMM, we made a stand-alone communications benchmark program. Based on the experience with this program, we made a new version of the CHARMM communications library which was later integrated with the rest of CHARMM.

The following table shows bandwidth results from the benchmark program running the old and new implementations of the library. Notice that typical vector lengths used in CHARMM are in the range 32-256 Kbytes.

TABLE 3.1
Performance of communications library operations on a 32-node CM-5E.

Operation	Vector length [bytes]	Bandwith [Mbytes/sec] Original	New
Distributed Vector Broadcast	4096	3.325	6.915
	8192	3.380	6.884
	16384	3.663	7.036
	32768	3.765	7.145
	65536	3.981	5.330
	131072	4.015	6.811
	262144	4.035	6.545
Vector Distributed Global Sum	1024	1.299	4.202
	2048	1.960	5.100
	4096	2.606	5.713
	8192	3.129	6.056
	16384	3.267	5.736
	32768	3.236	5.065
	65536	3.280	4.924
	131072	3.243	4.716
	262144	2.944	4.480
Vector Global Sum	1024	0.690	2.324
	2048	1.047	2.819
	4096	1.396	3.144
	8192	1.672	3.335
	16384	1.770	3.227
	32768	1.805	3.000
	65536	1.802	2.901
	131072	1.810	2.751
	262144	1.770	2.759

4. Improving the force computation: cache utilization, flop rates, and optimizing compilers. We performed a number of minor optimizations that produced a modest speed-up of the force subroutine. The code is executed on the CM-5 and CM-5E nodes which have Cypress and Viking SPARC processor sets respectively. To study cache utilization we made a self-contained kernel of the force calculation which runs isolated from CHARMM and reads the state of a simulation from a file (previously saved from a CHARMM run). We tried a few changes, including reordering and regrouping of instructions, to the innermost loops with the intent of improving superscalar performance. They produced only very small gains. The only change that produced a somewhat significant speed-up was the introduction of temporary arrays: at the beginning of the force subroutine we copied the value of the coordinates x, y, z, the electric charge and the four van der Waals pair coefficients to local tables in such way that when a given element is fetched the other variables, which will be required later in the loop, will already be loaded in cache memory. These temporary tables improved cache utilization and increased the speed of the force calculation by a factor of 1.18 by requiring a small increase in memory. The version included in release C23F4 did not include the use of these tables.

We also studied the flop rates inside the critical force loop. These are flop rates inside the innermost kernel of the non-bonded force subroutine (the loop over j neighbors for a given i particle). The following diagram shows the section of code analyzed.

```
Non-bonded forces subroutine (ENBFS8):

 Loop over Atom I
    Loop over (interacting) Atom J
         force calculation
    end do
    accumulation of results
 end do
```

The instruction count for the innermost loop, which accounts for 80 % of the time, is as follows:

Number of arithmetic instructions	=	59,
Number of indirections (table look-ups)	=	17,
Number of comparisons (if statements)	=	7,
Total	=	83,

and the times for one time step, i.e., one call to the subroutine with actual CHARMM data, are:

CM-5 node (compiler = f77 -O4)	=	4.52 secs,
CM-5 node (compiler = apf77 -O5)	=	3.04 secs,
CM-5E node (compiler = f77 -O4)	=	1.56 secs,
CM-5E node (compiler = apf77 -O5)	=	0.88 secs.

These numbers exclude the table optimization mentioned in the previous paragraph. Knowing the times and instruction counts we can compute the flop rates. The total number of instructions in one time step (209,066 passes over the j loop for this example) is $209,066 \times 59 = 12.33 \times 10^6$ for arithmetic operations and $209,066 \times 83 = 17.36 \times 10^6$ for all instructions. The flops rates are:

Mflop rates per node for the innermost loop

Machine/compiler	Mflops/node arithmetic	all
CM-5/f77 -O4	3.50	4.80
CM-5/apf77 -O5	4.10	7.10
CM-5E/f77 -O4	10.0	13.9
CM-5E/apf77 -O5	17.8	24.7

This table shows flop rates using the standard f77 (-O4) compiler and also Apogee's apf77 (-O5) optimizing compiler[7]. As can be seen in these numbers, the effect of using an aggressive optimizing compiler produces a very significant speed-up of about 1.8 in the force loop. The improvement of the full code using apf77 is about 1.4 with respect to the standard compiler f77.

5. Overall timings and performance. Our best times for the improved version are between 1.4 and 2 times faster than the original version. The new communications routines improve the overall performance by about 15-20%. The rest comes mainly from the use of the optimizing compiler. The following tables show new and old timings for one of the standard CHARMM benchmarks: Carboxy Myoglobin (MbCO) + 3830 water molecules (14026 atoms).

TABLE 5.1

Original Timings (C23F3) for Carboxy Myoglobin (MbCO) simulation 1000 steps, cutoff = 12-14 A.

Computer	nodes	CPU time [hours] Tot.	force	comm.	Memory [Mb/node]	μsecs per atom-update	Speed-up vs orig code	4 nodes
CM-5	4	12.2	93%	0.4%	24.6	3234	1	1
CM-5	32	1.74	87%	4.3%	16.6	447	1	7.2
CM-5	64	1.04	76%	7.7%	16.5	267	1	12.1
CM-5	128	0.61	71%	14%	16.5	155	1	20.9
CM-5	256	0.42	58%	19%	16.5	108	1	29.9
CM-5E	4	4.07	90%	0.8%	24.6	1162	1	1
CM-5E	32	0.59	80%	9%	16.6	151	1	7.7
CM-5E	64	0.35	68%	19%	16.5	89	1	13.1

TABLE 5.2

Improved version's timings for the Carboxy Myoglobin (MbCO) simulation 1000 steps, cutoff = 12-14 A. This version has the improved communications, improved force computation and uses the apogee apf77 (-O5) compiler.

Computer	node	CPU time [hours] Tot.	force	comm.	Memory [Mb/node]	musecs per atom-update	Speed-up vs orig code	4 nodes
CM-5	4	6.10	89%	0.6%	23.7	1566	2.0	1
CM-5	32	0.90	83%	7%	15.8	231	1.9	6.8
CM-5	64	0.63	81%	10%	15.3	160	1.7	9.8
CM-5	128	0.42	75%	16%	15.0	108	1.4	14.5
CM-5	256	0.30	64%	24%	15.0	77	1.4	20.3
CM-5E	4	2.24	85%	1%	23.7	575	2.0	1
CM-5E	32	0.32	77%	11%	15.8	83	1.8	6.9
CM-5E	64	0.20	71%	18%	15.3	52	1.7	11.1

Figure 5.1 shows the communication and total timings as a function of the number of processors.

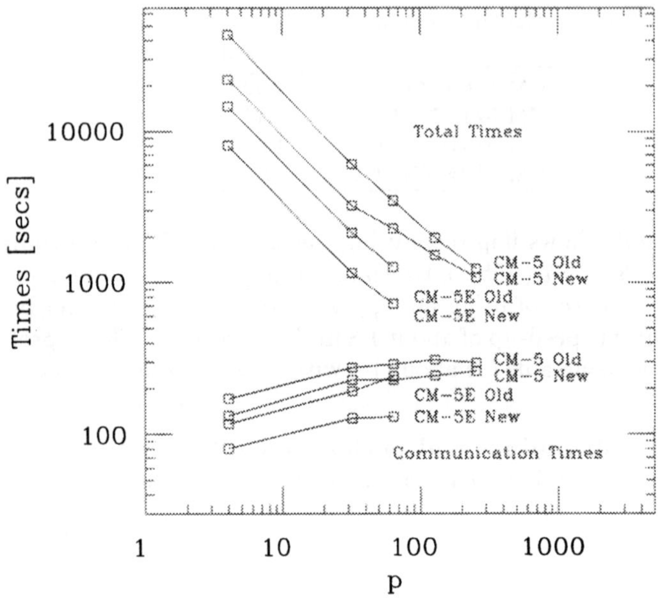

FIG. 5.1. *Communications and total timings for the original (old) and new versions of CHARMM on CM-5 and CM-5E machines.*

TABLE 5.3
CHARMM23 timings in hours for Carboxy Myoglobin (MbCO) simulation 1000 steps, cutoff = 12-14. (We thank Milan Hodoščék for making these numbers available to us).

Nodes	1	2	4	8	16	32	64	128	256	512
HP/735	2.93	1.49	0.78	0.48						
Intel Gamma	20.72	10.49	5.31	2.66	1.37	0.73	0.42	0.26		
Intel Delta									0.17	0.13
Intel Paragon	15.31	7.73	3.92	1.98	1.03	0.55	0.31	0.18		
IBM/SP1	6.55	3.42	1.78	0.96	0.59	0.28	0.22	0.19		
CM-5			6.10			0.90	0.63	0.42	0.30	
CM-5E			2.24			0.32	0.20			

Table 5.3 shows comparisons of parallel CHARMM on a variety of machines.

The conclusion is that in terms of cost performance a cluster of workstations is probably the best choice. However, if small absolute times are important then the MPP machines are able to provide reasonable speedups, for example a speed-up of 44 on 64 nodes for the CM-5E, and provide short turnaround times.

To get a better idea about the current performance and limitations of the new version we analyzed the average percentage of time spent on different sections of the program as is shown in table 5.4. These averages are computed over all processors and several runs with several models (input files). They don't sum to 100% because of averaging and rounding.

TABLE 5.4
Average Percentages for different parts of the code.

# Processors	External Terms	Internal Terms	Non-Bonded List	Communications	Waiting
32	38%	4%	45%	8%	3%
64	34%	6%	41%	15%	3%
128	29%	8%	36%	23%	3%
256	23%	11%	30%	30%	3%

Communications times are now fixed (as an absolute number) but they are still a significant factor with large numbers of processors. The non-bonded list generation scales well but it is a bigger factor than we expected. The computation of the internal terms does not scale as well and this deserves further study. The load imbalance appears small in these average numbers but some processors might still be under-utilized. Some of these are basic limitations of the current parallelization approach.

6. Conclusions. As a result of our study of parallelization issues of CHARMM on the CM-5 we managed to improved performance by a factor of 1.4-2 without making any major algorithmic changes. This is a relatively modest but significant improvement taking into account the extensive use of CHARMM throughout the academic community and supercomputer centers worldwide. The scaling of performance with increasing number of nodes is much better than in the original C23F3 version. In addition some of the results gave us insight into the most general problem of optimizing distributed Molecular Dynamics calculations.

After making the changes to produce the optimized version we have verified all the test cases which parallel CHARMM supports.

The forthcoming release of CHARMM (C23F4) has a CM-5 version with most of the improvements of the optimized program presented here.

We thank B. Brooks, M. Hodoščeк, M. Karplus and K. Schulten for discussions about Molecular Dynamics and CHARMM. We also thank Adam Greenberg for provided invaluable help with some of the debugging sessions.

REFERENCES

[1] B. R. Brooks, R. E. Bruccoleri, B. D. Olafson, D. J. States, S. Swaminathan and M. Karplus, *J. Comp. Chem.* **4**, 187 (1983).
[2] B. R. Brooks and M. Hodoščеk, *Chemical Design Automation News* **7**, 16 (1992).
[3] G. C. Fox, M. A. Johnson, G. A. Lyszerga, S. W. Otto, J. K. Salmon and D. W. Walker, *Solving Problems on Concurrent Processors, Vol. 1*, Prentice Hall, Englewod Cliffs, NJ (1988).
[4] R. A. van de Geijn, Technical Report CS-91-138 University of Tennessee.
[5] T. von Eicken, D. E. Culler, S. C. Glodstein and K. E. Schauser, in *Procs. of the 19th International Symposium on Computer Architecture.* ACM Press, Australia, May 1992.
[6] L. W. Tucker and A. Mainwaring, *Parallel Computing*, **20**, No 4, March 31, 1994.

[7] The apogee apf77 compiler (version 2.3) is a commercial product provided by Apogee Software Inc.

COMPUTATIONAL STUDIES OF SPATIALLY CONSTRAINED DNA

WILMA K. OLSON*, TIMOTHY P. WESTCOTT*, JENNIFER A. MARTINO*, AND GUO-HUA LIU*

1. Introduction. Closed loops of double stranded DNA are ubiquitous in nature, occurring in systems ranging from plasmids, bacterial chromosomes, and many viral genomes, which form single closed loops, to eukaryotic chromosomes and other linear DNAs, which appear to be organized into topologically constrained domains by DNA-binding proteins [1,2]. The topological constraints in the latter systems are determined by the spacing of the bound proteins along the contour of the double helix along with the imposed turns and twists of DNA in the intermolecular complexes [3,4]. As long as the duplex remains unbroken, the linking number *Lk*, or number of times the two strands of the DNA wrap around one another, is conserved [5-8]. If one of the strands is nicked and later re-sealed, the change in overall folding that accompanies DNA-protein interactions leads to a change in *Lk*. The supercoiling brought about by such protein action, in turn, determines a number of key biological events, including replication, transcription, and recombination [9].

The influence of protein binding and reactivity on the overall folding of DNA is analogous to the changes in topology seen in the manipulation of physical models, such as a rubber band or a guitar string. Sealing the ends of a model that is partially wrapped like a piece of thread around a spool generates figure-8 and interwound structures once the spool is removed. The resulting configurations are sensitive to the degree of wrapping and the topological state of the model upon closing (e.g., whether the rod is already supercoiled or "relaxed" to a twist-free state). Mechanical methods commonly used to analyze these elastic materials are at once applicable to the study of protein-bound double helical DNA.

In this article we survey and compare various elastic energy models and computer simulation techniques that are useful in studying protein-induced folding of DNA. The examples presented here correspond to nicked closed circular duplexes containing one or more chain fragments partially wrapped around a "phantom" cylindrical core of proteins. The models consist of two parts: rigid double helical segments wrapped around protein and unbound connecting DNA loops. The loops are flexible and have at least one single-stranded scission so that they can assume a relaxed "twist-free" state. Thus, the shapes of the loops at equilibrium are simply those of minimum bending energy (see below). The degree of wrapping around protein, the chain length of DNA, and the number of proteins are

* Department of Chemistry, Rutgers, the State University of New Jersey, New Brunswick, New Jersey 08903.

variables in the calculations. The chain representations take advantage of curve fitting methods and rod representations commonly used in computer aided design and engineering [10,11], while the minima are identified using stochastic, deterministic, and iterative techniques. The results are consistent with physical models and suggest new ways to think about how proteins add and remove supercoils to and from DNA. The findings bear relevance to DNA with natural base sequence dependent curvature as well as to protein-induced deformations of the double helix.

2. Elastic model

Inextensible Thin Rod

Smooth elastic deformations of an inextensible thin rod with circular cross section are normally described in terms of two bending contributions, κ_1 and κ_2, and the twist density, κ_3, at all points s along the contour of the rod:

$$E_{\text{elastic}} = \frac{A_1}{2}\oint(\kappa_1(s)-\kappa_1^o(s))^2 ds + \frac{A_2}{2}\oint(\kappa_2(s)-\kappa_2^o(s))^2 ds + \frac{C}{2}\oint(\kappa_3(s)-\kappa_3^o(s))^2 ds. \tag{2.1}$$

The angles are components of the vector, $\hat{\kappa}(s) = [\kappa_1, \kappa_2, \kappa_3]$, describing the angular rotation of local coordinate frames embedded in cross sections of the rod at s and $s + ds$. If the rod is divided into a set of discrete elements and the spacing between planar slabs, Δs, is taken equal to the 3.4 Å distance between neighboring base pairs in double helical B-DNA, κ_1 and κ_2 approximate the so-called roll and tilt angles, ρ and τ, between base pair planes, and κ_3 the base pair twist, Ω.

The elastic force constants (A_1, A_2, C) in Eq. 2.1 describing the energy associated with deformations of the rod from its equilibrium rest state, where $\hat{\kappa} = \hat{\kappa}^o = [\kappa_1^o, \kappa_2^o, \kappa_3^o]$, can then be defined in terms of the energy impeding the bending and twisting of DNA base pairs. We choose rest states equal to the mean roll, tilt, and twist angles in the currently available B-DNA crystal structures ($\langle\rho\rangle = 0°$, $\langle\tau\rangle = 0°$, $\langle\Omega\rangle = 36°$, respectively) and constants of rolling, tilting, and twisting such that the mean-square fluctuations in the crystal sample ($\langle\rho^2\rangle^{1/2} = 6°$, $\langle\tau^2\rangle^{1/2} = 3.5°$, $\langle\Omega^2\rangle^{1/2} = 6°$) correspond to angular deviations that raise the energy per base pair step by $0.5kT$, where k is the Boltzmann constant and T the absolute temperature [12-14]. The elastic energy of DNA thus reduces to a sum of contributions over the set of neighboring base pairs $\{bp\}$:

$$E_{\text{elastic}} = \sum_{\{bp\}} \frac{kT}{2}\left(\frac{(\rho-\langle\rho\rangle)^2}{\langle\rho^2\rangle} + \frac{(\tau-\langle\tau\rangle)^2}{\langle\tau^2\rangle} + \frac{(\Omega-\langle\Omega\rangle)^2}{\langle\Omega^2\rangle}\right). \tag{2.2}$$

Isotropic Rod

If the bending is isotropic and the rod naturally straight, i.e., $A_1 = A_2 = A$ and $\kappa_1^o = \kappa_2^o = 0$, the bending energy becomes a function of the curvature of the axis of the rod, $\kappa = \sqrt{\kappa_1^2 + \kappa_2^2}$. It is further known that the twist density, $\kappa_3(s)$, in such a rod is uniform, and when incorporated in a closed DNA, the twist energy can be expressed in terms of the total contour length of the chain L, the writhing number Wr (a measure of overall chain folding obtained with standard differential geometric formulas [8]), and the difference of the linking number, $\Delta Lk = Lk - Lk_0$, in the current and the twist-free equilibrium rest states. The linking number, an integer, is a topological invariant in the sense that it cannot be changed except by severing (and resealing) one or both of the strands that comprise the cyclic double helix [5]. In such cases, the elastic energy becomes a simple function of the axis of the rod given by [6]:

$$E_{\text{elastic}} = \frac{A}{2} \oint \kappa(s)^2 ds + \frac{2\pi^2 C}{L}(\Delta Lk - Wr)^2. \tag{2.3}$$

We define the overall bending stiffness, A, of an idealized naturally straight B-DNA with isotropic bending in terms of the average mean-square fluctuations in roll and tilt:

$$A = \left[\frac{2kT}{\langle \Delta\rho^2 \rangle + \langle \Delta\tau^2 \rangle}\right]. \tag{2.4}$$

As a result, when $\langle \Delta\rho^2 \rangle = \langle \Delta\tau^2 \rangle = \langle \Delta\Theta^2 \rangle$ and $\langle \rho \rangle = \langle \tau \rangle = 0$ in Eq. 2.2, the bending contribution reduces to a function of the net bending angle, $\Theta = \sqrt{\rho^2 + \tau^2}$, between adjacent base pair planes:

$$E_{\text{bend}} = kT\left(\frac{\Theta^2}{\langle \Theta^2 \rangle}\right). \tag{2.5}$$

This expression is equivalent to the energy of bending a thin, naturally straight, isotropic rod given in Eq. (2.3) if we define the curvature as a discretized parameter consistent with its differential geometric meaning, i.e., the magnitude of the rate of change of the tangent,

$$\kappa(s) = \left|\frac{\Delta \boldsymbol{t}}{\Delta s}\right| = \frac{|\boldsymbol{t}(s + \Delta s) - \boldsymbol{t}(s)|}{\Delta s}, \tag{2.6}$$

and further recognize that the net bending angle Θ between neighboring base pair planes is defined by the scalar product of successive tangents, i.e., $\cos\Theta = \boldsymbol{t}(s + \Delta s) \cdot \boldsymbol{t}(s)$. Thus, $\kappa(s) = \Delta\Theta/\Delta s$ and $A = 2kT\Delta s/\langle \Delta\Theta^2 \rangle$. The parameter Δs is the known separation between base pair centers.

3. Chain representation

Overview

The representation of closed chain molecules with ends confined to a fixed end-to-end separation and orientation is a long standing problem in polymer physical chemistry that can be attacked from several points of view. In one approach the configurations of unconstrained linear chains that meet certain spatial criteria (normally a set of distances and angles between chain ends) are collected through exhaustive simulation studies [15]. This method, however, is not practical for studies of the preferred three-dimensional arrangements and intrinsic flexibility of constrained chains. The probability of identifying specific configurations from the random sampling of an unconstrained chain is so low that it is difficult to accumulate a meaningful sample of appropriate states. The method, however, is very useful for simulations of the kinetics of chain cyclization (i.e., ring closure) or the loop formation of open chains [16-19].

A second way to study constrained polymer chains is to start with a configuration that meets the desired structural criteria and then allow the system to deform subject to some potential function. The major technical difficulty in such computer simulations is the preservation of the constraints on the displacement and orientation of chain ends during the simulation. Individual Cartesian coordinates must be moved in small concerted steps, or internal torsion and valence angles varied in highly correlated pairwise increments, to maintain the fixed configuration of chain ends [20-23]. Alternatively, one can introduce explicit energy terms that constrain the chain ends to a given positioning. Elastic potentials with no physical significance are typically employed in Cartesian simulations to keep one or more interatomic distances within a desired range [24-26]. As outlined below, one can find the force needed to keep the ends of a rodlike polymer in a given position by direct numerical solution of the equations of elastic equilibrium.

A third way to treat constrained polymer chains involves the use of curve fitting methods commonly employed in computer aided design and engineering. These representations automatically satisfy the end-to-end limitations on constrained DNA using a relatively limited number of variables. The number of points on the overall curve, however, is independent of the number of variables so that chains of different lengths can be generated by simply rescaling the variables.

Base Pair Positioning

Treatment of DNA as a closed elastic rod necessitates finding the locations of all constituent base pairs. One way that this can be achieved is with a coordinate transformation, $\boldsymbol{T}(\phi(s), \theta(s), \psi(s))$, that relates a local reference frame at position s, $[\boldsymbol{d}_1(s), \boldsymbol{d}_2(s), \boldsymbol{d}_3(s)]$, to the global coordinate system, $[\boldsymbol{e}_1, \boldsymbol{e}_2, \boldsymbol{e}_3]$, through the Euler angles $(\phi(s), \theta(s), \psi(s))$:

$$\boldsymbol{T}(\phi(s),\theta(s),\psi(s)) =$$

$$\begin{bmatrix} \cos\psi(s) & \sin\psi(s) & 0 \\ -\sin\psi(s) & \cos\psi(s) & 0 \\ 0 & 0 & 1 \end{bmatrix} \begin{bmatrix} 1 & 0 & 0 \\ 0 & \cos\theta(s) & \sin\theta(s) \\ 0 & -\sin\theta(s) & \cos\theta(s) \end{bmatrix} \begin{bmatrix} \cos\phi(s) & \sin\phi(s) & 0 \\ -\sin\phi(s) & \cos\phi(s) & 0 \\ 0 & 0 & 1 \end{bmatrix}$$

(3.1)

The Euler angles so defined and their derivatives with respect to arc length, $\{\phi'(s),\theta'(s),\psi'(s)\}$, are related to the two bending components and the twist density of the local frame [11].

$$\begin{aligned} \kappa_1(s) &= (\theta'(s)\cos\psi(s) + \phi'(s)\sin\theta(s)\sin\psi(s))ds, \\ \kappa_2(s) &= (-\theta'(s)\sin\psi(s) + \phi'(s)\sin\theta(s)\cos\psi(s))ds, \\ \kappa_3(s) &= (\phi'(s)\cos\theta(s) + \psi'(s))ds)). \end{aligned} \quad (3.2)$$

Constraints on the vectorial location of the ends of the rod, $\boldsymbol{r}$, can be introduced through a Lagrange multiplier $\boldsymbol{L} = [\lambda_1,\lambda_2,\lambda_3]$, which acts as the effective force needed to keep the chain ends at the desired end-to-end position. The value of $\boldsymbol{L}$ is initially unknown but can be determined along with the complete set of $\kappa_1(s),\kappa_2(s),\kappa_3(s)$ values that minimize the energy of the constrained rod through solution of three Euler-Lagrange equations of the form,

$$\frac{d}{ds}\frac{\partial \oint[\frac{A_1}{2}(\kappa_1-\kappa_1^o)^2 + \frac{A_2}{2}(\kappa_2-\kappa_2^o)^2 + \frac{C}{2}(\kappa_3-\kappa_3^o)^2 + \boldsymbol{L}\cdot\boldsymbol{t}(s)]ds}{\partial\xi'} =$$

$$\frac{\partial \oint[\frac{A_1}{2}(\kappa_1-\kappa_1^o)^2 + \frac{A_2}{2}(\kappa_2-\kappa_2^o)^2 + \frac{C}{2}(\kappa_3-\kappa_3^o)^2 + \boldsymbol{L}\cdot\boldsymbol{t}(s)]ds}{\partial\xi}$$

(3.3)

where $\xi = \phi,\theta$, and ψ, ξ' is the derivative with respect to arc length, and $\kappa_i = \kappa_i(s), \kappa_i^o = \kappa_i^o(s), i = 1,3$. This expression incorporates the fixed end-to-end vector $\boldsymbol{r}$ as the integral of successive tangent vectors, $\boldsymbol{t}(s)$, along the rod:

$$\boldsymbol{r} = \oint \boldsymbol{t}(s)ds = \oint[\sin\phi(s)\sin\theta(s), -\cos\phi(s)\sin\theta(s), \cos\theta(s)]ds. \quad (3.4)$$

If we assess the rod in terms of a sum of energy contributions and displacements over DNA base pairs, Eq. 3.3 leads to a system of $3N$ second-order nonlinear differential equations which can be solved numerically (T. P. Westcott, I. Tobias, and W. K. Olson, manuscript in preparation). The elastic equations can also be solved with finite element techniques [27,28].

Constrained Double Helical Axis

If DNA is modeled as a naturally straight isotropic rod, the configuration of the chain is completely determined by the bending and twisting of its double helical axis. For the purpose of computer simulations, the chain closure requirement can be met by various cyclic functions, such as piecewise B-spline curves [29-31] or the sine and cosine functions comprising a finite Fourier series [3,32]. The former representation is a local formulation in the sense that the change of a single variable, a so-called controlling point, alters only part of the closed pathway, whereas the latter formulation is global in that the choice of an individual variable affects the entire curve. The principal advantage of the cubic B-spline parameterization is the direct control of the chain trajectory provided by the choice of independent variables. Its major limitation, as will become evident below, is the approximate nature of the curve. The choice of controlling points for the initial state is somewhat arbitrary, and the constructed curve does not usually pass through the controlling points themselves. Fourier analysis corrects for this deficiency and provides a direct connection between experimental measurement and computer modeling. Virtually any set of coordinates can be transformed into a finite Fourier series, the simplicity of which can be exploited for structural manipulation and analysis.

Closed B-spline Curve

The B-splines are polynomial expressions that smoothly connect a sequence of arbitrary controlling points $\boldsymbol{p}$ [10]. The curves are regionally defined functions that are evaluated separately over different intervals of the closed pathway. The order four (i.e., cubic) cyclic B-spline,

$$\boldsymbol{r}_i(u) = F_1(u)\boldsymbol{p}_{i-1} + F_2(u)\boldsymbol{p}_i + F_3(u)\boldsymbol{p}_{i+1} + F_4(u)\boldsymbol{p}_{i+2}, \tag{3.5}$$

with continuous first and second derivatives is sufficient for the calculation of the topological and energetic parameters of an ideal isotropic DNA rod. The four basis functions in this expression,

$$\begin{aligned} F_1(u) &= \frac{(-u^3 + 3u^2 - 3u + 1)}{6}, \\ F_2(u) &= \frac{(3u^3 - 6u^2 + 4)}{6}, \\ F_3(u) &= \frac{(-3u^3 + 3u^2 - 3u + 1)}{6}, \\ F_4(u) &= \frac{u^3}{6}, \end{aligned} \tag{3.6}$$

assure that successive piecewise curve segments, $\boldsymbol{r}_i$, and their first and second derivatives are smoothly connected. The number of points along each segment is determined by the mesh parameter, u, which varies between zero and unity. The number of controlling points is a function of the complexity of the curve with more points needed for the representation of more convoluted pathways. Some of the controlling points can be fixed during the course of the computations to simulate the effects of local rigidity within the DNA [33]. In the calculations reported below, a set of points describing a superhelix of appropriate proportions is used to model the presence of a protein assembly rigidly bound to circular DNA.

Fourier Series Representations

A finite three-dimensional Fourier series of the form:

$$\Delta \boldsymbol{r}(u) = \sum_{m=1}^{M} \{\boldsymbol{a}_m \sin(2\pi m u) + \boldsymbol{b}_m(\cos(2\pi m u) - 1)\} + \boldsymbol{a}_{M+1} \sin(2\pi(M+1)u), \tag{3.7}$$

corresponds to the difference in configuration between a given starting structure and an arbitrary folded pathway of the chain axis. The vectorial coefficients $\boldsymbol{a}_m$ and $\boldsymbol{b}_m$ are optimized during the course of computer simulation with the contour parameter u ranging, as for the B-spline curves, between zero and unity. The restraints on $\boldsymbol{b}_m$ assure that the ends of the DNA coincide with those of the starting state, while the definition of the coefficient $\boldsymbol{a}_{M+1}$,

$$\boldsymbol{a}_{M+1} = \frac{-1}{M+1} \sum_{m=1}^{M} m \boldsymbol{a}_m, \tag{3.8}$$

further guarantees that the tangents at the ends of the DNA coincide with those of the starting state [3]. Starting configurations can include any arbitrary set of curve points. The sine and cosine functions in the Fourier series can be differentiated infinitely, so that unlike the cubic B-spline functions, the formalism does not need to be changed for more elaborate models of the chain.

4. Optimization

Monte Carlo Methods

Minimum energy forms of spatially constrained DNAs have been identified using stochastic (i.e., Monte Carlo), deterministic (i.e., direct minimization), and iterative computational methods. The Monte Carlo schemes entail random variation of an arbitrary B-spline controlling point or Fourier coefficient. The energy of the resulting chain representation is compared with that of the preceding state. Arrangements of lower energy are automatically accepted, while those of higher energy are accepted according to the standard Metropolis criterion [34]. That is, a move to a state of higher energy is allowed only if a random number (between zero and unity) is less than the Boltzmann factor of the energy difference ΔE between successive configurations, i.e., $\exp(-\Delta E/kT)$. The simplicity of the algorithm and the ease of programming are counterbalanced by the long times required to identify the global energy minimum of a particular chain. Monte Carlo simulations have been carried out at a fixed (high) temperature with the repetition of successful downhill moves to accelerate convergence [32], and more gradually over a series of lower temperatures in a simulated annealing scheme [30].

Deterministic Minimization

The direct minimization involves computation of the energy and its first and second derivatives with respect to the three components of the independent variables (i.e., B-spline controlling points or Fourier coefficients) [31,35-37]. The requisite programming is thus more demanding than the Monte Carlo method. The minimization problem is solved using TNPACK, a FORTRAN package containing the truncated-Newton algorithm [38,39]. This second derivative method combines the rapid convergence properties of Newton methods with the modest computational and storage requirements more typical of gradient methods. The method takes advantage of the separability of the potential energy function into terms of different complexity, using representative terms to obtain an approximate solution of the search direction at each step of the calculation. The positive definiteness of the Hessian at the minimum ensures location of a true local minimum. While the independent variables in the minimization procedure are the artificial parameters describing the pathway of the double helical axis (i.e., B-spline controlling points or Fourier coefficients), the energies and derivatives are determined from interactions involving all mesh points along the DNA curve.

Iterative Solution

The iterative scheme is applied only to the numerical solution of the Euler-Lagrange equations in Eq. 3.3. For a DNA made up of N base pairs, this system of nonlinear equations involves $3N+3$ unknowns corresponding to the set of Euler angles at each base pair, $\{\phi_i, \theta_i, \psi_i\}$ $i = 1, N$, plus the three components of the unknown force (represented by the Lagrange multiplier $\boldsymbol{L}$) that holds the ends of the chain in place. Each of the nonlinear expressions in Eq. 3.3, $f_j(\{\xi_i\}) = 0, i = 1, 3N+3$, can be reexpressed using the Newton-Raphson method as a linear equation of the form, $\partial f_j(\{\xi_i\})/\partial \xi_i|_{\xi_i=\xi_i^*}\, \Delta\xi_i = -f_j(\{\xi_i\})|_{\xi_i=\xi_i^*}$ involving the set of corrections $\{\Delta\xi_i\}$ which improve an initial approximation of the solution, $\{\xi_i^*\}$ [40]. The set of $3N+3$ linear expressions is solved by Gaussian elimination. The corrections are added to the initial guess and the process is reiterated until the corrections are less than a desired level of accuracy (here taken to be 0.0001%). Computer programming is straightforward once the Euler-Lagrange equations are used to derive the equations of equilibrium.

Chain Length Constraint

The simple torsionally relaxed system described below entails no long-range self-contacts that require monitoring by van der Waals' or other pairwise energy terms. A harmonic expression of the form, $E_L = (L - L_0)^2$, however, must be introduced in the Monte Carlo simulations and the deterministic energy minimization studies to constrain the chain length to a fixed value. The contour length is automatically fixed in the system of elastic equations considered here (Eq. 3.3).

5. Results

Nicked DNA Loops

To model the effects of protein binding on overall DNA structure, we divide the elastic chain model into two parts. The part associated with protein is restricted to a rigid left-handed superhelical pathway of 45 Å radius and 30 Å pitch. The rest of the chain, the protein-free double helical loop, is allowed to relax to a configuration that minimizes the energy and satisfies the boundary conditions (i.e., end positions and/or orientations) at the junction between the two sections. For simplicity, the DNA is modeled as a twist-free, naturally straight thin isotropic rod, so that only the bending energy is included in the optimization (i.e., Eq. 2.5). The global minimum thus corresponds to the equilibrium rest state of a torsionally relaxed (i.e., nicked) DNA loop. Each such configuration has a writhing number which depends on both the shape of the loop and the assumed shape of the rigid, protein-bound segment. If the nicked strand is sealed, the DNA will be characterized, following Eq. 2.3, by a linking number difference numerically equal to the writhing number (since $\Delta Lk - Wr = 0$ in the torsionally relaxed state). Upon release of the protein, the molecule will assume a new shape, consistent with the value of ΔLk so determined. At this point the DNA is no longer twist free, but converted to a new protein-free equilibrium state with a new writhing number Wr' and twist energy proportional to $(\Delta Lk - Wr')^2$.

Comparative Chain Folding

Figure 5.1 is a comparative plot of the change in writhing number found from independent computational studies of a 1000 bp nicked DNA with varying degrees of superhelical wrapping around the idealized protein core. The points obtained through numerical solution of the Euler-Lagrange equations describe a solid line (with irregularities that reflect the computed numerical values), while the data resulting from Monte Carlo optimization of the Fourier coefficients of the duplex axis correspond to the open circles. The former approach, involving no more than 50 iterations per computed equilibrium configuration, is significantly more efficient than the latter computation. Each Monte Carlo value is the result of $\sim 40,000$ randomly generated configurational moves [3]. The bending force constant is set at a value corresponding to a root-mean-square fluctuation of 5.8° per base pair at 298 K in both studies. Significantly, the overall structural picture provided by the two schemes is identical: the writhing number undergoes a series of sudden drops at roughly half integral superhelical turns. The unbound loop snaps at these points from an open contact-free state to a collapsed configuration with a point of long-range self-association. At low values of superhelical wrapping, the presence of protein has essentially

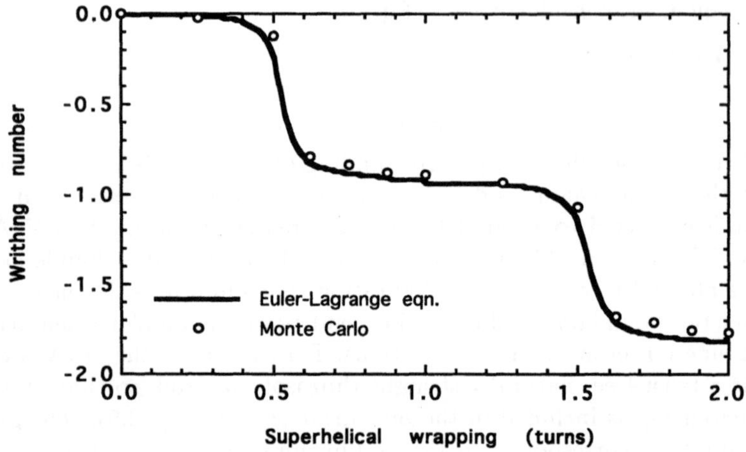

FIG. 5.1. *Comparative variation of the writhing number at the global energy minimum versus the number of superhelical turns of a torsionally relaxed* 1000 *bp DNA constrained to wrap in a superhelical pathway (radius* = 45 *Å; pitch* = 30 *Å) around an ideal cylindrical protein core. Curves generated with data obtained from numerical solution of the Euler-Lagrange equations at* 0.0125 *turn increments of wrapping (solid line with irregularities that reflect the computed values) and from high temperature Monte Carlo optimization at points spaced by* 0.125 *or* 0.25 *turns (open circles [3]).*

no effect on the writhing number, its value being near zero up to ~ 0.5 superhelical turns. The midpoint of the first configurational collapse, as measured by the change in Wr with increase in superhelical turns found from the solution of Eq. 3.3, occurs at 0.54 superhelical turns. Above this point and up to the second sharp transition at 1.54 superhelical turns, the average value of Wr is -0.93. The slight discontinuity in the slope at one complete turn of wrapping reflects a configurational transition which allows the protein-bound fragment to revolve from the outside to the inside of the free loop without significant change in Wr [3]. Beyond the second major configurational collapse at 1.54 turns, the writhing number levels off to a value of approximately -1.8.

Energy Variation

Plotted in Figure 5.2 are the elastic energies of the optimized protein-free DNA fragments in the minimum energy configurations identified by solution of the Euler-Lagrange equations. (The slight irregularities in the curve reflect the computed numerical values.) The variation of energy in the loops reveals that the large configurational changes noted above at ~ 0.5 and ~ 1.5 superhelical turns reflect a critical elastic transition where the imposed bending on the ends of the DNA becomes too severe. (There is an even greater build-up of bending energy at a complete helical turn of wrapping corresponding to the movement of the protein from the

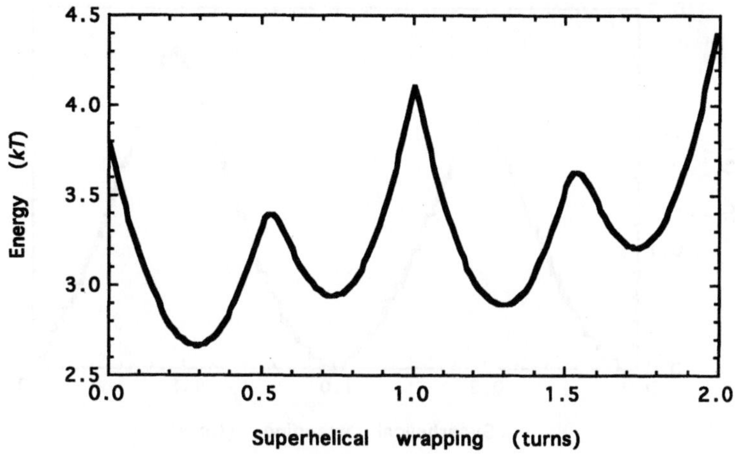

FIG. 5.2. *Variation of the bending energy, in units of kT, of the free unbound DNA in a torsionally relaxed chain of* 1000 *bp versus the degree of superhelical wrapping around an ideal cylindrical protein core. Data correspond to the equilibrium configurations in Figure 5.1 obtained from solution of the Euler-Lagrange equations.*

inside to the outside of the free DNA loop [3]). The data further show that open loops constrained by protein-bound fragments of ~ 0.3 turns are lower in energy than all other configurational states, including the protein-free nicked circle of maximum chain length. (The number of base pairs in the unbound DNA loops decreases by ~ 84 for every complete turn of superhelical wrapping so that the circle is the largest loop in the series of constrained configurations). Open loops resulting from ~ 1.3 turns of superhelical wrapping and collapsed loops resulting from ~ 0.7 or ~ 1.7 turns of wrapping are also energetically favored compared to the closed circle. These energy differences would be even more pronounced if the contour length of the free loop were fixed and the protein binding constraint simply altered the orientation and distance between chain ends.

The energy of the protein-bound DNA is not included in Figure 5.2. If the bound superhelical DNA were an ideal isotropic rod, its elastic energy contribution of $\sim 0.6kT$ per base pair would overwhelm that of the attached constrained loop. Presumably there are favorable (electrostatic, van der Waals', hydrogen bonding, etc.) contacts between protein and DNA that would compensate for the increase in bending energy associated with the tight superhelical wrapping of intrinsically straight nucleic acid around a cylindrical core. On the other hand, if the protein-bound DNA were naturally curved by $\sim 4.5^o$ per base pair, i.e., corresponding to the ~ 80 bp/turn found in our idealized example [41], the superhelical pathway would be a favored minimum energy state.

Figure 5.3 shows the magnitude of the work, in units of kT, needed to

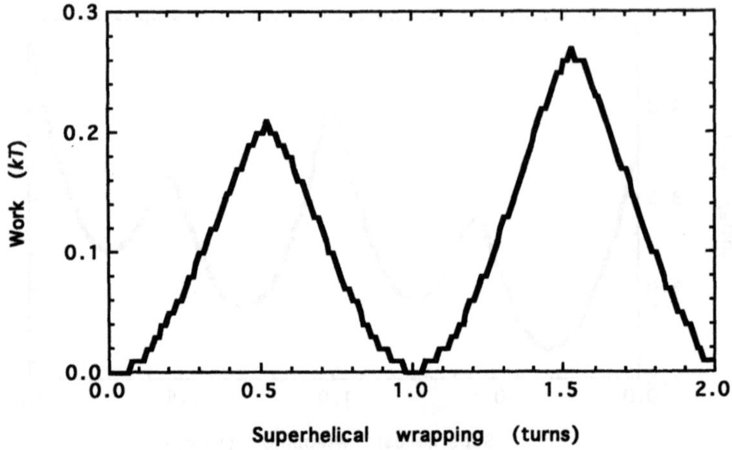

FIG. 5.3. *Magnitude of the work, in units of kT, needed to keep the ends of the free loop of a* 1000 *bp torsionally relaxed DNA at the distance and orientation specified by a given degree of superhelical wrapping around an ideal cylindrical protein. Data correspond to configurations in Figures 5.1 and 5.2 obtained by numerical solution of Eq. 3.3.*

keep the ends of the free DNA chain at the vectorial separation specified by a given degree of superhelical wrapping around protein. These data, which are an order of magnitude smaller than the elastic energy of the free DNA loop, correspond to the product of the distance between the ends of the free DNA loop and the magnitude of the Lagrange multiplier that minimizes the energy of the closed DNA (i.e., protein-bound and free segments where $\mathbf{r} = \mathbf{0}$). (The irregularities in the numerical values are magnified by the scale of the figure.) The maximum work occurs at the midpoints of the configurational transitions between open and collapsed loops and the minimum work at zero helical wrapping corresponding to the nicked planar circle.

Comparative Geometry

Table 5.1 and Figure 5.4 provide a more detailed comparison of the three-dimensional structures of protein-bound DNA loops identified by different computational approaches. Listed in the table for several independently determined structures (all with 1.75 superhelical turns of wrapping around protein) are the writhing numbers, geometric centers of the optimized free loops $\{< x >, < y >, < z >\}$, and root-mean-square (rms) displacements of points on the loops with respect to a common reference frame embedded in the idealized cylindrical protein (i.e., $\{\Delta x, \Delta y, \Delta z\}$, where $\Delta\zeta = \sqrt{< \zeta^2 > - < \zeta >^2}$, for $\zeta = x, y, z$). The optimized coordinates are superimposed in the figure. The minima correspond in two cases to chains of 500 bp and in three others to chains of 1000 bp. As evident from the table and figure, there are subtle differences in the three-dimensional

TABLE 5.1

*Comparative performance and predicted geometries of protein-bound DNA obtained by different computational methods**

Chain Representation (Independent Variables†)	Computational Technique	Performance (computed steps)	L (bp)	Wr	$<x>$ (Å)	$<y>$ (Å)	$<z>$ (Å)	Δx (Å)	Δy (Å)	Δz (Å)	L_{free} (Å)
B-spline (25 control pts)	Monte Carlo simulated annealing	~ 3,000,000	500	−1.44	302.4	27.6	40.4	153.6	28.6	96.9	1216.0
Fourier series (10 coefficients)	Monte Carlo downhill acceleration	~ 40,000	500	−1.71	340.9	−8.0	1.4	174.6	82.9	39.0	1267.2
Rod (3000 Euler angles +3 Lagrange coefficients)	Elastic eqns. iterated solution	50	1000	−1.75	640.2	2.2	−2.3	360.6	254.1	68.1	2907.0
Fourier series (14 coefficients)	Truncated Newton minimization	124	1000	−1.76	632.8	−1.5	−0.1	345.0	269.0	67.2	2901.0
Fourier series (10 coefficients)	Monte Carlo downhill acceleration	~ 40,000	1000	−1.71	534.7	2.0	−2.8	439.6	170.6	78.2	2934.7

* The protein-bound DNA is constrained to 1.75 turns of a superhelix of 30 Å pitch and 45 Å radius.

† The number of independent variables in the *B*-spline and Fourier series chain representations is 3 times the number of controlling points and Fourier coefficients listed.

structures of the optimized chains at the level of the precise pathway of the double helical axis. These differences are reflected, in part, in the writhing number which ranges between −1.44 and −1.76 in the five structures. The writhing number, however, is not a unique measure of chain folding, but simply a numerical assessment of the close contacts in the chain.

The one structure that stands out as most different from the other minimum energy states is the 500 bp cubic B-spline represented curve identified from four cycles of Monte Carlo simulated annealing (Figure 5.4(a), dark gray curve). The spline function described by Eqs. 3.5 and 3.6 places no positional or orientational constraints on the junction between free and bound parts of the DNA. The bound DNA is approximated by a set of 25 fixed controlling points, some of which influence the path of the free DNA (i.e., the three points closest to each terminus of the rigid protein core help to define the first three spline fragments at each end of the free DNA; see Eq. 3.5). The precise location and angular orientation of the junction between rigid and free DNA and, indeed, the length of the bound fragment are all poorly defined. Note, for example, the slight peeling of the free DNA loop away from the ends of the protein core in the optimized B-spline structure. It is therefore not surprising that the optimized B-spline curve differs from the other representations of the same protein-bound DNA system since the ends of the free loop do not adhere to the boundary restrictions placed upon the other curves (see discussion section below). The free loop of the B-spline curve is less symmetrically positioned and is rotated with respect to the fixed protein segment compared to the other structures, while the writhing number (−1.44) is noticeably less negative than those of the other states. As shown below, simulated annealing of a B-spline chain model, while approximate and computationally lengthy, provides a practical first step in examining the three-dimensional folding of DNA with more than one bound protein. The simplicity of the Monte Carlo algorithm together with the local control of chain structure allows for rapid program development and preliminary assessment of prototypical systems for further investigation.

The close agreement between the optimum configuration of a 1000 bp DNA identified by numerical solution of the equations of elasticity and that found upon deterministic minimization of a Fourier series represented chain axis with 14 sets of variable coefficients is particularly notable. The axes of the two curves coincide very closely (Figure 5.4(b), black and gray curves, respectively). The differences in the centers of gravity and rms displacements of the chains reported in Table 5.1 reflect not only the locations but also the different numbers and spacing of the computed points on the two curves. The 1000 bp chain configuration identified from Monte Carlo optimization of 10 sets of Fourier coefficients adopts a folding pattern similar to those of the other DNAs of this length, but the free loop is somewhat more elongated in overall shape (apparent by eye in Figure 5.4(b) and from the different values of Wr, Δx, and Δy in Table 5.1). As evident from the

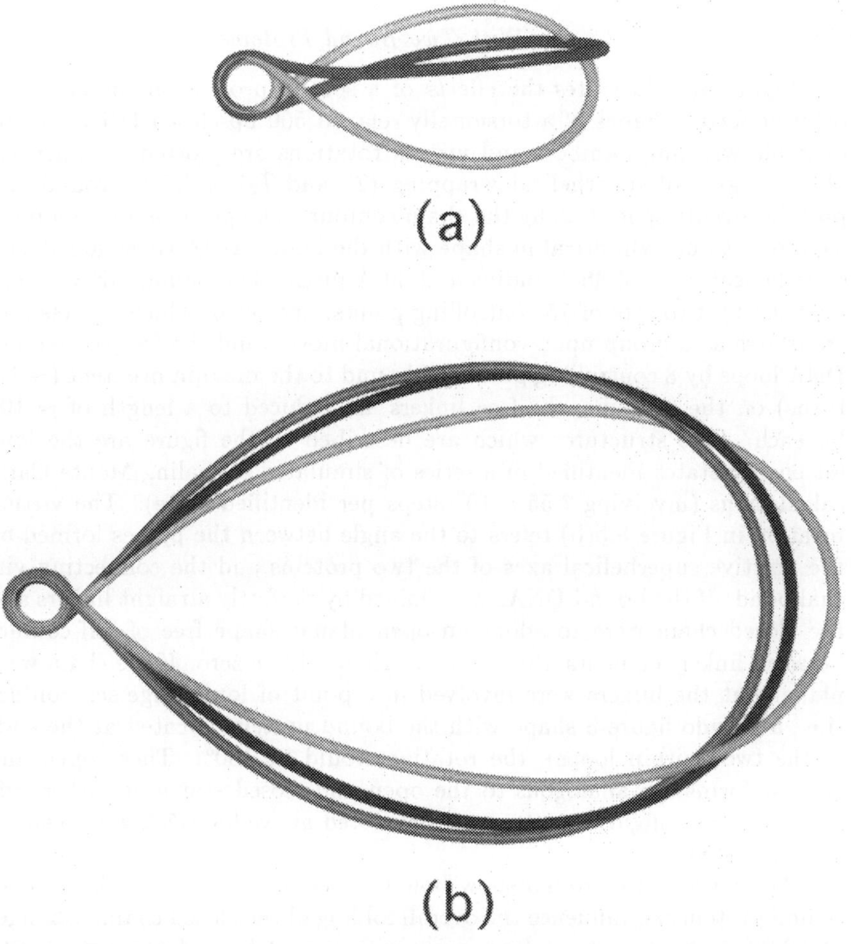

FIG. 5.4. *Comparative molecular representations of global energy minima of a torsionally relaxed DNA constrained to* 1.75 *turns along a superhelical pathway* (45 Å *radius and* 30 Å *pitch) around an idealized cylindrical protein: (a) Chains of* 500 *bp identified by Monte Carlo simulated annealing of a B-spline representation of the double helical axis (dark gray) and with constant temperature Monte Carlo searching using downhill acceleration with a Fourier series representation of the chain axis (light gray* [3,32]*); (b) chains of* 1000 *bp identified by numerical solution of the Euler-Lagrange equations (black), from direct minimization of energy as a function of the Fourier coefficients defining the chain axis (gray), and by Monte Carlo searching with downhill acceleration and a Fourier series representation of the chain axis (light gray* [3,32]*). To facilitate comparison between chains of different lengths, all structures are drawn in the same orientation and at the same scale. See Table 5.1 for further details.*

number of computer steps involved, the Monte Carlo method is a poor way to identify energy minima, although it is more efficient that the simulated annealing study carried out with a B-spline chain representation. The ensemble of configurations generated with Monte Carlo calculations, nevertheless, is very useful in developing a thermodynamic description of spatially constrained DNA [19,23,42].

Chains With Two Bound Proteins

Figure 5.5 illustrates the effects of a second protein on the energetically preferred shapes of a torsionally relaxed 500 bp closed DNA. In this example writhing numbers and virtual rotations are plotted as functions of the degree of superhelical wrapping (T_1 and T_2) of DNA around two proteins evenly spaced along the chain contour. The proteins are assumed, as above, to be cylindrical in shape with the bound DNA constrained to a superhelical path of 45 Å radius and 30 Å pitch. The bound DNA is represented by two sets of 15 controlling points, one set of which rotates and translates as a group upon configurational moves, and the free connecting DNA loops by 8 controlling points. If bound to the maximum extent ($\sim$ 1.8 turns) on the proteins, the free linkers are reduced to a length of $\sim$ 100 bp each. The structures which are described in the figure are the lowest energy states identified in a series of simulated annealing Monte Carlo calculations (involving 7.55×10^6 steps per identified state). The virtual rotation in Figure 5.5(b) refers to the angle between the planes formed by the positive superhelical axes of the two proteins and the connecting virtual bond. If the bound DNAs were joined by perfectly straight linkers and the closed chain were to adopt an open planar shape free of self-contact between linker segments, the virtual angle would be zero. If the chain were planar but the linkers were involved in a point of long-range self-contact (i.e., a pseudo figure-8 shape with the bound proteins located at the ends of the two hairpin loops), the rotation would be 180°. These open and crossed forms are analogous to the open and closed states on either side of the sharp configurational transitions noted above for DNA with a single bound protein.

The data in Figure 5.5 show how the peeling of DNA at the ends of bound protein can influence the overall folding of the closed chain. Changes of ± 0.2 turns in superhelical wrapping with respect to a reference state with $\sim$ 1.6 turns produce writhing numbers ranging between -2.8 and -1.8. The decrease in writhing number with increased wrapping around protein reflects a reorientation of the protein cores as measured by the virtual torsion. The computed values of Wr and the virtual rotation are strongly correlated with a linear correlation coefficient of 0.997. Identification of the precise degree of turning needed to bring about large-scale configurational collapse of the chain with two evenly spaced proteins, however, is uncertain due to the approximate nature of the B-spline function. The jump in Wr accompanying the global folding of the chain, nevertheless, is clearly more

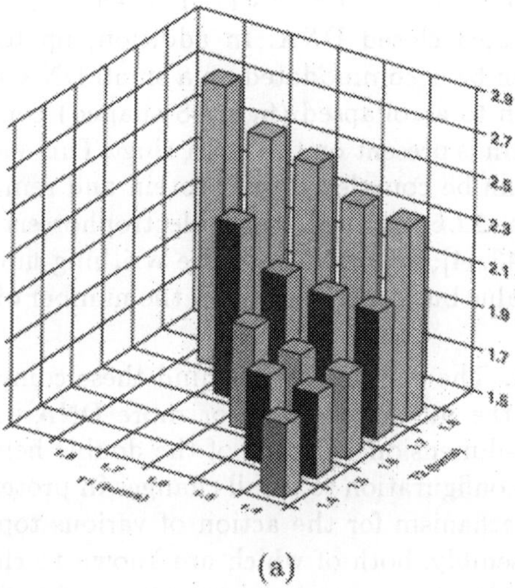

(a)

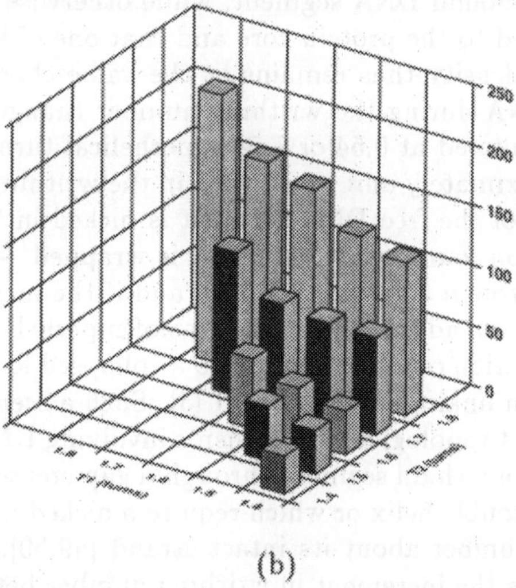

(b)

FIG. 5.5. *Variation of (a) the writhing number and (b) the virtual rotation of two symmetrically positioned proteins bound to DNA as a function of the degree of superhelical wrapping of a torsionally relaxed 500 bp chain about each protein (T_1 and T_2). Energy minima identified by Monte Carlo simulated annealing of a B spline representation of the chain axis. See text for further details.*

gradual and occurs at a greater degree of superhelical wrapping than the corresponding transition in a chain of the same length with a single bound protein of the same type. Up to $\sim$ 1.8 turns of DNA can be bound on each protein without decreasing the average writhing number per protein core to the value of -1.7 to -1.8 observed when a single protein binds to a torsionally relaxed closed DNA. In addition, up to $\sim$ 1.7 turns of DNA per protein can be accommodated on a small DNA without inducing a buckling transition to a collapsed (figure-8 shaped) configuration, when more than one protein is present on the DNA ring. This suggests that up to $\sim$ 140 bp of DNA can be complexed per protein and remain in agreement with the changes in ΔLk attained in gel electrophoresis studies of DNA minichromosomes [43,44]. The change in the writhing number per protein is thus not a fixed value but dependent upon the number of bound proteins.

6. Discussion. The driving force behind these computations is a desire to understand the influence of one or more DNA-wrapping proteins on the overall three-dimensional folding of the double helix. The observed sensitivity of DNA configuration to small changes in protein wrapping suggests a plausible mechanism for the action of various topoisomerases and the histone core assembly, both of which are known to change the linking number of closed molecules and to bend long segments of DNA [43,45-48]. The mechanism follows directly from the variation in writhing number with degree of superhelical wrapping reported in Figure 5.1. The computations assume that the unbound DNA segment, while otherwise free, has its two ends firmly anchored to the protein core and that one of its two strands is nicked. The twist density thus remains at the value characteristic of torsionally relaxed DNA during the writhing number change. The net effect of the transition centered at 0.54 or 1.54 superhelical turns in the figure is a decrease of approximately unit magnitude in the writhing number. If one of the two strands of the free DNA segment is nicked in the "open" equilibrium configuration found when the DNA is wrapped $\sim$ 0.4 superhelical turns around the protein core, for example, and if the nicked strand is ligated (resealed) after an additional $\sim$ 0.2 turns of superhelical wrapping and concomitant completion of the transition to a collapsed loop, the DNA will undergo a change in linking number of -1 [3]. Such a scenario differs from previously suggested topological mechanisms involving DNA which require controlled passage of a chain segment through a gap created by severing either strand of the double helix or which require a nicked molecule to rotate at fixed writhing number about its intact strand [49,50]. Here the increment in ΔLk equals the increment in writhing number between instants of nicking and ligation. That increment is determined by a small change in wrapping at points where the DNA increases or decreases its contact with protein.

The sudden re-orientation of DNA with increased levels of protein binding reflects a tug-of-war between the local geometric constraints of the protein binding site versus the overall bending energy of the protein-free hairpin loop. As evident from Figure 5.2, there is a build-up of bending energy (and hence local curvature) at the ends of the unbound DNA immediately prior to the sudden drop in writhing number. The transition is analogous (in the sense that a small change in a local parameter produces a large-scale change in structure) to the sharp configurational jump between the circle and figure-8 observed as a function of imposed ΔLk in ordinary closed circular DNA. Here, as noted above, the curves are all torsionally "relaxed" forms rather than closed circular states under topological stress. The increase in local bending energy in the loop becomes too great as the chain wraps more than a half superhelical turn around protein and thus induces the transformation to a collapsed configuration with a new crossover point of relatively close contact. Chemical agents which enhance or prevent the local binding of DNA to protein can thus control the overall tertiary structure of the elastic duplex. An agent which enhances the binding around the protein core can bring about the macromolecular collapse, while one that keeps the DNA off the protein can help to maintain an open state.

The degree of protein binding further determines which residues, normally distant along the polymer contour, are brought into direct interaction and also defines the orientation of nonbonded chain segments. This rather subtle long-range control is relevant to the action of other kinds of DNA-binding proteins, allowing communication between spatially separate elements on the chain. Geometric changes at the ends of the free DNA loop control the positioning of intervening residues with particular restrictions on the protein boundary, fixing specific nonbonded sites at characteristic distances and orientations (see, for example, Figure 5.4). It is thus easy to understand why the activity of a regulatory protein may be linked to the binding or reactivity of other molecular agents at different sites along the DNA. Both the angle and sequential position of contact are altered when the DNA is wrapped to a different extent around the protein core [3]. Thus, the specific geometry resulting from binding one protein to DNA can potentially affect the action of other molecular species.

Significantly, the computationally stable configurations of the DNA loops do not appear to be tied to either the form of chain representation or the simulation method. Several very different minimization and modeling approaches lead to the same structural conclusions. Moreover, the data are also consistent with recent extension of the theory of elastic rods which leads to explicit expressions for the equilibrium configuration of a constrained, torsionally relaxed DNA loop with boundary conditions of the type considered here [4]. According to the theory, the equilibrium configuration of a spatially constrained ideal DNA rod is a function of the distance and angle between the fixed end points together with the total length of the chain. The computations, however, go beyond the theory in being able to incorporate local base sequence-dependent structural features (e.g., bending anisotropy, natural curvature, etc.), long-range self contacts (e.g., electrostatic interactions), and external forces and moments (such as the random and viscous drag forces introduced in standard Langevin dynamics formulations) [13,32,35-37,51]. Here we have focused attention simply on the relative performance of different representations and computational treatments of a constrained, naturally straight, uncharged, inextensible, isotropic rod rather than examine effects of specific structural and environmental features or imposed supercoiling. We have further restricted the current analysis to the stable minimum energy equilibrium structures of the DNA rather than treat the full ensemble of macromolecular states associated with room temperature fluctuations of local structure.

Table 5.1 clearly shows that the most efficient ways to identify stable energy minima are through numerical solution of the equations of elasticity or by application of an efficient energy optimization package, such as TNPACK [38,39]. The former approach is critical for the correct treatment of bending and twisting in "real" DNA, where the ten common dimers are characterized by specific equilibrium roll, tilt, and twist angles and are subject to anisotropic sequence-dependent angular fluctuations [14,41]. The gradient and Hessian functions accumulated in the latter calculations are useful for molecular dynamics applications [31,35-37], but must be coupled in the future to the elastic treatment for a correct accounting of twisting and bending in supercoiled models, including the ideal isotropic rod. The Monte Carlo methods, while inefficient for the identification of global energy minima, are very useful for wide sampling of configuration space and are currently being combined with the elastic equations to examine the configurational features of long DNAs with several bound proteins (J. A. Martino & W. K. Olson, unpublished data). Full treatment of the electrostatic interactions of the charged phosphate groups along the DNA backbone, and thus proper accounting of the critical long-range contacts introduced upon supercoiling, requires the incorporation of new algorithms that speed-up the enumeration of pairwise energy terms. Application of a fast adaptive scheme to the analysis of electrostatic interactions in supercoiled DNA is under investigation (M. O. Fenley, K. Chua, A. H. Boschitsch, & W. K. Olson, unpublished data).

7. Acknowledgments. This research has been generously supported by the U.S. Public Health Service under research grant GM34809. Calculations were performed at the Rutgers Center for Computational Chemistry. Computer graphics were prepared with the dedicated assistance of Mr. Andrew Olson. J.A.M. is a Howard Hughes Medical Institute predoctoral fellow and T.P.W. a predoctoral trainee supported in part by grants from the U.S. Public Health Service (Molecular Biophysics Training Grant GM08319) and the U.S. Department of Education (National Needs in Chemistry Program).

REFERENCES

[1] Bates, A. D. & Maxwell, A., *DNA Topology*, IRL Press, Oxford, Chapter 6 (1993).

[2] Travers, A., *DNA-Protein Interactions*, Chapman & Hall, London, Chapter 7 (1993).

[3] Zhang, P., Tobias, I. & Olson, W. K., *Computer simulation of protein-induced structural changes in closed circular DNA*, J. Mol. Biol. 242, 271–290 (1994).

[4] Tobias, I., Coleman, B. & Olson, W. K., *Dependence of DNA tertiary structure on end conditions: Theory and implications for topological transitions*, J. Chem. Phys. 101, 10990–10996 (1994).

[5] White, J. H., *Self-linking and the Gauss integral in higher dimensions*, Amer. J. Math. 91, 693–728 (1969).

[6] Fuller, F. B., *The writhing number of a space curve*, Proc. Natl. Acad. Sci., USA 68, 815–819 (1971).

[7] Fuller, F. B., *Decomposition of the linking number of a closed ribbon: A problem from molecular biology*, Proc. Natl. Acad. Sci., USA 75, 3557–3561 (1978).

[8] White, J. H., *An introduction to the geometry and topology of DNA structure*, in *Mathematical Methods for DNA Sequences*, Waterman, M. S., Ed., CRC Press, Boca Raton, FL, pp. 225–253 (1989).

[9] Benjamin, H. W. & Cozzarelli, N. R., *DNA-directed synapsis in recombination: Slithering and random collision of sites*, Proc. R. A. Welch Found. Conf. Chem. Res. 29,107–126 (1986).

[10] Mortenson, M. E., *Geometric Modeling*, John Wiley & Sons, New York, Chapter 2 (1985).

[11] Dill, E. H., *Kirchhoff's theory of rods*, Archive for History of Exact Science 44, 1–23 (1992).

[12] Berman, H. M., Olson, W. K., Beveridge, D. L., Westbrook, J., Gelbin, A., Demeny, T., Hsieh, S.-H., Srinivasan, A. R. & Schneider, B., *The nucleic acid database: A comprehensive relational database of three-dimensional structures of nucleic acids*, Biophys. J. 63, 751–759 (1992).

[13] Olson, W. K., Babcock, M. S., Gorin, A., Liu, G.-H., Marky, N. L., Martino, J. A., Pedersen, S. C., Srinivasan, A. R., Tobias, I., Westcott, T. P. & Zhang, P., *Flexing and folding double helical DNA*, Biophys. Chem. 55, 7–29 (1995).

[14] Gorin, A. A., Zhurkin, V. B. & Olson, W. K. *DNA twisting correlates with base pair morphology*, J. Mol. Biol. 247, 34–48 (1995).

[15] Yoon, D. Y. & Flory, P. J., *Moments and distribution functions for polymer chains of finite length. II. Polymethylene chains*, J. Chem. Phys. 61, 5366–5380 (1974).

[16] Marky, N. L. & Olson, W. K., *Loop formation in polynucleotide chains. I. Theory of hairpin loop closure*, Biopolymers 21, 2329–2344 (1982).

[17] Hagerman, P. J., *Analysis of ring-closure probabilities of isotropic wormlike chains: Application to duplex DNA*, Biopolymers 24, 1881–1897 (1985).

[18] Levene, S. D. & Crothers, D. M., *Ring closure probabilities for DNA fragments by Monte Carlo simulation*, J. Mol. Biol. 189, 61–72 (1986).

[19] Vologodskii, A. V., Levene, S. D., Frank-Kamenetskii, M. D. & Cozzarelli, N. R., *Conformational and thermodynamic properties of supercoiled DNA*, J. Mol. Biol. 227, 1224–1243 (1992).

[20] Frank-Kamenetskii, M. D., Lukashin, A. V. & Vologodskii, A. V., *Statistical mechanics and topology of polymer chains*, Nature (London) 258, 398–402 (1975).

[21] Vologodskii, A. V., Anshelevich, V. V., Lukashin, A. V. & Frank-Kamenetskii, M. D., *Statistical mechanics of supercoils and the torsional stiffness of the DNA double helix*, Nature (London) 280, 294–298 (1979).

[22] Frank-Kamenetskii, M. D. & Vologodskii, A. V., *Topological aspects of the physics of polymers: The theory and its biophysical applications*, Sov. Phys. Usp. (Eng. ed.) 24, 679–696 (1981).

[23] Klenin, K. V., Vologodskii, A. V., Anshelevich, V. V., Dykhne, A. M. & Frank-Kamenetskii, M. D., *Computer simulation of DNA supercoiling*, J. Mol. Biol.

217, 413–419 (1991).

[24] Tan, R. K.-Z. & Harvey, S. C., *Molecular mechanics models of supercoiled DNA*, J. Mol. Biol. 205, 573–591 (1989).

[25] Tan, R. K.-Z. & Harvey, S. C., *Succinct macromolecular models: Application to supercoiled DNA* in *Theoretical Biochemistry and Molecular Biophysics Volume 1: DNA*, Beveridge, D. L. & Lavery, R., Eds., Adenine Press, Schenectady, NY, pp. 125–137 (1990).

[26] Malhotra, A., Tan, R. K.-Z. & Harvey, S. C., *Modeling large RNAs and ribonucleoprotein particles using molecular mechanics techniques*, Biophys. J. 66, 1777–1795 (1994).

[27] Yang, Y., Tobias, I. & Olson, W. K., *Finite element analysis of DNA supercoiling*, J. Chem. Phys. 98, 1673–1686 (1993).

[28] Bauer, W. R., Lund, R. A. & White, J. H., *Twist and writhe of a DNA loop containing intrinsic bends*, Proc. Natl. Acad. Sci., USA 90, 833–837 (1993).

[29] Hao, M.-H. & Olson, W. K., *Modeling DNA supercoils and knots with B-spline functions*, Biopolymers 28, 873–900 (1989).

[30] Hao, M.-H. & Olson, W. K., *Searching the global equilibrium configurations of supercoiled DNA by simulated annealing*, Macromolecules 22, 3292–3303 (1989).

[31] Schlick, T. & Olson, W. K., *Supercoiled DNA energetics and dynamics by computer simulation*, J. Mol. Biol. 223, 1089–1119 (1992).

[32] Zhang, P., Olson, W. K. & Tobias, I., (1991) *Accelerated record keeping Fourier series Monte Carlo simulations of an isotropic elastic rod model of DNA*, Comp. Polymer Sci. 1, 3–17 (1991).

[33] Olson, W. K. & Zhang, P., *Computer simulation of DNA supercoiling*, Methods in Enzymology 203, 403–432 (1991).

[34] Metropolis, N., Rosenbluth, A. W., Rosenbluth, M. N., Teller, A. & Teller, E., *Equation of state calculations by fast computing machines*, J. Chem. Phys. 21, 1087–1092 (1953).

[35] Schlick, T., Olson, W. K., Westcott, T. & Greenberg, J. P., *On higher buckling transitions in supercoiled DNA*, Biopolymers 34, 565–597 (1994).

[36] Schlick, T., Li, B. & Olson, W. K., *The influence of salt on the structure and energetics of supercoiled DNA*, Biophys. J. 67, 2146–2166 (1994).

[37] Liu, G., Olson, W. K. & Schlick, T., *Application of Fourier analysis to computer simulation of supercoiled DNA*, Comp. Polymer Sci. 5, 7–27 (1995).

[38] Schlick, T. & Overton, M., *A powerful truncated method for potential energy minimization*, J. Comp. Chem. 8, 1025–1039 (1987).

[39] Schlick, T. & Fogelson, A., *TNPACK - A truncated Newton minimization package for large-scale problems: I. Algorithm and usage, and II. Implementation example*, ACM Trans. Math. Soft. 18, 46–70 and 71–111 (1992).

[40] Press, W. H., Flannery, B. P., Teukolsky, S. A. & Vetterling, W. T., *Numerical Recipes*, Cambridge University Press, Cambridge, Chapter 9 (1986).

[41] Olson, W. K., Marky, N. L., Jernigan, R. L. & Zhurkin, V. B., *Influence of fluctuations on DNA curvature. A comparison of flexible and static wedge models of intrinsically bent DNA*, J. Mol. Biol. 232, 530–554 (1993).

[42] Rybenkov, V. V., Cozzarelli, N. R. & Vologodskii, A. V., *Probability of DNA knotting and the effective diameter of the DNA double helix*, Proc. Natl. Acad. Sci., USA 90, 5307–5311 (1993).

[43] Germond, J. E., Hirt, B., Oudet, P., Gross-Bellard, M. & Chambon, P., *Folding of the DNA double helix in chromatin-like structures from simian virus 40*, Proc. Natl. Acad. Sci., USA 72, 1843–1847 (1975).

[44] Zivanovic, Y., Goulet, I., Revet, B., Le Bret, M. & Prunell, A., *Chromatin reconstitution on small DNA rings II. DNA supercoiling on the nucleosome*, J. Mol. Biol. 200, 267–290 (1988).

[45] Moore, C. L., Klevan, L., Wang, J. C. & Griffith, J. D., *Gyrase·DNA complexes visualized as looped structures by electron microscopy*, J. Biol. Chem. 258, 4612–4617 (1983).

[46] Richmond, T. J., Finch, J. T., Rushton, B., Rhodes, D. & Klug, A., *Structure of the nucleosome core particle at 7 Å resolution*, Nature (London) 311, 532–537 (1984).
[47] Klug, A., Finch, J. T. & Richmond, T. J., *Crystallographic structure of the octamer histone core of the nucleosome*, Science 229, 1109–1110 (1985).
[48] Bates, A. D. & Maxwell, A. *DNA gyrase can supercoil DNA circles as small as 174 base pairs*, EMBO J. 8, 1861–1866 (1989).
[49] Champoux, J. J., *Mechanistic aspects of type-I topoisomerases*, in *DNA Topology and Its Biological Effects*, Cozzarelli, N. R. & Wang, J. C., Eds., Cold Spring Harbor Laboratory Press, Cold Spring Harbor, NY, pp. 217–242 (1990).
[50] Hsieh, T.-S., *Mechanistic aspects of type-II DNA topoisomerases* in *DNA Topology and Its Biological Effects*, Cozzarelli, N. R. & Wang, J. C., Eds., Cold Spring Harbor Laboratory Press, Cold Spring Harbor, NY, pp. 243–263 (1990).
[51] Fenley, M. O., Olson, W. K., Tobias, I. & Manning, G. S., *Electrostatic effects in short superhelical DNA*, Biophys. Chem. 50, 255–271 (1994).

PURSUING LAPLACE'S VISION ON MODERN COMPUTERS

TAMAR SCHLICK*

Abstract. This contribution is an informal essay based on a talk delivered at the Institute for Mathematics and its Applications (IMA) in Minneapolis, under the summer program in molecular biology, July 18–22, 1994. I exclude many technical details, which can be found elsewhere, and instead focus on the basic ideas of molecular dynamics simulations, with the goal of conveying to students and non-specialists the key concepts of the theory and practice of large-scale simulations. Following a description of the basic idea in molecular dynamics, I discuss some of the practical details involved in simulations of large biological molecules, the numerical timestep problem, and approaches to this problem based on implicit-integration techniques. I end with a perspective of open challenges in the field and directions for future research.

1. Introduction. Consider the following statement by the celebrated French mathematician Pierre Simon de Laplace (1749–1827) [1]:

> *Une intelligence qui, pour un instant donné, connaîtrait toutes les forces dont la nature est animée et la situation respective des êtres qui la composent, si d'ailleurs elle était assez vaste pour soumettre ces données à l'Analyse, embrasserait dans la même formule les mouvements des plus grands corps de l'univers et ceux du plus léger atome: rien ne serait incertain pour elle, et l'avenir, comme le passé, serait présent à ses yeux. L'esprit humain offre, dans la perfection qu'il a su donner à l'Astronomie, une faible esquisse de cette intelligence.*

> *An intelligence which could, at any moment, comprehend all the forces by which nature is animated and the respective positions of the beings of which it is composed, and moreover, if this intelligence were far-reaching enough to subject these data to analysis, it would encompass in that formula both the movements of the largest bodies in the universe and those of the lightest atom: to it nothing would be uncertain, and the future, as well as the past, would be present to its eyes. The human mind offers us, in the perfection which it has given to astronomy, a faint sketch of this intelligence.*

* New York University and the Howard Hughes Medical Institute, Chemistry Department and Courant Institute of Mathematical Sciences, 251 Mercer Street, New York, NY 10012, schlick@nyu.edu

Although these words were written about 175 years ago — at a time when the significant effect of high-speed computers on the solution of physical problems could not have been imagined — all the ingredients of molecular dynamics (MD) simulations, as performed today, are contained: mathematical construction of the suitable force field, design of appropriate numerical integration tools, and long-time propagation of the equations of motion. These aspects are briefly discussed below, with a focus on the multiple-timescale problem and how it might be resolved. Of course, even Newtonian mechanics taken to its extreme cannot predict with certainty the future motion of all bodies. As became evident by the work of Poincaré less than a century after Laplace's statement, the solar system is chaotic even though the underlying laws can be clearly expressed. This understanding, however, should not deter us from pursuing Laplace's dream; rather, it should stimulate us to explore as deeply as possible the consequences of Newtonian physics.

2. The basic idea of MD. MD is the science of simulating motion of a system of particles according to Newton's classical equations of motion [2,3]. It is an exciting field, not only in the interface of theory and experiment, but also at the crossroads of mathematics, biology, chemistry, physics, and computer science. As Laplace indicated, the capability to analyze and predict motion — be it of the solar system or biological systems — provides the link between the past and the future. Clearly, a static picture of a biomolecule as obtained from X-ray crystallography — while extremely valuable — is still insufficient for understanding a wide range of biological activity. It only provides an average, frozen view of a complex system, but molecules are live entities, with their constituent atoms continuously interacting among themselves and with their environment. Thus, MD is necessary to explain the wide range of thermally accessible states of a system and to connect structure to function [4]. By following the dynamics of a molecular system in time, we can obtain a rich amount of information concerning mean atomic fluctuations, heat capacities, rates of configurational changes, enzyme-substrate binding, various types of concerted motion and, ultimately perhaps, protein folding [5,6].

MD is simple in principle. We simulate motion under the influence of a specified force field by following configurations in time according to Newton's equation of motion. We write these equations for a system of N atoms as the following pair of first-order differential equations:

$$\mathbf{M}\dot{V}(t) = -\nabla E(X(t)) + \dots \tag{2.1a}$$

$$\dot{X}(t) = V(t). \tag{2.1b}$$

In these equations, $X \in R^{3N}$ denotes the collective Cartesian vector of the system (i.e., the x, y, and z components of each atom are listed in turn);

V is the collective velocity vector; $\mathbf{M}$ is the diagonal mass matrix (i.e., the masses of each atom are repeated three times in the diagonal array of length $3N$); and the dot superscripts denote differentiation with respect to time, t. The force in the right-hand-side of eq. (2.1a) is composed of the systematic force, which is the negative gradient (vector of first partial derivatives)* of the potential energy E and, possibly, additional terms that mimic the environment.

For example, in its simplest form, the *Langevin* equation of motion (a special case of the *generalized* Langevin equation) represents the interactions between the solute (e.g., biomolecule) and the solvent environment through an effective frictional term that depends on the damping constant, γ (also known as *collision frequency*), and a stochastic restoring force:

$$\text{(2.2a)} \qquad \mathbf{M}\dot{V}(t) = -\nabla E(X(t)) - \gamma \mathbf{M} V(t) + R(t).$$

The random force R is a Gaussian process, whose purpose is to balance frictional dissipation with fluctuations that may increase the energy, and therefore maintain thermal equilibrium at a temperature T [7]:

$$\text{(2.2b)} \qquad \langle R(t) \rangle = 0, \quad \langle R(t)R(t')^T \rangle = 2\gamma k_B T \mathbf{M} \delta(t - t'),$$

where k_B is Boltzmann's constant, and δ is the usual Dirac delta symbol.

The potential energy E is a simple, semi-empirical function that has been parameterized for small systems (for which structure and thermodynamic data are available) and extended to large systems composed of the same basic chemical subgroups. The functions used today for biomolecules are basically the outgrowth of *molecular mechanics* [8,9,10,11], or force-field development and energy minimization efforts, that began in the 1960s in several leading laboratories across the world.

In molecular mechanics, the molecule is considered as a *mechanical system*, in which particles are atoms and springs are bonds. Simple, pairwise-additive potentials are then constructed to express how the composite atoms stretch, vibrate, and rotate about the bonds in response to intra- and intermolecular forces. Essentially, this energy is parameterized on the basis of three main sources: quantum mechanics (where available), experimental measurements (vibrational spectra, molecular conformations, heats of formation, thermodynamic functions, dipole moments, crystal properties, etc.), as well as trial and error, to reproduce observed properties of model systems.

Although there are many variations in the potential energy expressions that govern nucleic-acid and protein simulations today [12,13,14], the basic components include *nonbonded* and *bonded* terms. In the first cat-

* Each gradient component i, $i = 1, \ldots 3N$, is given by: $\nabla E(X)_i = \partial E(X)/\partial x_i$, where x_i denotes an x, y, or z component of an atom.

egory, Lennard-Jones (E_{LJ}) and Coulombic (E_{COUL}) potentials are typically used and, in the second, bond-length (E_{BOND}), bond-angle (E_{BANG}), and dihedral-angle (E_{TOR}) terms, as follows.

$$E = E_{LJ} + E_{COUL} + E_{BOND} + E_{BANG} + E_{TOR} \tag{2.3a}$$

$$E_{LJ} = \sum_{i,j \in S_{NB}} \left(\frac{-A_{ij}}{r_{ij}^6} + \frac{B_{ij}}{r_{ij}^{12}} \right) \tag{2.3b}$$

$$E_{COUL} = \sum_{i,j \in S_{NB}} \left(\frac{Q_i Q_j}{\epsilon(r_{ij})\, r_{ij}} \right) \tag{2.3c}$$

$$E_{BOND} = \sum_{i,j \in S_B} S_{ij} \left(r_{ij} - \bar{r}_{ij} \right)^2 \tag{2.3d}$$

$$E_{BANG} = \sum_{i,j,k \in S_{BA}} K_{ijk} (\cos\theta_{ijk} - \cos\bar{\theta}_{ijk})^2 \tag{2.3e}$$

$$E_{TOR} = \sum_{ijk\ell \in S_{DA}} \sum_n \left(\frac{V n_{ijk\ell}}{2} \left[1 + \cos(n \tau_{ijk\ell}) \right] \right) \tag{2.3f}$$

In these expressions, the symbols S_B, S_{BA}, and S_{DA} denote the sets of all bonds, bond angles, and dihedral angles, respectively. The nonbonded set, S_{NB}, typically includes all (i, j), $i < j$, atom pairs separated by three bonds or more. The parameters A, B, Q, S, K, and V are energy constants. The symbols r, θ, and τ denote the internal variables of bond lengths, bond angles, and dihedral angles, respectively, with bar symbols ($\bar{\theta}$, $\bar{r}$) indicating reference values. The integer n in eq. (2.3f) denotes the rotational periodicity of a certain dihedral angle (typically 2,3,6). A bond angle θ_{ijk} formed by a bonded triplet of atoms i–j–k is computed by the inner product of the two vectors (j to k and j to i). A dihedral angle $\tau_{ijk\ell}$ defines the rotation of bond i–j about the central j–k bond with respect to k–l, and is computed from the dot product of the two unit normals $\mathbf{n}_{ijk}$ and $\mathbf{n}_{jkl}$ with respect to the planes spanned by atoms i–j–k and j–k–l, respectively [15]. The Coulombic potential above (eq. 2.3c) includes a possible modification by a distance-dependent dielectric function, $\epsilon(r)$. Additional terms may be used (e.g., improper-torsion and hydrogen-bond potentials) [16,17]. For further details, the reader is invited to consult the literature cited above and a recent volume of *Chemical Reviews* (November 1993), which surveys several aspects of the broad field of molecular mechanics and modeling [18]. Many *macroscopic* models (i.e., not an all-atom level) use similar functions or potentials derived from the functions above.

Once we have established this approximate energy function, minimization gives information on favorable regions in configuration space (molecular mechanics or statics), and integration of the equations of motion reveals the dynamics of the system under the influence of the associated force field. Thus, the Newton equations of motion are fundamental because they relate three key quantities: time, configuration, and energy. Besides providing continuous dynamics information, MD is also used today as a conformational search tool — to bypass the multiple-minimum problem — and as a vehicle to refine low-resolution X-ray or NMR (nuclear magnetic resonance) data.

3. The practical difficulties of MD. Although the basic ideas are simple, MD is challenging in practice. The practical difficulties arise from the following components that enter into simulations: appropriate initial conditions, appropriate energy functions, suitable numerical integrators, careful simulation protocol, high-performance computers, careful analysis of the results, and powerful visualization.

Initial conditions. Initial coordinates and velocities are necessary to begin a dynamics simulation. Even when initial coordinates are available from experiment (e.g., crystal structure), the starting vector may not correspond to a minimum in the potential energy function used, and hence minimization (further refinement) is needed to relax strained contacts. When an experimental structure is not available, a ***build-up*** technique may be used to construct a structure on the basis of the known building blocks, and minimization again is required. The initial velocity vector is typically set from a Gaussian distribution so that the total kinetic energy of the system corresponds to the expected value at the target temperature T. (According to the classical equipartition theorem, each mode of degree of freedom has $(k_BT)/2$ energy at thermal equilibrium). Thus, an initial round of ***equilibration*** is necessary to initiate a molecular dynamics trajectory. In Langevin dynamics, this task is simpler because the random forces quickly lead the system to thermal equilibrium, even with a zero initial velocity vector.

Energy function. As mentioned above, the development of suitable potential energy functions for biomolecules is an ongoing effort [8,18]. Uncertainties are well recognized in the representation of solvent [19] and electrostatic interactions [13,14] and in the functional form of the local potentials (i.e., lack of anharmonic [20] and cross terms [21]). Particularly promising are efforts to combine quantum and molecular mechanics calculations and to include atomic polarizabilities. Clearly, the reliability of the results simulated by MD can only be as good as the governing force field.

Numerical integrator. As for numerical integrators, many of the well-known schemes, such as Verlet [22] and Runge-Kutta, have been used for a long time for solving ordinary differential equations that arise in chemistry (e.g., in quantum mechanics and MD applications). However, improved methods with greater stability are needed for the multiple timescale, large-scale systems of coupled harmonic oscillators present in MD of polymers and biological molecules. Indeed, both the increasing availability and speed of computers and the desire to extend simulation times to longer timeframes have stimulated exciting developments in algorithms for MD. These advances include symplectic [23, for example] and implicit integration methods, multiple timescale approaches, constrained dynamics formulations, internal-coordinate MD, and various strategies based on separating frameworks. Several of these will be detailed below. The ultimate goal is to develop integration schemes suitable for conservative systems that can simulate macromolecular systems reliably over long time frames at a significant computational gain over current ("brute-force") methods. At present, however, approaches that can extend to greater timescales generally sacrifice resolution of the local detail; moreover, computational time tends to be large.

Simulation protocol. A careful simulation protocol is essential for MD to ensure equilibration, proper enforcement of the boundary conditions (when, for example, a biomolecule is placed in a box of water molecules), proper positioning of solvent and salt molecules, computation of the nonbonded terms, monitoring of the kinetic temperature and the energetic fluctuations (for the possibility of systematic drifts or abrupt changes), and so on. For example, special improved techniques have been offered for truncating smoothly the nonbonded terms (to minimize artificial effects) [24] or, alternatively, for computing all the nonbonded terms in linear complexity by fast particle methods [25,26].

Computers. High-performance computers are essential for performing the computationally-intensive MD simulations [27]. The molecular models are large for biomolecules (thousands of atoms), and hence the energy and derivative computations are expensive. Further, millions of integration steps are required to enter the nanosecond range by typical methods. Thus, although MD simulations of condensed systems were performed early in this century, simulations of biomolecules only gained momentum in the mid-to-late 1980s with the advent of high-speed computing. Thus, the field of biomolecular simulations can be considered as relatively young, with many practical and algorithmic issues still to be resolved. In Table 3.1, representative calculations are shown for a perspective. Clearly, there has been a tendency to increase system size rather than the simulation time. Dynamics over longer times and larger spatial scales can only be approximated through macroscopic models of proteins and DNA.

Analysis and visualization. In addition to all of the above issues, careful analysis of the results and powerful visualization techniques are essential components of biomolecular simulations today. With the increasing ease and accessibility of generating computer trajectories, the challenge remains of carefully analyzing the voluminous data to distill the essential findings. While many scalar and vector functions can be computed from a

TABLE 3.1
*A Perspective on Molecular Mechanics and Dynamics Simulations**

Period	System and Size	Application	Simulated Time
Early 1970s	ATP (around 45 atoms)	MM	—
Late 1970s	BPTI (58 residues)	MD	10 ps
Early 1980s	GnRH (linear decapeptide)	MD	15 ps
Early 1980s	DNA (12 and 24 base-pair systems in vacuum)	MD	100 ps
Mid 1980s	DNA (5 base pairs with solvent and salt)	MD	100 ps
Late 1980s	Phospholipid Micelle (85 LPE and 1591 water molecules, roughly 7,000 atoms)	MD	100 ps
1994	Lipid bilayer system (27,000 atoms)	MD	250 ps

* The examples shown for each period are only representative. In the Application column, "MM" and "MD" denote molecular mechanics and molecular dynamics, respectively. The first system, ATP, is adenosine triphosphate, and in the early 1970s it was considered large for finding all minima by MM [10]. (Global optimization remains a difficult problem!). BPTI is the small protein Bovine Pancreatic Trypsin Inhibitor, on which a pioneering MD simulation was performed by McCammon, Gelin & Karplus [71]. GnRH denotes a gonadotropin-releasing hormone, a linear decapeptide, which triggers LH and FSH hormones; the simulation above was performed by Hagler *et al.* in 1983 [72]. The simulations of the DNA 12 and 24 base-pair fragments were performed in vacuum by Levitt [73] without electrostatics, and that of the DNA pentamer system, with 830 water molecules and 8 sodium ions and full electrostatics, by Kollman and co-workers [74]. The phospholipid aggregate simulations by Wendoloski *et al.* [75] incorporated a hydrated lysophosphatiadylethanolamine (LPE) micelle, containing 85 LPE and 1591 water molecules. (A micelle is a spherical aggregate of phospholipid molecules). The lipid bilayer simulation performed by Schulten and co-workers included about 5500 water molecules [76].

long series of configurations, understanding the dynamics requires a combination of sophisticated analysis and visualization tools.

Molecular graphics was more of a problem in the early days of molecular mechanics than it is now; before the surge of graphics innovations, researchers relied more heavily on mechanical models and plain intuition. Indeed, many companies have made an industry of molecular modeling and simulations and offer sophisticated tools to both experimentalists and computational/theoretical chemists. The dazzling capability of computer graphics today to render and animate a large, complex three-dimensional image — often so "real" in appearance that the source may be obscured — has made biological interpretation much easier, but one still has to know exactly where and for what to look.

4. The complexity of biomolecules. As mentioned above, while MD was applied to condensed systems much earlier this century, macromolecular applications were made possible only in the recent decade, with the advent of high-speed computing. There are many reasons why this lag occurred, and the issue at heart is best captured by the following statement by Frauenfelder and Wolynes [28]:

> *"Whatever complexity means, most people agree that biological systems have it."*

Energy landscape. Indeed, the energy landscape is complex for biomolecules [29]. A protein is a long chain of amino acids which can fold back on itself in many ways; similarly, DNA and RNA are long polymers made up of nucleotides which can fold up in three dimensions into many intricate patterns. In the folded forms, many segments that are far apart in the linear sequence can come into close contact. These contacts — be they hydrogen bonds, disulfide bonds, or noncovalent interactions like stacking and favorable electrostatics — are difficult to predict *a priori*. Thus, the multidimensional potential energy surface that governs biomolecular structure has many maxima, minima, and saddle points. Further, the distributions about each favorable or unfavorable state are highly anisotropic, with the width depending on the entropy associated with that state.

Asymmetry. Biomolecules are also asymmetric in comparison to simple systems, such as homogeneous liquid clusters, which were successfully simulated long ago. Certainly, there are symmetries in many aspects of protein and nucleic acid structure (e.g., many proteins are dimers, and the "idealized" DNA double helix has an axis of symmetry), but in realistic environments there are many sequence-specific motifs and binding interactions with other biomolecules in the environment that induce local structural variations in macromolecules. These local trends can produce profound global effects.

Complex motion. The motion of biomolecules is also more complex than that of small or homogeneous systems. The collective motion is a superposition of many fundamental motions ("normal modes") that characterize the dynamics of a biomolecule: stretches, bends, torsions, and combinations of those. Although these motions have associated frequencies of disparate magnitudes, the overall motion of a biomolecule is highly cooperative. That is, small local fluctuations can trigger a chain of events that will lead to a large global rearrangement. Energy transfer among vibrational modes may also be involved. In this sense, biomolecular systems can be considered chaotic; a unique trajectory cannot be expected from an MD simulation, but a 'representative' trajectory is sought.

Solvent and salt. In addition to the complexity of the energy landscape and the motion and the overall asymmetry involved, the solvent and ions in the environment influence the structure and dynamics of biomolecules profoundly. Solvent molecules surround and interpenetrate proteins and nucleic acids and damp many characteristic motions that might be present in vacuum. Similarly, ions such as sodium, calcium, and magnesium influence structure and dynamics significantly, especially of the polyelectrolytic nucleic acids [30, for example, in reference to supercoiled DNA]. Thus, for physical reliability, these long-range electrostatic interactions are essential to consider in MD simulations of macromolecules, in particular to accurately describe kinetic processes.

Multiple timescales. Another property associated with biomolecules that makes them both fascinating and complex is the *extraordinary* range of timescales involved. Relevant frequencies range from the rapid, high-frequency vibrations of characteristic times of 10^{-14} s (e.g., bond stretching) to protein folding and other global rearrangements, in the second range, to evolutionary timescales in the terasecond regime. In particular, there are many coupled characteristic motions in the 10^{-14} to 10^{2} s range (e.g., global bending and twisting of DNA, hinge-bending motion of proteins), so the problem is not easily amenable to a simple separating framework.

Statistical mechanics. Another important factor in simulating biomolecules, a product of all the above, is the need to consider a global framework for analysis, namely that of statistical mechanics. That is, *ensemble* properties are of interest, and this more global framework is necessary to analyze and put into perspective results generated today by single-molecule trajectories at relatively short times. Even for small molecules, it is difficult to sample all relevant conformations and, moreover, obtain the correct expected Boltzmann distribution [31,32]. For biomolecules, we expect to sample only a small region of the feasible energy surface, and this lack of ergodicity demands a global framework for trajectory analysis.

Most error analyses are performed for simple systems for which the trajectory is known in advance and is well defined [33]. Appropriate mathematical theories of error-analyses and long-time stability for ordinary differential equations of the type that arise in chemistry — multidimensional, highly nonlinear, frequently chaotic systems — are not yet available. Indeed, long-time stability analysis is quite at its infancy as a field [34]. In the context of biomolecular MD, we are interested in both the local (detailed kinetics) and global (sampling) features of our systems. In other words, from a practical point of view, we would like to merge both qualitative and quantitative behavior of biological systems. Therefore, one can imagine that different models, in combination with different integration or propagation methods, could be designed to address different aspects of dynamics problems for macromolecules. Accordingly, the appropriateness of an algorithm should be tightly coupled to the scientific questions that are being addressed with it. Accurate, small timestep methods may be excellent for understanding certain rapid dynamic correlations, while they are likely to fail at giving correct Boltzmann statistics. Similarly, enhanced sampling techniques may reveal more configurational states, but this additional information may come at a loss of continuous dynamics information.

5. The timestep problem. As mentioned above, the sampling problem is one of the two most urgent in the field, the other being the development of improved force fields. The sampling problem has two components: *spatial* and *temporal*. The multiple-minimum problem is associated with the former [35,36]. None of the currently available techniques — Monte Carlo (MC), global optimization, Brownian, Langevin and molecular dynamics — can guarantee finding the global minimum of a multivariate function associated with a biomolecule, though many promising methods are under development [35,37, and references cited therein]. The timestep problem is associated with the temporal component of the sampling problem. In principle, MD simulations can combine both the spatial and temporal features but, at present, significant progress is required to extend our results beyond the nanosecond horizon to capture the many thermally-accessible conformation states and transitional pathways of biomolecules.

The timestep problem arises from the large range of characteristic timescales associated with internal motions of biomolecules. These times span more than 10 orders of magnitude. The high-frequency, localized, and small-amplitude motions reside at the 10^{-15}s end, and many characteristic motions exist all the way up to the slow-frequency, global and large-amplitude processes near the second range. But to resolve the high frequencies, typical integration schemes require the timestep of $\Delta t = 1$ fs for numerical stability and, hence, one million steps for a nanosecond. The nanosecond timescale is thus the typical approachable range today for standard all-atom simulations, even on high-speed computers.

Let us examine the general discretization problem of an ordinary differential equation of the general form $y' = f(y)$ where y is a vector. (For example, y consists of X and $\dot{X}$ for MD). We specify a timestep Δt and then choose a numerical integration (or quadrature) scheme that balances desired accuracy, stability, and complexity [38,39]. Two broad classes of integration schemes are available: *explicit* and *implicit*. In explicit schemes, the solution at the end of the timestep is obtained by performing operations on the variables at the beginning of each timestep. Symbolically, we write $y^{n+1} = f(y^n, \Delta t, \ldots)$, where f is some nonlinear function, y^n is the difference-equation approximation to the solution y at time $n\Delta t$, and the dots indicate other parameters or previous solutions (i.e., y^{n-1}, y^{n-2}). With implicit integrators, the final solutions are functions of both the initial and final variables: $y^{n+1} = f(y^{n+1}, y^n, \Delta t, \ldots)$. This generally requires solution of a coupled nonlinear equation at each timestep to propagate the trajectory. The explicit versions generally involve simple algorithms that (for propagation only) use modest memory, while implicit methods involve more complex algorithms but are often more powerful for treating systems with disparate timescale dynamics.

A harmonic oscillator example. Consider, for example, two different discretizations of the equation governing the dynamics of a harmonic oscillator with friction:

$$\ddot{x} + \gamma \dot{x} + \omega^2 x = 0. \tag{5.1}$$

In this equation, ω denotes the natural frequency of an oscillator of unit mass.

Explicit Discretization. A typical explicit discretization, which corresponds to the well-known leap-frog [2] formulation in the case of $\gamma = 0$, gives the recursion relation for the position:

$$x^{n+1} - 2x^n \left[1 - \frac{\gamma\Delta t}{2} - \frac{(\omega\Delta t)^2}{2}\right] + x^{n-1}(1-\gamma\Delta t) = 0 \,. \tag{5.2}$$

To analyze the stability of this formulation, we let z represent the complex solution of the above equation so that $z^n = x^n$. To simplify the expressions, we set $\epsilon \equiv \omega\,\Delta t$ and obtain the following quadratic equation for z:

$$z^2 - 2z\left(1 - \frac{\gamma\,\Delta t}{2} - \frac{\epsilon^2}{2}\right) + (1-\gamma\,\Delta t) = 0 \,. \tag{5.3}$$

The solution for z can be written for an underdamped oscillator ($\gamma < 2\omega$), with $i = \sqrt{-1}$, as

$$z = a \pm \epsilon\, i\sqrt{b}, \tag{5.4}$$

where

$$a = 1 - \left(\frac{\epsilon^2}{2} + \frac{\gamma\,\Delta t}{2}\right), \tag{5.5a}$$

and

$$b = 1 - \left(\frac{\epsilon^2}{4} + \frac{\gamma\,\Delta t}{2} + \frac{\gamma^2\Delta t^2}{4\epsilon^2}\right), \tag{5.5b}$$

as long as the discriminant above is nonnegative. This condition requires:

$$\Delta t \leq \frac{2\omega - \gamma}{\omega^2} \,. \tag{5.6}$$

Thus, the timestep must be sufficiently small to guarantee numerical stability in the explicit discretization. In particular, when $\gamma = 0$, we have the familiar condition $\Delta t \leq 2/\omega$.

We are also interested in the behavior of the solution z as $n \to \infty$. This limit of $|z|^n$ is of interest when $t = n\,\Delta t \to \infty$ with Δt fixed. (Here $|\cdot|$ denotes the magnitude of the complex quantity). We obtain from eqns. (5.4, 5.5a,b) that

$$|z|^2 = 1 - \gamma\,\Delta t \,, \tag{5.7}$$

and so

$$|z|^n = (1-\gamma\,\Delta t)^{n/2} \approx e^{-(\gamma/2)t} \,. \tag{5.8}$$

Thus, the difference-equation solution realizes only the ***physical***, frictional damping effect, at an amplitude decay rate of $\gamma/2$. In particular, when $\gamma = 0$, $|z| = 1$, so the solution magnitude is conserved from step to step.

Implicit Discretization. Implicit schemes are designed to relax the severe timestep constraint associated with explicit methods. To illustrate, we discretize eq. (5.1) by the implicit-Euler (IE) scheme, the simplest known implicit scheme that has high stability [39]. The IE discretization for $y' = f(y)$ is $(y^{n+1} - y^n)/\Delta t = f(y^{n+1})$. After algebraic rearrangements, we obtain in analogy to eq. (5.2) the relation:

$$x^{n+1}[1 + \gamma\,\Delta t + (\omega\Delta t)^2] - x^n(2 + \gamma\,\Delta t) + x^{n-1} = 0\,. \tag{5.9}$$

In the form of eq. (5.4), the solution for z can be written with the coefficients:

$$a = \frac{1 + \left(\frac{\gamma\,\Delta t}{2}\right)}{(1 + \gamma\,\Delta t + \epsilon^2)}\,, \tag{5.10a}$$

$$b = \frac{1 - \left(\frac{\gamma}{2\omega}\right)^2}{(1 + \gamma\,\Delta t + \epsilon^2)^2}\,. \tag{5.10b}$$

To determine the solution behavior in time, we derive, as above, the relation:

$$|z|^2 = (1 + \gamma\,\Delta t + \epsilon^2)^{-1} \tag{5.11}$$

from which

$$|z|^n = (1 + \gamma\,\Delta t + \epsilon^2)^{-n/2} \approx e^{-[\gamma/2 + (\omega^2\Delta t)/2]t}\,. \tag{5.12}$$

Here, the second source of damping is a *numerical* one, with an intrinsic amplitude decay rate of $(\omega^2\Delta t)/2$. Thus, the IE damping becomes more severe as ω and/or Δt increase, while the physical damping due to friction only depends on γ.

The stability of the IE scheme over all Δt can also be seen from the analysis above. Since the magnitude of z is less than one for all positive values of γ, ω, and Δt (eq. 5.11), the solution will not grow in magnitude over time. (This notion of stability is known as "unconditional stability" or "A-stability"). In fact, we see from eq. (5.12) that even in the absence of friction the solution magnitude will decrease from step to step; as $t \to \infty$, $|z| = (1+\epsilon^2)^{-1/2} \to 0$. Thus, straightforward IE integration cannot be used for MD, and some mechanism (e.g., Langevin formulation) is necessary to counteract numerical damping in some way.

Numerical Illustration. In Figure 5.1, numerical results are shown for the example above. Four timesteps (0.05, 0.1, 1.0, and 2.0) and three values of γ (0, 0.1, and 0.5) are used (the units are arbitrary) for an oscillator with $\omega = 1$ (frequency 2π radians). The motion is propagated by the two schemes above, with $x^0 = 0.8$, $v^0 = 0.5$, and x^1 set to $x^0 + (\Delta t)v^0$: *filled*

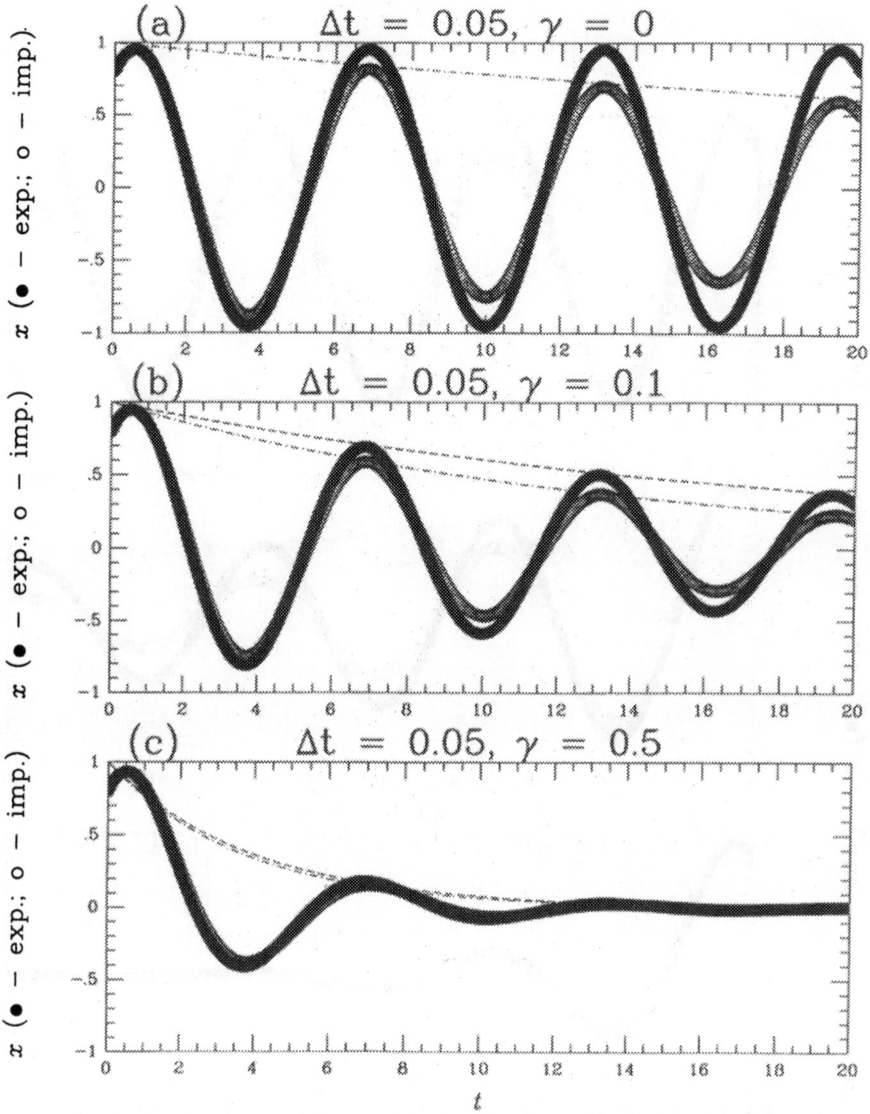

FIG. 5.1. (a,b,c). The Motion of a Harmonic Oscillator with Friction Simulated by Explicit and Implicit Schemes. *The dynamics is propagated by the explicit scheme described in eq. (5.2) of the text (filled circles) and the implicit-Euler (IE) scheme of eq. (5.9) (hollow circles) for various values of timesteps (Δt) and damping constants (γ), as indicated. For $\Delta t = 0.05$ and 0.1, the magnitude formulas of eq. (5.8) (dashed curve) for nonzero γ and eq. (5.11) (dot-dash) are shown for the explicit and implicit schemes, respectively. Note the different ranges in positions for various timesteps.*

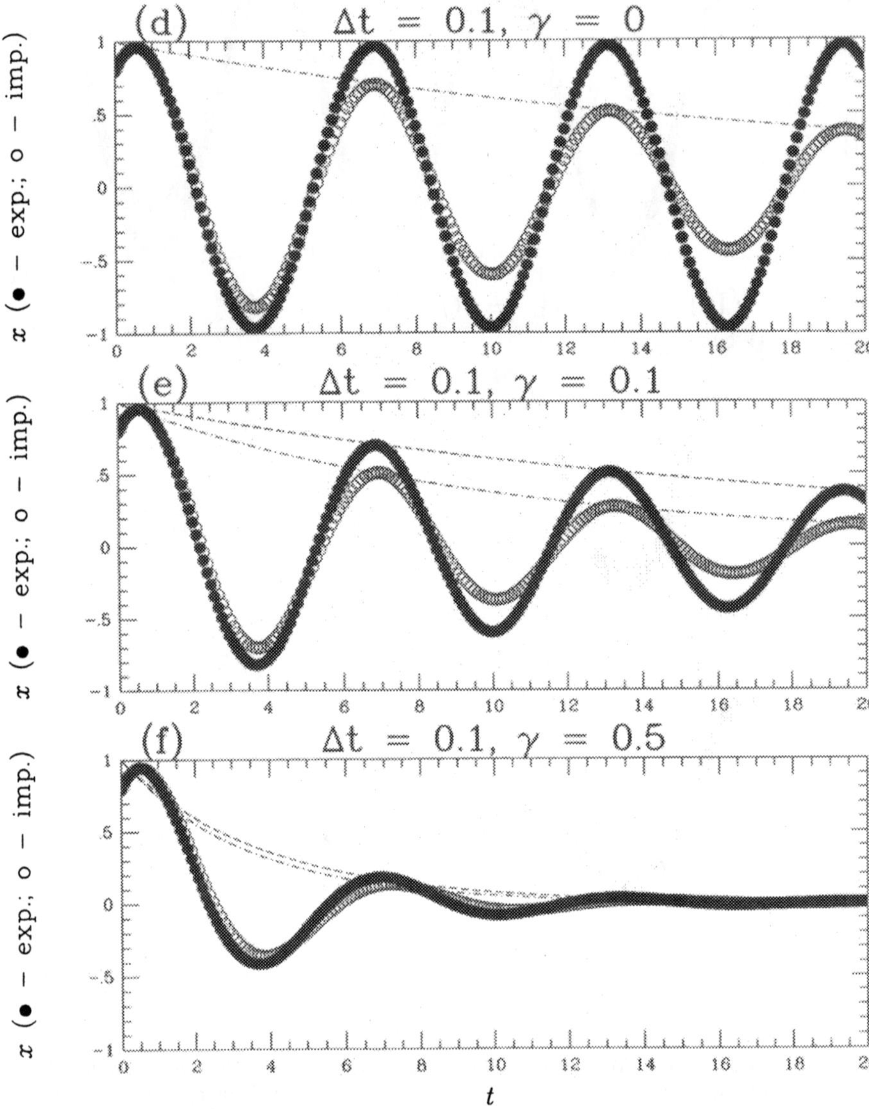

FIG. 5.1. (d,e,f).

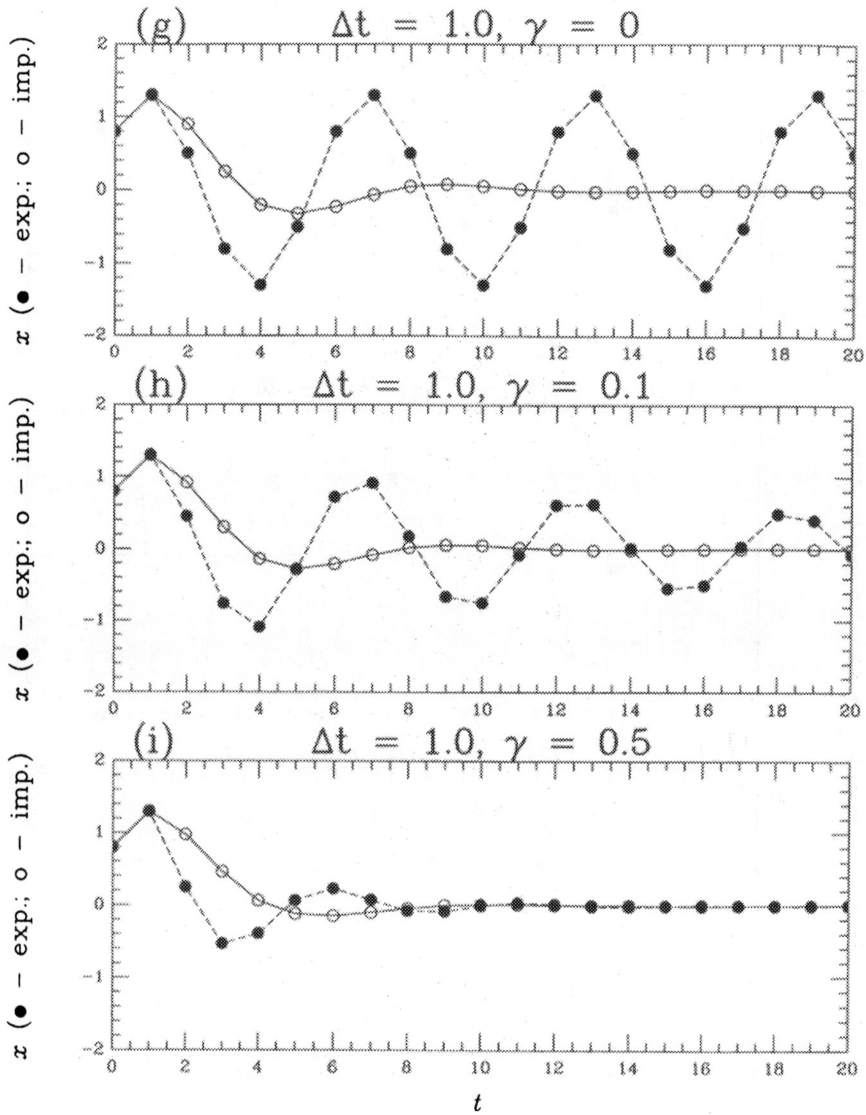
(g) Δt = 1.0, γ = 0
(h) Δt = 1.0, γ = 0.1
(i) Δt = 1.0, γ = 0.5
x (● – exp.; o – imp.)
t
2
1
0
−1
−2
0
2
4
6
8
10
12
14
16
18
20

FIG. 5.1. (g,h,i).

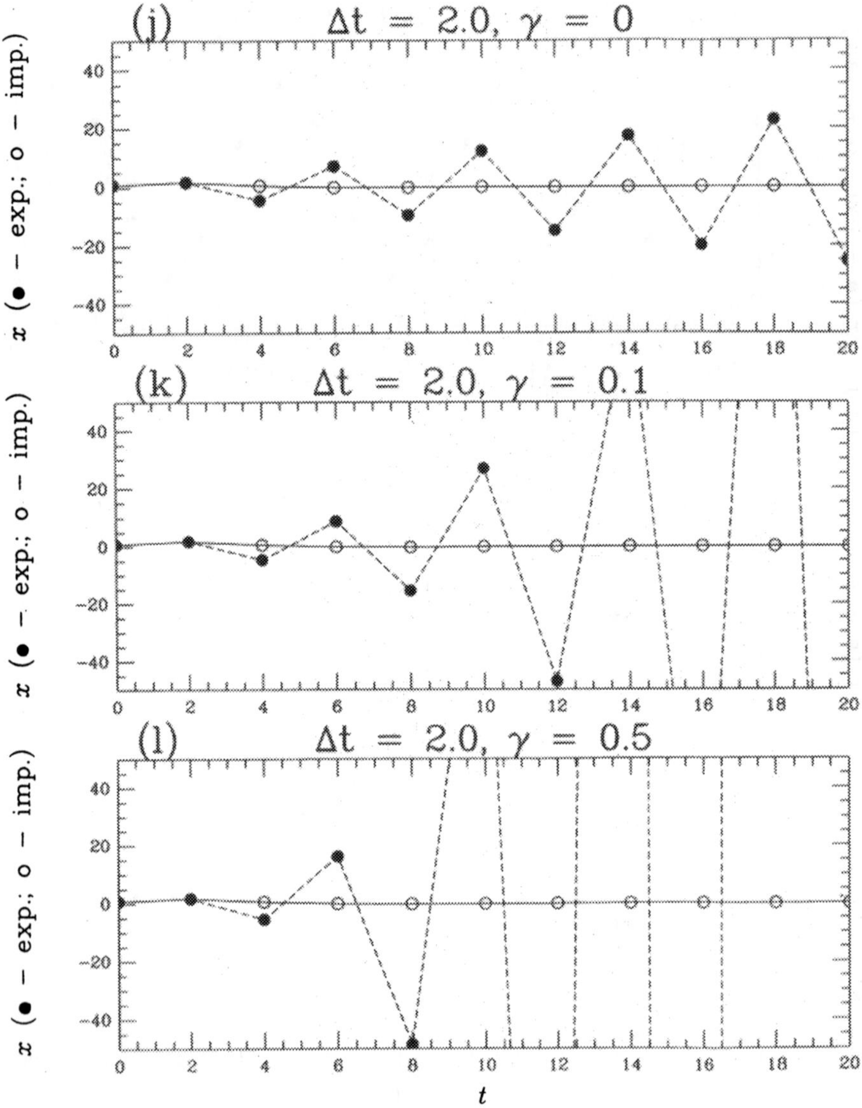
(j) Δt = 2.0, γ = 0
(k) Δt = 2.0, γ = 0.1
(l) Δt = 2.0, γ = 0.5
x (● – exp.; o – imp.)
40
20
0
−20
−40
0 2 4 6 8 10 12 14 16 18 20
t

FIG. 5.1. (j,k,l).

circles for the explicit scheme (eq. 5.2) and *hollow* circles for IE (eq. 5.9). In addition, for the smaller two timesteps the magnitude formulas of eq. (5.8) (for nonzero γ) and eq. (5.11) are drawn, with *dashed* curves representing the explicit scheme and *dot-dash* curves corresponding to the implicit scheme. We clearly note from these views the intrinsic damping of the IE scheme, increasing in severity with the timestep. For larger γ, however, the physical damping dominates so the two schemes produce similar results for small timesteps. As the timestep is increased, the explicit scheme clearly becomes unstable, with divergence occurring more rapidly with larger γ (compare panels j,k,l). Note that the scales for position range from -1 to 1 for the two timestep values $\Delta t = 0.05$ and 0.1; from -2 to 2 for $\Delta t = 1$; and from -50 to 50 for $\Delta t = 2$.

6. Approaches to large timesteps. In developing novel approaches to long-time integration for biomolecular MD, we must accept that "accurate" trajectories cannot be expected in the sense of detailed agreement with small-timestep methods. We basically seek "correct" behavior in a larger context, allowing if necessary some compromise between enhanced sampling and retention of kinetics. Three basic approaches are possible.

(a) **Resolving the high-frequency modes**

This is the approach taken by explicit integrators (e.g., Verlet) today. For example, for the Langevin equation (2.2a,b), a generalization of the Verlet for MD known as BBK [40] can be written as follows:

(6.1a)

$$X^{n+1} = X^n + V^n \Delta t \left[\frac{1 - \frac{\gamma \Delta t}{2}}{1 + \frac{\gamma \Delta t}{2}} \right] + \frac{\Delta t^2}{1 + (\gamma \Delta t)/2} \mathbf{M}^{-1} \left[\mathbf{R}^n - \nabla E(X^n) \right],$$

(6.1b)

$$V^{n+1} = \frac{(X^{n+1} - X^n)}{\Delta t},$$

where

(6.1c)

$$\langle R^n \rangle = 0, \qquad \langle R^n (R^m)^T \rangle = 2 \gamma k_B T \mathbf{M} (\delta_{nm} / \Delta t).$$

This second-order accurate scheme limits the timestep to the femtosecond range, as in standard MD. Certainly, such explicit approaches for MD are important when detailed kinetics are the focus. However, these approaches will have to rely on machine speed alone to enter longer timeframes.

(b) **Suppressing the high frequencies**

Since the fluctuations of the high frequencies are local and small in amplitude, it is reasonable to suppress their motion in an attempt to allow larger timesteps and hence longer simulation times. However, because of the cooperative nature of biomolecular motion, as well as the energy transfer among vibrational modes, an issue of reliability of the results naturally arises: How will the freezing of the high-frequency motion affect the overall dynamics? Clearly, some of the dynamic details will be altered and barriers may be more difficult to surpass in this representation.

Standard techniques of effectively freezing the fast vibrational modes by a constrained formulation [41,42,43,44] (typically the bond lengths only) increase the timestep by a small factor such as two, still with added complexity at each step. In addition, suppressing the flexibility of bond angles is known to affect the detailed dynamics [45].

"LI": The Langevin/Implicit-Euler Scheme. An alternative to these approaches relies on the intrinsic damping of the IE scheme, as described above, to "quench" the high-frequency motion [46,47]. By exploiting frictional and numerical damping, we can effectively quench vibrational modes with frequencies $\omega \gg \omega_c$ (ω_c = *cutoff frequency*), while fully activating modes $\omega \ll \omega_c$. LI can be described by the following pair of first-order differential equations:

$$\text{(6.2a)} \qquad \mathbf{M}\left[\frac{V^{n+1}-V^{n}}{\Delta t}\right] = -\nabla E(X^{n+1}) - \gamma \mathbf{M} V^{n+1} + R^{n+1},$$

$$\text{(6.2b)} \qquad \frac{X^{n+1}-X^{n}}{\Delta t} = V^{n+1} \, .$$

The random-force R is computed according to eq. (6.1c). Since a simple analytical relation is not possible as above (for harmonic oscillators), we must determine X^{n+1} by solving the nonlinear system above or, equivalently, minimizing the "dynamics" function Φ:

$$\text{(6.3a)} \qquad \Phi(X) = \frac{1}{2}(1+\gamma\Delta t)(X - X_0^n)^T\,\mathbf{M}(X - X_0^n) + (\Delta t)^2\,E(X) \, .$$

In this function, the quantities known from previous timesteps are collected in the vector

$$\text{(6.3b)} \qquad X_0^n = X^n + \frac{\Delta t}{1+\gamma\Delta t}\left[V^n + \Delta t\,\mathbf{M}^{-1}\,R^{n+1}\right] .$$

Following minimization, V^{n+1} is obtained directly from (6.2b). The function Φ contains "kinetic" and "potential" terms and thus at large timesteps energy minimization dominates and the scheme is stable. The sparsity of Φ's Hessian is identical to that of the potential energy E, and rapid minimization of Φ is possible with the truncated-Newton package tailored for molecular mechanics, TNPACK [48,49], in combination with a good initial guess (e.g., X_0^n).

The approach of LI yields a far more rigid dynamics, and hence unrealistic, description of all-atom systems than expected [50,51,32]. However, LI is effective in terms of both computational time and reliability for macroscopic models where the high frequencies are absent, such as supercoiled DNA (see Figure 6.1) [52,53,54]. For standard all-atom models, the activities and associated energies of the high-frequency modes must clearly be retained to capture interesting global conformational changes successfully. A variant of LI, termed LIN (Langevin/Implicit-Euler/Normal Modes) has begun to address this damping issue [55] (see below), and a new "dynamics driver" algorithm was also developed to counteract numerical damping and enhance sampling [32]. Scheraga and co-workers have also used the framework of LI with a counteracting mechanism to study the folding/unfolding of the small protein BPTI [78].

(c) **Approximating the high-frequency modes**

Since suppressing the high-frequency modes is certain to alter the dynamics of macromolecules in general, strategies for approximating their contribution in some way are reasonable to consider. In this category of schemes, various multiple-timestep methods have been developed, in which the slow and fast forces are updated at different frequencies, as needed [56,57,25,58]. These approaches certainly provide additional speedup, though there has been discussion about their stability [59]. Modifications may also be possible to "constrained MD" (as described above) for small, but significant, CPU gains.

"LIN": The Langevin/Implicit-Euler/Normal Mode Scheme. Our alternative LIN is an extension of LI which combines normal mode analysis (NM) with IE. The basic idea in LIN is to obtain a large timestep approximation to the solution which captures most of the oscillations within that timestep by using normal mode analysis (NM) and then correct it by a large timestep integration by IE. This aims at fully activating vibrational modes for all-atom systems while still maintaining the large timestep stability of LI. In this way, the rapidly-varying motion (X_h) is captured in the first component, while the slowly-varying motion (Z) is resolved in the second ($X = X_h + Z$).

In brief, we describe the method as follows. A harmonic approximation to E at a reference position X_r produces for the Langevin equation (2.2a,b):

$$(6.4)\quad \mathbf{M}\dot{V}_h = -\nabla E_h - \mathbf{H}_h(X_h - X_r) - \gamma \mathbf{M} V_h + R, \qquad \dot{X}_h = V_h,$$

where ∇E_h is a vector, and $\mathbf{H}_h$ is a Hessian approximation (see below). At the beginning of timestep $(n+1)$, system (6.4) is solved for the "harmonic" coordinate and velocity vectors, X_h^{n+1} and V_h^{n+1}, by standard NM techniques [55]. This involves the determination of an orthogonal matrix $\mathbf{T}$ that diagonalizes $\mathbf{H}' \equiv \mathbf{M}^{-\frac{1}{2}}\mathbf{H}\mathbf{M}^{-\frac{1}{2}}$ and the application of several transformations that produce a decoupled set of differential equations for the normal modes, whose solutions are known.

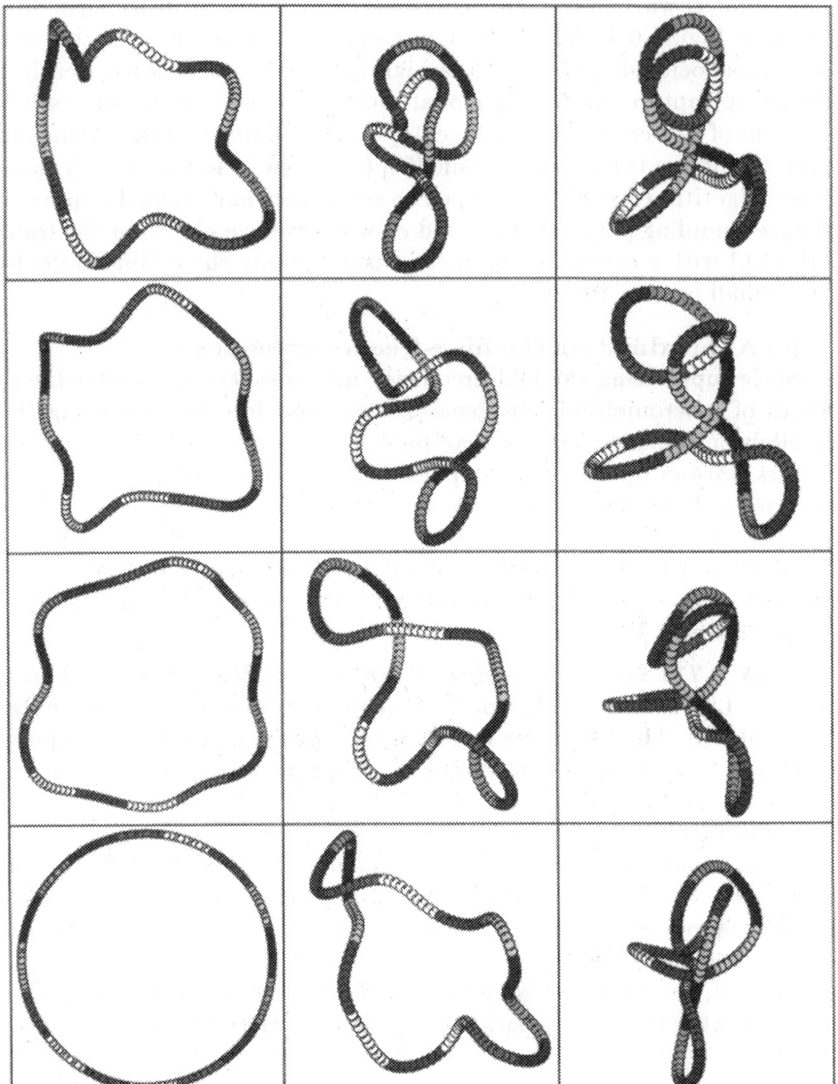

FIG. 6.1. **Snapshots of Supercoiled DNA Dynamics.** *The frames shown (ordered left to right, top to bottom, beginning at the top left corner) are selected from a molecular dynamics trajectory obtained by the Langevin/Implicit-Euler scheme for a closed double-stranded DNA system subjected to torsional stress. Images are rendered by the program MOLSCRIPT [77] through an interface written by Constantine Kreatsoulas. The DNA chain of 2000 base pairs is modeled by B-splines, and an elastic and electrostatic energy function is used. Typical salt concentrations and superhelical densities are used here to suggest how a DNA plasmid will supercoil and branch in response to torsional stress. Note how rapidly kinks are formed along the chain, leading to a folded interwound structure. The DNA continues to bend and twist fluidly, forming a branched form, as commonly seen in electron microscopy photographs. For further details on the model and applications, see references [30,52,53,54].*

To resolve the remaining component of the motion, $Z(t) = X(t) - X_h(t)$, we reformulate the discretized equations of motion in terms of $X(t)$ rather than $Z(t)$ and apply the implicit-Euler discretization. We then translate the resulting nonlinear system into an optimization problem so that X can be obtained as a minimum of the "dynamics" function $\Phi(X)$ of eq. (6.3a) where, instead of eq. (6.3b),

$$\text{(6.5)} \quad X_0^n = X_h^{n+1} + \frac{(\Delta t)^2}{(1+\gamma\Delta t)}\mathbf{M}^{-1}[\nabla E_h^{n+1} + \mathbf{H}_h^{n+1}(X_h^{n+1} - X_r)].$$

(The superscripts $(n+1)$ for ∇E_h and $\mathbf{H}_h$ above denote the values used in the harmonic approximation of eq. 6.4). This minimization, handled as LI by TNPACK [48], is significantly accelerated since X_h^{n+1} provides an excellent initial guess. V^{n+1} is then obtained by eq. (6.2b).

Results to date demonstrate that for liquid butane the high-frequency modes can be fully activated by LIN [55] and that computational gain can be obtained for *dense* decomposition implementations only beyond 250 fs [60]. However, energetic fluctuations are large, and work is needed to reduce them. We also hope to improve computational performance for biomolecular systems with significant linear-algebra work for sparse matrices. This involves careful implementations of the NM details. Since the approximate Hessian in the linearized Langevin equation is system dependent, systematic procedures for protein and nucleic acid systems must be developed regarding: (i) choice of the approximate Hessian in the linearized Langevin equation, (ii) frequency of the normal-mode decomposition, and (iii) efficient decompositions for large, sparse systems.

Computationally, the approximate Hessian for the normal-mode analysis must be sparse and fast to decompose. One possibility is a block-diagonal Hessian H_h. Another is a sparse molecular preconditioner (from the local potential terms), which has nearly a band structure [49] and can be subjected to reordering schemes for sparse systems to obtain block-diagonal or banded forms [61,62]. Such sparse structures can be exploited to perform parallel decompositions for the blocks, or to apply block-Lanczos or Givens decomposition techniques [63,64]. Lanczos techniques are attractive for large systems since they are iterative and cheap per step, like their related Conjugate Gradient methods (though they are sensitive to round-off error). Givens orthogonalization techniques are especially efficient for banded systems, as small rotation matrices are repeatedly applied to obtain a tridiagonal form, which must then be reduced to diagonal.

The frequency of the NM decomposition depends on the approximate Hessian used and the system in question. Preliminary analysis has suggested to us that a 1 ps interval is reasonable for retaining the same NM description [60]. However, clearly normal-mode analyses adequately approximate the system as long as it is oscillating in the basin of a local energy minimum; this approximation may break down when an energy barrier is crossed. This situation will have to be monitored in practice, and the NM description updated as necessary.

Finally, we hope that efficient implementations for high-performance platforms will make LIN competitive for biomolecular simulations. LIN offers opportunities for parallelization in the spectral decompositions (e.g., block Lanczos) and in the minimization (e.g., function evaluations for the line search). Further, optimization of the energy and derivative loops might result in greater speedup in the context of LIN since such calculations are required more extensively.

Other avenues can also be pursued with LIN in the hope of making computational time competitive with small-timestep simulations. One possibility is to obtain the normal-mode information as the simulation by Fourier transform and filtering techniques [65]. It might also be interesting to consider a reversed strategy of LIN, in which the normal mode information is used to resolve the slow frequencies rather than the high frequencies. This information may be computed from dynamic correlation functions (e.g., covariance matrix of positional fluctuations) accumulated during the dynamics trajectory [66]. These functions also incorporate anharmonic and solvent effects. In this combination, a suitable strategy will have to be developed for the high-frequency component (e.g., a constrained formulation). Another variation on LIN is to replace the IE integration component by a nondamping scheme, such as implicit midpoint; this might also remove the requirement for the additional Langevin terms.

7. Future prospects. As Laplace suggested, formulation and applications of MD hold the key to many important biological phenomena of macromolecules. Nevertheless, many challenging issues must still be resolved. Even on modern computers, innovative combinations of algorithmic and programming approaches are needed to extend our simulation scope significantly in order to capture a glimpse of the future. What makes these multiple-timescale integration problems particularly challenging is the fact that solutions demand much more than straightforward application of standard mathematical techniques. At this point it appears that the optimal algorithms for MD will require a combination of methods and strategies discussed above, including symplectic and implicit numerical integration schemes that have minimal intrinsic damping, and correct resolution of the subdynamics of the system by some other techniques (e.g., normal-mode analysis).

The implicit-midpoint scheme. In this category, interesting issues remain to be resolved with respect to the implicit-midpoint (IM) scheme. In the same framework of LI, the IM discretization for the Langevin equation can be written as follows:

(7.1a)

$$\mathbf{M}\left[\frac{V^{n+1}-V^{n}}{\Delta t}\right] = -\nabla E\left(\frac{X^{n+1}+X^{n}}{2}\right) - \gamma\mathbf{M}\left(\frac{V^{n+1}+V^{n}}{2}\right) + R^{n},$$

(7.1b)
$$\frac{X^{n+1} - X^n}{\Delta t} = \frac{V^{n+1} + V^n}{2} .$$

(The random-force R is computed according to eq. (6.1c)). The corresponding "dynamics" function Φ to be minimized is:

(7.2a)
$$\Phi(X) = \frac{\alpha}{2}(X - X_0^n)^T \mathbf{M}(X - X_0^n) + (\Delta t)^2 \, E\left(\frac{X + X^n}{2}\right) ,$$

with

(7.2b)
$$X_0^n = X^n + \frac{V^n \Delta t}{\alpha} + \frac{(\Delta t)^2 \mathbf{M}^{-1} R^n}{2\alpha} ,$$

where $\alpha = 1 + (\gamma \Delta t)/2$. The new velocity, V^{n+1}, is computed after X^{n+1} is known from the relation:

(7.3)
$$V^{n+1} = 2\left(\frac{X^{n+1} - X^n}{\Delta t}\right) - V^n .$$

This symplectic scheme [67,68] is nondamping and known to conserve energy in the linear regime for very large timesteps. The symplecticness can be seen by applying IM to the harmonic oscillator system (eq. 5.1) with $\gamma = 0$ to obtain:

(7.4)
$$\begin{pmatrix} v^{n+1} \\ \omega x^{n+1} \end{pmatrix} = \mathbf{Q}(\theta) \begin{pmatrix} v^n \\ \omega x^n \end{pmatrix} ,$$

where $\mathbf{Q}(\theta)$ is the rotation matrix

(7.5)
$$\mathbf{Q}(\theta) = \begin{pmatrix} \cos\theta & -\sin\theta \\ \sin\theta & \cos\theta \end{pmatrix}, \quad \theta = \tan^{-1}\left[\frac{\epsilon}{1 - \epsilon^2/4}\right] ,$$

(or equivalently $\theta = 2\tan^{-1}[\epsilon/2]$). Recall that $\epsilon = \omega \, \Delta t$ so the rotation angle above is timestep dependent. The solutions for x^n can be shown to be of the form $A\lambda_1^n + B\lambda_2^n$, where $\lambda_{1,2} = e^{\pm i\theta}$ [68], and hence there is no numerical damping. However, in the nonlinear regime, interesting "resonance" phenomena have been observed which limit the confidence by which IM can be applied to biomolecular MD [67,69]. In particular, we have observed very large errors or instability at particular timesteps (Figure 7.1). This resonance is due to a particular relation between the timestep and the approximated frequency of the system by the IM scheme. If these phenomena are better understood, a counteracting mechanism may be devised so that IM might be applied to biomolecular systems at moderate timesteps to be both reliable and competitive (when an efficient minimizer is used) with small-timestep approximations. At this point, we are pessimistic. However, resonance problems are also relevant to the Verlet integrator [69] and hence deserve further attention.

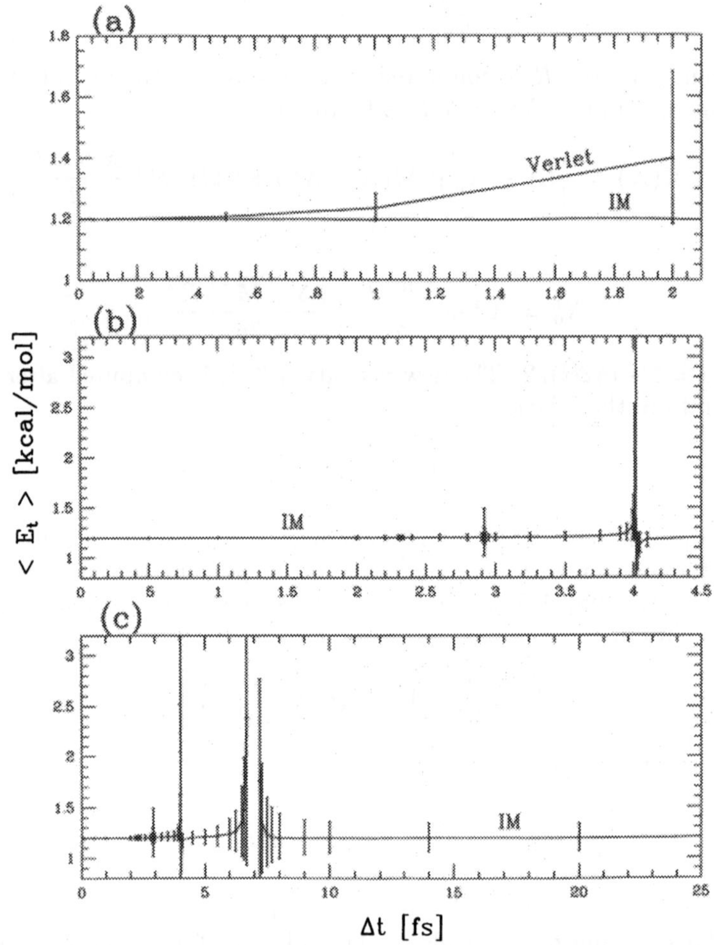

FIG. 7.1. The Energy Envelopes Obtained for a System Governed by a Morse Potential by Explicit (Verlet) and Implicit-Midpoint (IM) Integration. *The Morse potential is a function of bond length r: $D_e[1-\exp(-a(r-r_0))]^2$, where D_e is the dissociation energy, a is a constant, and r_0 is the equilibrium bond length. The resulting trajectory-averaged energy ranges of a one-dimensional system clearly increase rapidly with Δt for the explicit Verlet scheme (panel a) but are indicative of resonance problems for the IM scheme (e.g., near $\Delta t = 2.9$, 4, 6.9 fs, panels b,c). Significantly, at very large timesteps (i.e., beyond 10 fs) the envelopes are bound for the IM trajectories (panel c). See discussion in [69].*

In addition to these interesting issues with the IM scheme, of interest is the establishment of an organized theory for simulation evaluation, including both local error analysis (the main focus today), long-time behavior, finite-time behavior, and a statistical-mechanical framework for evaluation of both simulation results and new integrators and conformational propagators.

Global error analysis. A more rigorous analytical framework for global error analysis should focus on the question of ensemble versus trajectory error. This is important particularly in the Langevin framework. How does a "trajectory error" relate to an "ensemble error"? What kind of conclusions can we draw from results of one trajectory, or a finite ensemble of trajectories, regarding various static and dynamic properties that are of interest?

Relevance of the limit $\gamma \to 0$. How relevant is the limit $\gamma \to 0$ for practical simulation? Is it appropriate to compare results of Langevin simulations to their corresponding molecular dynamics equations where the stochastic terms are zero? For example, harmonic oscillator phase diagrams from MD and LD are qualitatively very different [68].

Relevance of the limit $\Delta t \to 0$. Furthermore, since we are dealing with integrations where the timestep Δt is not necessarily small, how relevant are results in the limit $\Delta t \to 0$? Are such asymptotic results relevant to practical problems? That is, how does finite time influence our results, and what can be derived as guidelines for practical simulations that are routinely performed today for very large systems?

Sensitivity of results to numerical schemes. Another question in this spirit deals with the sensitivity of specific measurements to numerical schemes. Can we pair specific static and dynamic properties with specific Langevin integrators? That is, which properties are best/poorly reproduced by certain schemes and why? This question is particularly of interest for novel integration and propagation schemes [70, for example]. How should those methods be evaluated consistently? The standard approach of comparing results to small-timestep simulation is inadequate in many situations, and alternatives could be offered.

These problems, which we have recently begun to explore [79], and many related issues should keep computational chemists, applied mathematicians, biologists, and computer scientists quite busy during the next decade. Indeed, it is my belief that significant progress will emerge from software developments in the coming years, just as the hardware developments have sparked activities in macromolecular MD in the past decade.

Acknowledgments. I am grateful to Klaus Schulten, Jill Mesirov, and DeWitt Sumners for inviting me to participate in the IMA workshop and to the IMA staff, as well as other participants, for making the week enjoyable and productive. I thank Yaël and Shulamith Schlick and Philippe

Derreumaux for assisting with the French translation, and Gomathi Ramachandran and Margaret Mandziuk for providing Figures 6.1 and 7.1, respectively. This research was generously supported by the National Science Foundation, the National Institutes of Health, and the Alfred P. Sloan Foundation. T.S. is an investigator of the Howard Hughes Medical Institute.

REFERENCES

[1] P. S. de Laplace. *Oeuvres Complètes de Laplace. Théorie Analytique des Probabilités*, volume VII. Gauthier-Villars, Paris, France, third edition, 1820.

[2] J. A. McCammon and S. C. Harvey. *Dynamics of Proteins and Nucleic Acids.* Cambridge University Press, Cambridge, MA, 1987.

[3] M. P. Allen and D. J. Tildesley. *Computer Simulation of Liquids.* Oxford University Press, New York, New York, 1990.

[4] C. L. Brooks III, M. Karplus, and B. M. Pettitt. *Proteins: A Theoretical Perspective of Dynamics, Structure, and Thermodynamics*, volume LXXI of *Advances in Chemical Physics.* John Wiley & Sons, New York, New York, 1988.

[5] F. M. Richards. The protein folding problem. *Sci. Amer.*, 264:54–63, 1991.

[6] H. S. Chan and K. A. Dill. The protein folding problem. *Physics Today*, 46:24–32, 1993.

[7] R. W. Pastor. Techniques and applications of Langevin dynamics simulations. In G. R. Luckhurst and C. A. Veracini, editors, *The Molecular Dynamics of Liquid Crystals*, pages 85–138. Kluwer Academic Publishers, Dordrecht, The Netherlands, 1994.

[8] M. Vásquez, G. Némethy, and H. A. Scheraga. Conformational energy calculations on polypeptides and proteins. *Chemical Reviews*, 94:2183–2239, 1994.

[9] J. Phillip Bowen and N. L. Allinger. Molecular mechanics: The art and science of parameterization. In K. B. Lipkowitz and D. B. Boyd, editors, *Reviews in Computational Chemistry*, volume II, pages 81–97. VCH Publishers, New York, New York, 1991.

[10] U. Burkert & N. L. Allinger. *Molecular Mechanics*, volume 177 of *American Chemical Society Monograph.* ACS, Washington, D. C., 1982.

[11] S. Lifson. Potential energy functions for structural molecular biology. In D. B. Davies, W. Saenger, and S. S. Danyluk, editors, *Methods in Structural Molecular Biology*, pages 359–385. Plenum Press, London, 1981.

[12] I. K. Roterman, M. H. Lambert, K. D. Gibson, and H. A. Scheraga. A comparison of the CHARMM, AMBER and ECEPP potentials for peptides. *J. Biomol. Struct. Dyn.*, 7:391–452, 1989.

[13] P. A. Kollman and K. A. Dill. Decisions in force field development: An alternative to those described by Roterman *et al. J. Biomol. Struct. Dyn.*, 8:1103–1107, 1991.

[14] K. B. Gibson and H.A. Scheraga. Decisions in force field development: Reply to Kollman and Dill. *J. Biomol. Struct. Dyn.*, 8:1109–1111, 1991.

[15] T. Schlick. *Modeling and Minimization Techniques for Predicting Three-Dimensional Structures of Large Biological Molecules.* PhD thesis, New York University, Courant Institute of Mathematical Sciences, New York, New York, October 1987.

[16] B. R. Brooks, R. E. Bruccoleri, B. D. Olafson, D. J. States, S. Swaminathan, and M. Karplus. CHARMM: A program for macromolecular energy, minimization, and dynamics calculations. *J. Comp. Chem.*, 4:187–217, 1983.

[17] S. J. Weiner, P. A. Kollman, D. T. Nguyen, and D. A. Case. An all atom force field for simulations of proteins and nucleic acids. *J. Comp. Chem.*, 7:230–252, 1986.

[18] Molecular mechanics and modeling, November 1993. Special issue of *Chemical Reviews* (Volume 93, Number 7).

[19] C. A. Schiffer, J. W. Caldwell, P. A. Kollman, and R. M. Stroud. Protein structure prediction with a combined solvation free energy-molecular mechanics force field. *Mol. Sim.*, 10:121–149, 1993.

[20] J. R. Maple, M.-J. Hwang, T. P. Stockfisch, U. Dinur, M. Waldman, C. S. Ewing, and A. T. Hagler. Derivation of class II force fields. I. Methodology and quantum force field for the alkyl functional group and alkane molecules. *J. Comp. Chem.*, 15:162–182, 1994.

[21] P. Derreumaux and G. Vergoten. Influence of the spectroscopic potential energy function SPASIBA on molecular dynamics of proteins: Comparison with the AMBER potential. *J. Mol. Struct.*, 286:55–64, 1993.

[22] L. Verlet. Computer 'experiments' on classical fluids: I. Thermodynamical properties of Lennard-Jones molecules. *Physical Review*, 159(1):98–103, July 1967.

[23] M. P. Calvo and J. M. Sanz-Serna. The development of variable-step symplectic integrators, with application to the two-body problem. *SIAM J. Sci. Comput.*, 14:936, 1993.

[24] P. J. Steinbach and B. R. Brooks. New spherical-cutoff methods for long-range forces in macromolecular simulation. *J. Comp. Chem.*, 15:667–683, 1994.

[25] H. Grubmuller, H. Heller, A. Windemuth, and K. Schulten. Generalized Verlet algorithm for efficient molecular dynamics simulations with long-range interactions. *Mol. Sim.*, 6:121–142, 1991.

[26] J. A. Board Jr., J. W. Causey, T. F. Leathrum Jr., A. Windemuth, and K. Schulten. Accelerated molecular dynamics simulations with the parallel fast multiple algorithm. *Chem. Phys. Lett.*, 198:89–94, 1992.

[27] J. A. Board Jr., L. V. Kalé, K. Schulten, R. D. Skeel, and T. Schlick. Modeling biomolecules: Larger scales, longer durations. *IEEE Computational Science & Engineering*, 1:19–30, Winter 1994.

[28] H. Frauenfelder and P. G. Wolynes. Biomolecules: Where the physics of complexity and simplicity meet. *Physics Today*, 47:58–64, 1994.

[29] H. Frauenfelder, S. G. Sligar, and P. G. Wolynes. The energy landscapes and motions of proteins. *Science*, 254:1598–1603, 1991; P.G. Wolynes, J.N. Onuchic, and D. Thirumalai, Navigating the Folding Routes, *Science*, 267:1619–1620, 1995.

[30] T. Schlick, B. Li, and W. K. Olson. The influence of salt on DNA energetics and dynamics. *Biophys. J.*, 67:2146–2166, 1994.

[31] W. F. van Gunsteren and P. K. Weiner, editors. *Computer Simulation of Biomolecular Systems.* ESCOM, Leiden, The Netherlands, 1989.

[32] P. Derreumaux and T. Schlick. Long-time integration for peptides by the dynamics driver approach. *Proteins, Structure, Function and Genetics*, 21:282–302, 1995.

[33] R. W. Pastor, B. R. Brooks, and A. Szabo. An analysis of the accuracy of Langevin and molecular dynamics algorithms. *Mol. Phys.*, 65:1409–1419, 1988.

[34] A. M. Stuart and A. R. Humphries. Model problems in numerical stability theory for initial value problems. *SIAM Review*, 36:226–257, 1994.

[35] H. A. Scheraga. Predicting three-dimensional structures of oligopeptides. In K. B. Lipkowitz and D. B. Boyd, editors, *Reviews in Computational Chemistry*, volume III, pages 73–142. VCH Publishers, New York, New York, 1992.

[36] Mathematical challenges from theoretical/computational chemistry, National Research Council Report, National Academy Press, Washington, D.C., 1995.

[37] Z. Wu. The effective energy transformation scheme as a special continuation approach to global optimization with application to molecular conformation. *SIAM J. Opt.*, 6, 1996.

[38] E. Hairer, S. P. Nørsett, and G. Wanner. *Solving Ordinary Differential Equations I. Nonstiff Problems*, volume 8 of *Springer Series in Computational Mathematics*. Springer-Verlag, New York, New York, second edition, 1993.

[39] E. Hairer and G. Wanner. *Solving Ordinary Differential Equations II. Stiff and Differential-Algebraic Problems*, volume 14 of *Springer Series in Computational Mathematics*. Springer-Verlag, New York, New York, 1991.

[40] A. Brünger, C. B. Brooks, and M. Karplus. Stochastic boundary conditions for molecular dynamics simulations of ST2 water. *Chem. Phys. Lett.*, 105:495–500, 1982.

[41] W. F. van Gunsteren. Constrained dynamics of flexible molecules. *Mol. Phys.*, 40:1015–1019, 1980.

[42] W. F. van Gunsteren and H.J.C. Berendsen. Algorithms for macromolecular dynamics and constraint dynamics. *Mol. Phys.*, 34:1311–1327, 1977.

[43] J. P. Ryckaert, G. Ciccotti, and H. J. C. Berendsen. Numerical integration of the Cartesian equations of motion of a system with constraints: Molecular dynamics of n-alkanes. *J. Comp. Phys.*, 23:327–341, 1977.

[44] S. Miyamoto and P. A. Kollman. SETTLE: An analytical version of the SHAKE and RATTLE algorithm for rigid water models. *J. Comp. Chem.*, 13:952–962, 1992.

[45] W. F. van Gunsteren and M. Karplus. Effect of constraints on the dynamics of macromolecules. *Macromolecules*, 15:1528–1543, 1982.

[46] C. S. Peskin and T. Schlick. Molecular dynamics by the backward Euler's method. *Comm. Pure App. Math.*, 42:1001–1031, 1989.

[47] T. Schlick and C. S. Peskin. Can classical equations simulate quantum-mechanical behavior? A molecular dynamics investigation of a diatomic molecule with a Morse potential. *Comm. Pure App. Math.*, 42:1141–1163, 1989.

[48] T. Schlick and A. Fogelson. TNPACK — A truncated Newton minimization package for large-scale problems: I. algorithm and usage. *ACM Trans. Math. Softw.*, 14:46–70, 1992.

[49] P. Derreumaux, G. Zhang, B. Brooks, and T. Schlick. A truncated-Newton method adapted for CHARMM and biomolecular applications. *J. Comp. Chem.*, 15:532–552, 1994.

[50] T. Schlick, S. Figueroa, and M. Mezei. A molecular dynamics simulation of a water droplet by the implicit-Euler/Langevin scheme. *J. Chem. Phys.*, 94:2118–2129, 1991.

[51] A. Nyberg and T. Schlick. Increasing the time step in molecular dynamics. *Chem. Phys. Lett.*, 198:538–546, 1992.

[52] T. Schlick and W. K. Olson. Supercoiled DNA energetics and dynamics by computer simulation. *J. Mol. Biol.*, 223:1089–1119, 1992.

[53] T. Schlick and W. K. Olson. Trefoil knotting revealed by molecular dynamics simulations of supercoiled DNA. *Science*, 257:1110–1115, 1992.

[54] G. Ramachandran and T. Schlick. Solvent effects on supercoiled DNA dynamics explored by Langevin dynamics simulations. *Phys. Rev. E*, 51:6188–6203, 1995.

[55] G. Zhang and T. Schlick. LIN: A new algorithm combining implicit integration and normal mode techniques for molecular dynamics. *J. Comp. Chem.*, 14:1212–1233, 1993.

[56] W. B. Streett, D. J. Tildesley, and G. Saville. Multiple time step methods in molecular dynamics. *Mol. Phys.*, 35:639–648, 1978.

[57] M. E. Tuckerman and B. J. Berne. Molecular dynamics in systems with multiple time scales: Systems with stiff and soft degrees of freedom and with short and long range forces. *J. Comp. Chem.*, 95:8362–8364, 1992.

[58] M. Watanabe and M. Karplus. Dynamics of molecules with internal degrees of freedom by multiple time-step methods. *J. Chem. Phys.*, 99:8063–8074, 1993.

[59] J. J. Biesiadecki and R. D. Skeel. Dangers of multiple-time-step methods. *J. Comp. Phys.*, 109:318–328, 1993.

[60] G. Zhang and T. Schlick. The Langevin/implicit-Euler/Normal-Mode scheme (LIN) for molecular dynamics at large time steps. *J. Chem. Phys.*, 101:4995–5012, 1994.

[61] J. O'Neil and D. B. Szyld. A block ordering method for sparse matrices. *SIAM J. Sci. Stat. Comp.*, 11:811–823, 1990.

[62] I. S. Duff, A. M. Erisman, and J. K. Reid. *Direct Methods for Sparse Matrices.* Oxford University Press, New York, New York, 1986.

[63] G. H. Golub and C. F. van Loan. *Matrix Computations.* John Hopkins University Press, Baltimore, MD, second edition, 1986.

[64] Z. Zlatev. *Computational Methods for General Sparse Matrices.* Kluwer Academic Publishers, Dordrecht, The Netherlands, 1991.

[65] P. Dauber-Osguthorpe and D. J. Osguthorpe. Partitioning the motion in molecular dynamics simulations into characteristic modes of motion. *J. Comp. Chem.*, 14:1259–1271, 1993.

[66] A. Amadei, A. B. M. Linssen, and H. J. C. Berendsen. Essential dynamics of proteins. *Proteins, Structure, Function and Genetics*, 17:412–425, 1993.

[67] J. C. Simo, N. Tarnow, and K. K. Wong. Exact energy-momentum conserving algorithms and symplectic schemes for nonlinear dynamics. *Computer Methods in Applied Mechanics and Engineering*, 100:63–116, 1991.

[68] G. Zhang and T. Schlick. Implicit discretization schemes for Langevin dynamics. *Mol. Phys.*, 84:1077–1098, 1995.

[69] M. Mandziuk and T. Schlick. Resonance in chemical-system dynamics simulated by the implicit-midpoint scheme. *Chem. Phys. Lett.*, 237:525–535, 1995.

[70] N. Grønbech-Jensen and S. Doniach. Long-time overdamped Langevin dynamics of molecular chains. *J. Comp. Chem.*, 15:997–1012, 1994.

[71] J. A. McCammon, B. R. Gelin, and M. Karplus. Dynamics of folded proteins. *Nature*, 267:585–590, 1977.

[72] R. S. Struthers, J. Rivier, and A. T. Hagler. Theoretical simulation of conformation, energetics, and dynamics in the design of GnRH analogs. *Transactions of the American Crystallographic Association*, 20:83–96, 1984. Proceedings of the Symposium on Molecules in Motion, University of Kentucky, Lexington, Kentucky, May 20–21, 1984.

[73] M. Levitt. Computer simulation of DNA double-helix dynamics. *Cold Spring Harbor Symp. Quant. Biol.*, 47:251–275, 1983.

[74] G. L. Seibel, U. C. Singh, and P. A Kollman. A molecular dynamics simulation of double-helical B-DNA including counterions and water. *Proc. Natl. Acad. Sci. USA*, 82:6537–6540, 1985.

[75] J. J. Wendoloski, S. J. Kimatian, C. E. Schutt, and F. R. Salemme. Molecular dynamics simulation of a phospholipid micelle. *Science*, 243:636–638, 1989.

[76] H. Heller, M. Schaefer, and K. Schulten. Molecular dynamics simulation of a bilayer of 200 lipids in the gel and in the liquid-crystal phases. *J. Phys. Chem.*, 97:8343–8360, 1993.

[77] P. J. Kraulis. Molscript: A program to produce both detailed and schematic plots of protein structures. *J. App. Crystallogr.*, 24:946–950, 1991.

[78] M.-H. Hao, M.R. Pincus, S. Rackovsky, and H.A. Scheraga. Unfolding and refolding of the native structure of bovine pancreatic trypsin inhibitor studied by computer simulations. *Biochemistry*, 32:9614–9631, 1993.

[79] B. Mishra and T. Schlick. Error analysis in numerical integration of the Langevin equation: 1. Linear analysis for five explicit and implicit schemes. *Preprint*, 1995.

IMA SUMMER PROGRAMS

1987 Robotics
1988 Signal Processing
1989 Robustness, Diagnostics, Computing and Graphics in Statistics
1990 Radar and Sonar (June 18 - June 29)
New Directions in Time Series Analysis (July 2 - July 27)
1991 Semiconductors
1992 Environmental Studies: Mathematical, Computational, and Statistical Analysis
1993 Modeling, Mesh Generation, and Adaptive Numerical Methods for Partial Differential Equations
1994 Molecular Biology
1995 Large Scale Optimizations with Applications to Inverse Problems, Optimal Control and Design, and Molecular and Structural Optimization
1996 Emerging Applications of Number Theory
1997 Statistics in Health Sciences

SPRINGER LECTURE NOTES FROM THE IMA:

The Mathematics and Physics of Disordered Media
Editors: Barry Hughes and Barry Ninham
(Lecture Notes in Math., Volume 1035, 1983)

Orienting Polymers
Editor: J.L. Ericksen
(Lecture Notes in Math., Volume 1063, 1984)

New Perspectives in Thermodynamics
Editor: James Serrin
(Springer-Verlag, 1986)

Models of Economic Dynamics
Editor: Hugo Sonnenschein
(Lecture Notes in Econ., Volume 264, 1986)

The IMA Volumes in Mathematics and its Applications

Current Volumes:

23 **Signal Processing Part II: Control Theory and Applications of Signal Processing** L. Auslander, F.A. Grünbaum, J.W. Helton, T. Kailath, P. Khargonekar, and S. Mitter (eds.)

24 **Mathematics in Industrial Problems, Part 2** A. Friedman

25 **Solitons in Physics, Mathematics, and Nonlinear Optics** P.J. Olver and D.H. Sattinger (eds.)

26 **Two Phase Flows and Waves** D.D. Joseph and D.G. Schaeffer (eds.)

27 **Nonlinear Evolution Equations that Change Type** B.L. Keyfitz and M. Shearer (eds.)

28 **Computer Aided Proofs in Analysis** K. Meyer and D. Schmidt (eds.)

29 **Multidimensional Hyperbolic Problems and Computations** A. Majda and J. Glimm (eds.)

30 **Microlocal Analysis and Nonlinear Waves** M. Beals, R. Melrose, and J. Rauch (eds.)

31 **Mathematics in Industrial Problems, Part 3** A. Friedman

32 **Radar and Sonar, Part I** R. Blahut, W. Miller, Jr., and C. Wilcox

33 **Directions in Robust Statistics and Diagnostics: Part I** W.A. Stahel and S. Weisberg (eds.)

34 **Directions in Robust Statistics and Diagnostics: Part II** W.A. Stahel and S. Weisberg (eds.)

35 **Dynamical Issues in Combustion Theory** P. Fife, A. Liñán, and F.A. Williams (eds.)

36 **Computing and Graphics in Statistics** A. Buja and P. Tukey (eds.)

37 **Patterns and Dynamics in Reactive Media** H. Swinney, G. Aris, and D. Aronson (eds.)

38 **Mathematics in Industrial Problems, Part 4** A. Friedman

39 **Radar and Sonar, Part II** F.A. Grünbaum, M. Bernfeld, and R.E. Blahut (eds.)

40 **Nonlinear Phenomena in Atmospheric and Oceanic Sciences** G.F. Carnevale and R.T. Pierrehumbert (eds.)

41 **Chaotic Processes in the Geological Sciences** D.A. Yuen (ed.)

42 **Partial Differential Equations with Minimal Smoothness and Applications** B. Dahlberg, E. Fabes, R. Fefferman, D. Jerison, C. Kenig, and J. Pipher (eds.)

43 **On the Evolution of Phase Boundaries** M.E. Gurtin and G.B. McFadden

44 **Twist Mappings and Their Applications** R. McGehee and K.R. Meyer (eds.)

45 **New Directions in Time Series Analysis, Part I** D. Brillinger, P. Caines, J. Geweke, E. Parzen, M. Rosenblatt, and M.S. Taqqu (eds.)

46 **New Directions in Time Series Analysis, Part II**
D. Brillinger, P. Caines, J. Geweke, E. Parzen, M. Rosenblatt, and M.S. Taqqu (eds.)

47 **Degenerate Diffusions**
W.-M. Ni, L.A. Peletier, and J.-L. Vazquez (eds.)

48 **Linear Algebra, Markov Chains, and Queueing Models**
C.D. Meyer and R.J. Plemmons (eds.)

49 **Mathematics in Industrial Problems, Part 5** A. Friedman

50 **Combinatorial and Graph-Theoretic Problems in Linear Algebra**
R.A. Brualdi, S. Friedland, and V. Klee (eds.)

51 **Statistical Thermodynamics and Differential Geometry of Microstructured Materials**
H.T. Davis and J.C.C. Nitsche (eds.)

52 **Shock Induced Transitions and Phase Structures in General Media** J.E. Dunn, R. Fosdick, and M. Slemrod (eds.)

53 **Variational and Free Boundary Problems**
A. Friedman and J. Spruck (eds.)

54 **Microstructure and Phase Transitions**
D. Kinderlehrer, R. James, M. Luskin, and J.L. Ericksen (eds.)

55 **Turbulence in Fluid Flows: A Dynamical Systems Approach**
G.R. Sell, C. Foias, and R. Temam (eds.)

56 **Graph Theory and Sparse Matrix Computation**
A. George, J.R. Gilbert, and J.W.H. Liu (eds.)

57 **Mathematics in Industrial Problems, Part 6** A. Friedman

58 **Semiconductors, Part I**
W.M. Coughran, Jr., J. Cole, P. Lloyd, and J. White (eds.)

59 **Semiconductors, Part II**
W.M. Coughran, Jr., J. Cole, P. Lloyd, and J. White (eds.)

60 **Recent Advances in Iterative Methods**
G. Golub, A. Greenbaum, and M. Luskin (eds.)

61 **Free Boundaries in Viscous Flows**
R.A. Brown and S.H. Davis (eds.)

62 **Linear Algebra for Control Theory**
P. Van Dooren and B. Wyman (eds.)

63 **Hamiltonian Dynamical Systems: History, Theory, and Applications**
H.S. Dumas, K.R. Meyer, and D.S. Schmidt (eds.)

64 **Systems and Control Theory for Power Systems**
J.H. Chow, P.V. Kokotovic, R.J. Thomas (eds.)

65 **Mathematical Finance**
M.H.A. Davis, D. Duffie, W.H. Fleming, and S.E. Shreve (eds.)

66 **Robust Control Theory** B.A. Francis and P.P. Khargonekar (eds.)

67 **Mathematics in Industrial Problems, Part 7** A. Friedman

68 **Flow Control** M.D. Gunzburger (ed.)

69 **Linear Algebra for Signal Processing**
A. Bojanczyk and G. Cybenko (eds.)

70 **Control and Optimal Design of Distributed Parameter Systems**
J.E. Lagnese, D.L. Russell, and L.W. White (eds.)

71 **Stochastic Networks** F.P. Kelly and R.J. Williams (eds.)

72 **Discrete Probability and Algorithms**
D. Aldous, P. Diaconis, J. Spencer, and J.M. Steele (eds.)

73 **Discrete Event Systems, Manufacturing Systems, and Communication Networks**
P.R. Kumar and P.P. Varaiya (eds.)

74 **Adaptive Control, Filtering, and Signal Processing**
K.J. Åström, G.C. Goodwin, and P.R. Kumar (eds.)

75 **Modeling, Mesh Generation, and Adaptive Numerical Methods for Partial Differential Equations** I. Babuska, J.E. Flaherty, W.D. Henshaw, J.E. Hopcroft, J.E. Oliger, and T. Tezduyar (eds.)

76 **Random Discrete Structures** D. Aldous and R. Pemantle (eds.)

77 **Nonlinear Stochastic PDEs: Hydrodynamic Limit and Burgers' Turbulence** T. Funaki and W.A. Woyczynski (eds.)

78 **Nonsmooth Analysis and Geometric Methods in Deterministic Optimal Control** B.S. Mordukhovich and H.J. Sussmann (eds.)

79 **Environmental Studies: Mathematical, Computational, and Statistical Analysis** M.F. Wheeler (ed.)

80 **Image Models (and their Speech Model Cousins)**
S.E. Levinson and L. Shepp (eds.)

81 **Genetic Mapping and DNA Sequencing**
T. Speed and M.S. Waterman (eds.)

82 **Mathematical Approaches to Biomolecular Structure and Dynamics**
J.P. Mesirov, K. Schulten, and D. Sumners (eds.)

FORTHCOMING VOLUMES

1993-1994: *Emerging Applications of Probability*
- Mathematical Population Genetics
- Stochastic Models in Geosystems
- Classical and Modern Branching Processes

1994-1995: *Waves and Scattering*
- Computational Wave Propagation
- Wavelet, Multigrid and Other Fast Algorithms (Multiple, FFT) and Their Use in Wave Propagation
- Waves in Random and Other Complex Media
- Inverse Problems in Wave Propagation
- Singularities and Oscillations
- Quasiclassical Methods
- Multiparticle Quantum Scattering with Applications to Nuclear, Atomic, and Molecular Physics

1995 Summer Program: *Large Scale Optimization with Applications to Inverse Problems, Optimal Control and Design, and Molecular and Structural Optimization*

1995-1996: *Mathematical Methods in Materials Science*
- Mechanical Response of Materials from Angstroms to Meters
- Phase Transformations, Composite Materials, and Microstructure
- Disordered Materials
- Particulate Flows: Processing and Rheology
- Interface and Thin Films
- Nonlinear Optical Materials
- Numerical Methods for Polymeric Systems
- Topology and Geometry in Polymer Science
- Mathematics in Industrial Problems, Part 9

MIX
Papier aus verantwortungsvollen Quellen
Paper from responsible sources
FSC® C105338

If you have any concerns about our products,
you can contact us on
ProductSafety@springernature.com

In case Publisher is established outside the EU,
the EU authorized representative is:
Springer Nature Customer Service Center GmbH
Europaplatz 3, 69115 Heidelberg, Germany

Printed by Libri Plureos GmbH
in Hamburg, Germany